Processing of Heavy Crude Oils - Challenges and Opportunities

Edited by Ramasamy Marappa Gounder

Published in London, United Kingdom

IntechOpen

Supporting open minds since 2005

Processing of Heavy Crude Oils - Challenges and Opportunities
http://dx.doi.org/10.5772/intechopen.74912
Edited by Ramasamy Marappa Gounder

Contributors
Samuel Sojinu, Onome Ejeromedoghene, Abdelaziz Nasr El-Hoshoudy, Saad M. Desouky, Emad Soliman, Sayeed Rushd, Rasel A Sultan, Shahriar Mahmud, Tarek Ganat, Cs Khor, Hazlina Husin, Mysara Mohyaldinn, Pradip Mandal, Nurul Aini Amran, Shafirah Samsuri, Mohd Afnan Ahmad, Feyisayo Victoria Adams, Chikaodili Chukwuneke, Bolade Agboola, Mahshad Pazouki, Faith Uchenna Uchenna Babalola, Alfred Susu, Ching Thian Tye, Ramasamy Marappa Gounder

First published in London, United Kingdom, 2019 by IntechOpen
IntechOpen is the global imprint of INTECHOPEN LIMITED, registered in England and Wales, registration number: 11086078, 7th floor, 10 Lower Thames Street, London, EC3R 6AF, United Kingdom
Printed in Croatia

British Library Cataloguing-in-Publication Data
A catalogue record for this book is available from the British Library

Additional hard and PDF copies can be obtained from orders@intechopen.com

Processing of Heavy Crude Oils - Challenges and Opportunities
Edited by Ramasamy Marappa Gounder
p. cm.
Print ISBN 978-1-83968-409-8
Online ISBN 978-1-83968-410-4
eBook (PDF) ISBN 978-1-83968-411-1

An electronic version of this book is freely available, thanks to the support of libraries working with Knowledge Unlatched. KU is a collaborative initiative designed to make high quality books Open Access for the public good. More information about the initiative and links to the Open Access version can be found at www.knowledgeunlatched.org

For EU product safety concerns:
IN TECH d.o.o., Prolaz Marije Krucifikse Kozulić 3, 51000 Rijeka, Croatia, info@intechopen.com or visit our website at intechopen.com.

Meet the editor

Dr. M. Ramasamy is currently an Associate Professor in the Chemical Engineering Department, Universiti Teknologi PETRONAS (UTP). He has been in teaching and research for more than 30 years. He is specialized in process simulation, optimization and control. His current research interests include advanced process control, heat exchanger fouling, thermal management of crude oil pipelines and energy management in oil and gas industries. He has supervised 7 PhD and 5 M.Sc. students. He has also published several research articles in reputed international journals, book chapters and a book. He has also delivered several invited lectures and short courses. He obtained his B.Tech. degree in Chemical Engineering from Madras University, India; his M.Tech. from the Indian Institute of Technology, Kharagpur and his PhD from the Indian Institute of Technology, Madras. He is a Chartered Engineer (Engineering Council, UK), Senior Professional Engineer (Engineering Council, India) and a Certified Energy Manager (GreenTech, Malaysia).

Contents

Preface

I am delighted to present this book titled *Processing of Heavy Crude Oils - Challenges and Opportunities* in an attempt to present the current status and technologies in this field.

It is envisaged that heavy crude oils will become a major energy source in the near future, replacing the conventional light crude oils, the reserves of which are depleting steadily over time. Heavy crude oils present several technological challenges in extraction, pumping, pipeline transportation and processing.

Very high viscosity impedes the mobility of the crude oil thereby introducing difficulties in production and transportation. Chapter 1 is an introductory chapter. Chapter 2 presents methods for enhancing the recovery of heavy crude oils. Chapter 3 discusses the laboratory analysis of PVT properties of black oils. Chapters 4 to 9 deal with pumping and pipeline transportation techniques.

Pre-treatment and removal of impurities/metals from heavy crude oils are presented in Chapters 10 to 12. Catalysts for hydrocracking and hydrotreating of heavy crude oils are discussed in Chapter 13. Chapter 14 presents the optimization of petroleum refinery configuration to process heavy crude oils. Processing of heavy crude oils generates different types of toxic organic and inorganic pollutants and Chapter 15 describes the potential environmental challenges that arise due to these pollutants.

I am very grateful to the editorial team at IntechOpen for their excellent and continuous support from the beginning to the end of this book project.

I am deeply indebted to my wife Chandragandhi and children, Ramadevi and Hemanand, for their unwavering support in all my activities.

Ramasamy Marappa Gounder
Universiti Teknologi Petronas,
Malaysia

Chapter 1

Introductory Chapter: Heavy Crude Oil Processing - An Overview

Ramasamy Marappa Gounder

1. Introduction

The demand for crude oil as an energy source is continuously increasing driven by the ever-growing population and economic development. Globally, over 80 million barrels per day of crude oil is produced currently out of which about 11 million barrels are classified as heavy crude oils. It has been well recognized that reservoirs of light conventional crude oils are decreasing and heavy crude oils are replacing them in the pool of crude oil feedstock in refineries which process the crude oils into various value-added products such as gasoline, diesel, liquefied petroleum gas, petrochemical feedstock, etc. The contribution of heavy crude oils in the crude oil pool is predicted to increase over the years from the current ~10%. An oil industry expert, Professor Emeritus Peter Odell, suggested that by 2100, the oil industry will be larger than in 2000 but up to 90% dependent on unconventional crude oils (high-acid crude oils, heavy crude oils, bitumen, and oil sand) [1].

Heavy crude oil reserves are proven to be abundant, and the active heavy oil-producing countries include Canada, Venezuela, United States, Mexico, Brazil, Russia, Indonesia, China, Colombia, Ecuador, Iraq, Kuwait, Saudi Arabia, Chad, and Angola. Heavy oils are relatively cheaper than light conventional crude oils, and, therefore, they present an economic opportunity to petroleum refiners to take advantage of their low cost to increase the refiner's profit margins. Heavy crude oils represent an important energy resource and require an unusually high levels of effort and expertise to economically produce and process. But as the energy demands grow, the economics of heavy oil production and processing will improve. The production and processing of heavy oils will be influenced by the effectiveness of available technologies.

Heavy crude oils are defined as any liquid petroleum with an API gravity in the range of 10–22.3° [2]. The fundamental characteristics of heavy crude oils are high viscosity and density, chemical complexity, high acidity, high sulfur, increased level of metal and heteroatom impurities, high asphaltenes content, low American Petroleum Institute (API) gravity, and low H/C ratio [3]. Due to these unfavorable characteristics, heavy crude oils are termed low-quality crude oils, and they pose many technological challenges in their production, transportation, and processing to both the producers and the refiners. Extracting and processing of heavy crude oil require large amounts of water and energy and can potentially cause extreme damages to surrounding environment.

This chapter will provide an overview on the challenges posed by heavy crude oils due to their unfavorable characteristics. Extraction of heavy oils is presented in Section 2. Section 3 will discuss in the flow assurance issues due to the high viscosity of heavy crude oils and techniques. Section 3 discusses the technological challenges that are encountered while processing the heavy crude oils in the refineries. Section 4 provides a summary of the issues discussed in this chapter. These issues are dealt in more details in the subsequent chapters of the book.

2. Extraction

Cold extraction and surface mining would be preferred for the recovery of heavy crude oils. However, as the depth of the reservoir increases, and due to the very high viscosities, recovery becomes extremely difficult. Water flooding becomes ineffective due to their huge differences in their viscosities. In order to increase the recovery factor, enhanced oil recovery methods are to be followed. As oil viscosity decreases with the increase of temperature, external thermal sources can be introduced to heat the oil and reduce the viscosity. Steam is generally used as a thermal source. Once the viscosity gets reduced, recovery techniques using chemicals and gases can be implemented to improve the displacement efficiency. Cyclic steam stimulation, steam flooding, steam assisted gravity drainage, in situ combustion, chemical injection, gas injection, and in situ catalytic upgrading are some of the typically applied methods to enhance the recovery of heavy crude oils. These methods require large volumes of water and energy on the wellheads to successfully implement. In situ catalytic recovery of heavy oil has many advantages over the steam, gas, and chemical injections. Milder operating conditions, reduced waste products with higher extend of upgrading, minimization of surface upgrading facilities, and reduced environmental impacts are some of the advantages [4].

3. Flow assurance

The viscosity of heavy crude oils is in the range of 100–10,000 cp, while the extra-heavy crude oils have viscosities greater than 10,000 cp. As the viscosity increases, flow characteristics of heavy crude oils dramatically change to very low mobility. The mobility of heavy crude oils reduces to such an extent that conventional pumping techniques fail to achieve any flow from the well bottom to the wellhead or in the transportation pipelines from the platforms to the receiving centers. In addition, wax and asphaltene depositions on the walls of the pipelines as temperature drops also hinder the flow through the pipelines. Such depositions can lead to multiphase flow, clogging of pipelines, high pressure drops, and occasional pipeline stoppages. Flow assurance is a technique through which the crude oil is transported from the wellbore to the processing facilities while mitigating all the risks and guaranteeing a manageable and profitable flow. The flow assurance techniques are generally classified into three approaches, namely, viscosity reduction, drag minimization, and in situ crude oil upgrading [5].

3.1 Viscosity reduction techniques

There are several methods in practice to reduce the viscosity of heavy crude oils in an effort to increase their mobility. These methods include (i) heating, (ii) dilution with light liquid hydrocarbons, (iii) oil-in-water emulsion formation, and (iv) depression of pour point [5, 6].

Heating is one of the common methods employed for pipeline transportation of heavy oils. Viscosity of heavy oils are highly dependent on temperature, and an increase in temperature produces a significant drop in viscosity. The pipelines are generally well insulated, and the temperature is maintained by providing external heating. Thermal management of production and transportation pipelines is very important, and there are several approaches to provide external heat in the pipelines, e.g., bundled pipelines in which a heating fluid is circulated through a pipeline within a carrier pipe and electrically heated subsea pipelines. For high wax crude oils, in addition to thermal management through heating/insulation, chemical inhibitor injection also helps in reducing crystallization of wax and thus viscosity.

Dilution is another method of reducing the viscosity of the heavy oil. It involves the addition of lighter liquid hydrocarbons such as condensates from natural gas production, conventional lighter crude oils, and product streams such as kerosene from the refinery. In some cases, up to 30 wt.% of kerosene is required to sufficiently reduce the viscosity. The disadvantage is that this uses up large quantities of valuable commercial products, and, moreover, the added product is to be processed again through the refinery along with the heavy crude oil. This, in turn, reduces the plant production capacity and at the same time reduces the plant energy efficiency. The addition of lighter liquid hydrocarbons may also affect the stability of wax and asphaltenes leading to pipeline clogging. Sometimes, the dilution is also necessary to meet the API requirements of the existing refineries.

Emulsion of oil-in-water is an alternative to the method of dilution to reduce the viscosity of heavy oils. When the heavy oil is dispersed in water, the flow characteristics improve tremendously. However, additional substances such as surfactants or stabilizing agents are added to ensure the stability of the emulsions.

Pour point depressants, also known as wax crystal modifiers, are the chemical additives that prevent nucleation and crystallization of paraffin. They reduce the viscosity and yield stress of the crude oil appreciably enabling the transportation of waxy crude oils.

3.2 Drag reduction techniques

Developing a core annular flow of heavy oils with a thin film of water or an aqueous solution between the core and the wall of the pipe reduces the pressure drop due to friction in the pipeline. This is akin to lubricating the inner core of the heavy oils and thus reducing the pressure drops. But, it is quite difficult to maintain the stability of the core through the length of the commercial pipelines.

3.3 In situ crude oil upgrading

Physicochemical upgrading of heavy oils produces a synthetic oil or syncrude with higher API gravity and low viscosity. Upgrading of crude oil involves in situ production of a solvent through separation, distillation, and thermal cracking, a part of the heavy oil to produce lighter fractions. The lighter fractions can be used to dilute the heavier fractions to upgrade their mobility.

4. Heavy crude oil refining

The processing of heavy oils in refineries, especially in the existing refineries, presents many difficulties due to their unfavorable characteristics. Due to very low H/C ratios, the yields of straight-run gasoline, kerosene, and diesel are very poor due to high levels of heavier components such as asphaltenes. A major fraction of the heavy crude oil after fractionation in an atmospheric distillation column is expected to report to the residue stream as the feed to the vacuum distillation unit.

The economic value of the residue stream is very low, and therefore, conversion techniques are required to upgrade them to maximize the production of value-added streams. The upgrading of the residue streams is generally done in one of the two methods, namely, hydrogen addition (e.g., hydrotreating or hydrocracking) or carbon rejection (e.g., deasphalting, delayed coking, visbreaking, and thermal/catalytic cracking) [7–9]. Fluidized catalytic cracking (FCC) and residue fluidized catalytic cracking (RFCC) convert heavy oils into more valuable gasoline and lighter products. These technologies enable the effective utilization of crude oils. FCC and RFCC process technologies have become the heart of modern refineries for upgrading heavy oils by upgrading atmospheric and vacuum residue streams. Refineries require a high flexibility in residue processing facilities like FCC and RFCC to cater to the highly varying nature of feedstocks.

There are several operational difficulties that arise in these upgrading processes. For example, very high operating temperatures in the range of 500–1000°C in thermal cracking units favors extensive coking [7, 10]: large amounts of expensive solvents for separating high-molecular-weight components in solvent deasphalting with relatively low yields of recovered deasphalted oil [11]. Catalytic deactivation caused by coking and metals (e.g., Ni, V, Ca, and Fe) and requirements of expensive hydrogen are major drawbacks of the hydrogen addition methods [11].

Except the heavy oil reserves from Africa, the rest of the heavy oil reserves have been characterized to contain high levels of sulfur. Desulfurization is therefore necessary in order to (i) meet the fuel specifications on sulfur content (e.g., 10 ppm limits on gasoline and diesel according to Euro 5 standards) and (ii) to protect catalysts from getting poisoned by sulfur. Hydrotreating removes materials such as sulfur, nitrogen, and metals harmful to the catalysts, and hydrotreaters are required to be located before the reformer, hydrocracker, and FCC units. Heavy metals also poison the catalysts used for the conversion techniques mentioned above and, thus, makes it imperative to pretreat crude oils to remove the metal contaminants almost completely.

Higher fractions of asphaltenes in crude oils contribute to heavy fouling in crude preheat train which affects the delicate balance of heat integration and stable operation of the atmospheric distillation column. Fouling is also pronounced with the presence of sulfur and naphthenic acid in the crude oil. High-acid crude oils require the removal of acid contents below certain threshold levels to safeguard the assets from corrosion effects.

5. Conclusion

Heavy crude oils will steadily replace the conventional light crude oils as energy sources to meet the increasing energy demands. The characteristics of heavy oils render the existing extraction and refining processes ineffective. This chapter enlisted some of the technological challenges faced by the oil producers and refiners. Flow assurance is a much challenging issue in producing and transporting heavy crude oils. Due to the low quality of the crude oil, the refining processes face enormous challenges. Issues such as fouling, corrosion, excessive coking, catalyst poisoning, increased contaminants in end products, the requirement of large hydrogen resources, environmental pollutions, high operating and maintenance costs, etc. require to be addressed with adequate and effective technologies. The synthesis of new catalysts for refining processes which are tolerant to the impurities present in heavy oils is also very crucial for the improved economics of refineries.

Author details

Ramasamy Marappa Gounder[1,2]

1 Department of Chemical Engineering, Universiti Teknologi PETRONAS, Seri Iskandar, Perak D.R, Malaysia

2 Centre for Process Systems Engineering, Institute of Autonomous Systems, Universiti Teknologi PETRONAS, Seri Iskandar, Perak D.R, Malaysia

*Address all correspondence to: marappagounder@utp.edu.my

References

[1] Sperling D, Gordon D. Two Billion Cars: Driving Towards Sustainability. Oxford, England, UK: Oxford University Press; 2009. p. 130

[2] Huc AY. Heavy Crude Oils: From Geology to Upgrading: An Overview. Paris, France: Editions Technip; 2011

[3] He L, Lin F, Li X, Sui H, Xu Z. Interfacial sciences in unconventional petroleum production: From fundamentals to applications. Chemical Society Reviews. 2015;**44**:5446-5494

[4] Guo K, Li H, Yu Z. In-situ heavy and extra-heavy oil recovery: A review. Fuel. 2016;**185**:886-902

[5] Palou RM, Mosqueira ML, Rendón BZ, Juárez EM, Huicochea CB, Clavel-López JC, et al. Transportation of heavy and extra-heavy crude oil by pipeline: A review. Journal of Petroleum Science and Engineering. 2011;**75**:274-282

[6] Santos RG, Loh W, Bannwart AC, Trevisan OV. An overview of heavy oil properties and its recovery and transportation methods. Brazilian Journal of Chemical Engineering. 2014;**31**(3):571-590

[7] Rana MS, Samano V, Ancheyta J, Diaz J. A review of recent advances on process technologies for upgrading of heavy oils and residua. Fuel. 2007;**86**:1216-1231

[8] Khan MK, Kwek W, Kim J. Upgrading heavy crude oils and extra heavy fractions in supercritical methanol. Energy & Fuels. 2017;**31**:12054-12063

[9] Canıaz RO, Erkey C. Process intensification for heavy oil upgrading using supercritical water. Chemical Engineering Research and Design. 2014;**92**:1845-1863

[10] Ebrahimi S, Moghaddas JS, Razavi Aghjeh MK. Study on thermal cracking behavior of petroleum residue. Fuel. 2008;**87**:1623-1627

[11] Speight J. Heavy and Extra-Heavy Oil Upgrading Technologies. 1st ed. UK: Gulf Professional Publishing; 2013

Chapter 2

Methods for Enhancing Recovery of Heavy Crude Oil

Mohd Afnan Ahmad, Shafirah Samsuri and Nurul Aini Amran

Abstract

The methods of enhancing recovery of heavy crude oil explore the importance of enhanced oil recovery and how it has grown in recent years due to the increased needs to locate unconventional resources such as heavy oil, shale, and bitumen. Unfortunately, petroleum engineers and managers are not always well-versed in the enhancement methods available when needed or the most economically viable solution to maximize their reservoir's productivity. Various recovery methods have been explored to extract heavy oil from deep reservoirs or oil spills. This chapter summarizes the details of methods, namely nanoparticle technology, carbon dioxide injection, thermal recovery and chemical injection, which include the methodology as well as the findings.

Keywords: enhancing oil recovery, nanoparticle, carbon dioxide injection, thermal recovery, gas injection

1. Introduction

The production of oil is classified into three phases; primary, secondary, and tertiary. First, the primary recovery involves the extraction of hydrocarbon which naturally rises to the surface. Then, for the second phase, water and gas are injected into the well to push oil to the surface [1]. After the second phase is done, there is still about 60–80% of oil left inside the well [2]. Thus, the implementation of enhanced oil recovery (EOR) during the last phase which is the tertiary phase can contribute up to 30% of original oil in place (OOIP) that can be extracted. Therefore, EOR can be represented by a few techniques namely nanoparticle technology, carbon dioxide injection, thermal recovery and chemical injection.

Heavy oils have the American Petroleum Institute (API) gravity of between 10 and 20 API and a viscosity greater than 100 cP with the characteristics of being asphaltic, dense and viscous. More energy demands are required for the elevated viscosity and the density of these crude oils in their production, and upgraded as well for transportation. Recovery of heavy oil is expected to make an important contribution towards environmental protection as well as energy and resource conservation.

2. Nanoparticle technology

Nanotechnology is one of the methods which attracts great attention nowadays in enhancing oil recovery because it is cost-effective and environmentally friendly [3].

Commonly, the size of nanoparticles for oil recovery is in a range of 1 to 100 nm. The size may slightly differ from any other international organization. First and foremost, the metallic oxide nanoparticles explaining the nature of the metal element which has low ionization potential and low electronegativity shows that it is a reactive and unstable element. The metal element can easily lose an electron, and form a stable state when in contact or react with oxygen. There are a few examples of metal oxide nanoparticles that have been studied lately such as aluminum oxide, copper(II) oxide, iron oxide, nickel oxide, magnesium oxide, tin oxide, titanium oxide and zinc oxide [4].

In enhancing oil recovery (EOR) with nanoparticles, the most influential factor is the interfacial tension (IFT). This parameter contributes to decreasing the capillary force, thereby increasing the oil recovery. Several studies show that IFT reduction between the oil and aqueous phase when mixed with nanofluids increase oil recoveries [5–7]. The trapped oil droplets may have deformed whilst the IFT between the oil and aqueous phase reduced, and it may pass the pore throats easily [5, 8]. Another parameter is the wettability, as it is measured by the complex interface boundary conditions acting within the pore space of sedimentary rocks [9, 10]. The alteration via wettability happens if nanoparticles are absorbed on the surface of grains. The most recommended metal oxide nanoparticles as enhancing oil recovery agent for heavy oil reservoirs is aluminum oxide (Al_2O_3) nanofluid. It can decrease the oil-brine IFT and oil viscosity. Spontaneous imbibition recovery in sandstone cores shows the highest recovery when the Al_2O_3 nanoparticles are dispersed in diesel [11]. Other than that, an experiment conducted by researchers found that Al_2O_3 nanoparticles can de-stabilize water drops which reduce the water in oil emulsion. This case indicates that Al_2O_3 may decrease the emulsion viscosity [12, 13]. However, it concludes that higher concentration of nanoparticles can block pore throats due to the aggregation of particles around the pores, and this may lead to prevention of oil recovery. This proves that in the study by Alomair et al., [12] the lowest concentration of 38.5% of oil recovery is obtained due to the IFT reduction and emulsion viscosity. Since iron oxide has a unique magnetism nature and low toxicity, iron oxide particles can reduce the viscosity of crude oil [14]. As iron oxide spreads in brine, it can be a good oil recovery agent in sandstone reservoirs. For unprompted imbibition in sandstone rocks, it shows that when diesel is selected as a dispersing agent, iron oxide can act as a better oil recovery candidate with instances of 82.5% of total oil recovery. Researchers have experimented with iron oxide to coat polymer in the separation of water and oil [15]. Polyvinylpyrrolidone (PVP) is the polymer used and coated with this nanoparticle, and results in near 100% of oil recovery due to the PVP that has the tendency to absorb both aliphatic and aromatic components of oil component, and the iron oxide acts as a structural support which allows magnetic separation from aqueous phase easily.

On the other hand, nickel oxide (Ni_2O_3) nanoparticles also show the same nature as Al_2O_3 nanoparticles. The effects of dispersed nanoparticles in heavy oil show the recovery of up to 85% of the asphaltenes in the original solution. According to Ogolo et al. [4], spontaneous imbibition and core flood experiments are seen on sandstone rock samples at room condition. The observation for spontaneous imbibition experiments which result in aluminum, nickel, and iron oxides are best found in oil recovery agents especially when diesel is selected as the dispersing agent. In the case of the core flooding experiments, nickel oxide nanoparticles are found to increase oil recovery when injected into sandstone cores after waterflooding. The recovery factor is found to be higher particularly when brine is used as the dispersant for nickel oxide particles. The study claims that the nickel oxide nanoparticles can increase the viscosity of the displacing fluid, and decrease the viscosity of the displaced oil.

Furthermore, magnesium oxide (MgO) and zinc oxide (ZnO) nanoparticles are used during core flood tests which spread in brine or ethanol. It can cause permeability impairment in sandstone rocks. It is found that soaking the rock samples in ethanol and magnesium oxide nanoparticle solution could significantly reduce the oil viscosity. For ZnO, the investigation of the applications of this inorganic compound in enhancing oil recovery processes are very limited. As stated by Ogolo et al. [4], similar to magnesium oxide, when ZnO is used as an enhancing oil recovery agent in sandstone, it shows a negative effect on the permeability of the samples used. The study claims that the problem initiated by agglomeration of the zinc oxide nanoparticles at the injection point can block the pores.

Other than that, Zirconium oxide (ZrO_2) nanoparticles are rarely used in the oil and gas industry and in enhancing the oil recovery process. Ogolo et al. [4] inject metal oxide as an enhancing oil recovery agent at room temperature into a sandstone core sample. It results in a small increase in oil recovery compared to the injection of distilled water alone. When brine or ethanol is used as dispersing agents, it reduces the recovery factor to less than that achieved in the absence of nanoparticles [4]. Tin oxide (SnO_2) nanoparticles are investigated by Naje et al. [16]. SnO_2 nanoparticles have recently attracted a lot of attention from researchers in various fields. Generally, SnO_2 is not used in oil recovery processes extensively. Studies are done by Ogolo et al. [4] on the potential of SnO_2 as oil recovery agent. The results obtained by these researchers show that SnO_2 performs like zirconium oxide and increases oil recovery in sandstone cores while spread in distilled water [16].

For TiO_2 nanoparticles, an analysis using these nanoparticles for water flooding are done, and 80% of oil recovery from oil-wet Berea sandstone in the EOR process comes out. After that, the test is done again, but with the absence of nanoparticles which show a result of 49% in amount [17]. However, the tendency of these particles to aggregate and precipitate results in a milky solution and impossible to measure the IFT [11]. They also conduct a coreflood experiment with TiO_2 and achieve 76% points of original oil in place (OOIP) with 0.05%wt of concentration by using povidone as dispersant since it reduces the particles plugged at inlet points [11].

Instead of the use of metal oxide nanoparticles, researchers also found organic and inorganic nanoparticles that may contribute to the EOR system. For organic nanoparticles, a study found the use of Multiwall Carbon Nano-tubes (MWNT) potential fluid for EOR agent in a high-temperature condition and high-pressure reservoirs [18]. There are two results which are in the absence and presence of electromagnetic waves. For the absence of electromagnetic waves, it shows 36% of oil recovery after the injection of the MWNT nanofluids, while in assistance of electromagnetic fields, it shows almost double the recovery. The higher results have been directly related to the oil viscosity reduction associated with the electromagnetic field. Also, the application of these nanotubes has been reported to increase the efficiency of drilling fluids [19].

In inorganic nanoparticles, the prominent element used is silica. The SiO_2 nanoparticles as proposed by Ogolo et al., [4] shows that the application of SiO_2 in water-wet sandstone reservoirs with this type of nanoparticles can be considered as a suitable EOR agent for this type of rock. Researches reveal that the specific surface area of the SiO_2 powders almost have no change even when it is heated to various temperature of up to 65°C, and proven with good thermal stability [20]. It also does not need a stabilizer compared to metal oxide by forming a more stable emulsion in 3%wt NaCl brine, and achieving higher oil-brine IFT compared to a mixture of brine and stabilizer on its own, resulting in higher oil recovery from Berea sandstone [9].

Researchers investigate that SiO_2 nanoparticles on the bubble surface enhance the foam stability against film rupture and Ostwald ripening [21]. The bubbles

Nanoparticles	Findings	References
Aluminum oxide (Al_2O_3) nanofluid	• Mostly used nanoparticles in enhancing oil recovery (EOR) process. • Al_2O_3 decreases oil brine interfacial tension (IFT). • Total recovery due to nanoparticles (distilled water as dispersing agent) is 12.5%. • Total recovery due to nanoparticles (brine as dispersing agent) is 5.0%. • Al_2O_3 decreases oil viscosity.	[4, 11, 12]
Iron oxide	• Has a unique magnetism characteristic. • Can reduce oil viscosity. • Total recovery due to nanoparticles (distilled water as dispersing agent) is 9.2%. • Diesel as a dispersing agent reached 82.5% of oil recovery.	[4, 14, 15]
Nickle oxide (Ni_2O_3)	• Has the same characteristics as Al_2O_3. • Total recovery due to nanoparticles (distilled water as dispersing agent) is 2.0%. • Total recovery due to nanoparticles (brine as dispersing agent) is 1.7%. • The oil recovery reached up to 85%.	[3, 4]
Magnesium oxide (MgO)	• Cause permeability impairment in sandstone rocks. • Reduce oil viscosity when soaking the rock sample in ethanol with MgO. • Total recovery due to nanoparticles (distilled water as dispersing agent) is 1.7%.	[3, 4]
Zinc oxide (ZnO)	• Very limited use in EOR. • These nanoparticles can block the pores, showed a negative result. • Total recovery due to nanoparticles (distilled water as dispersing agent) is 3.3%.	[3, 4]
Zirconium oxide (ZrO_2)	• Rarely used in EOR. • Show small recovery of oil compared to distilled water alone. • Total recovery due to nanoparticles (distilled water as dispersing agent) is 4.2%.	[4, 16]
Tin oxide (SnO_2)	• Same characteristics as zirconium oxide. • Increases oil recovery while spreading in distilled water. • Total recovery due to nanoparticles (distilled water as dispersing agent) is 3.3%.	[4, 16]
Titanium dioxide (TiO_2)	• Recover 80% of the oil from oil wet Berea sandstone. • Reduce the oil brine IFT. • Achieved higher oil recovery in wet formation compared to SiO_2.	[11, 17,]
Multiwall carbon nanotubes (MWNT)	• Absence of electromagnetic (EM) wave shows 36% of oil recovery. • The presence of EM waves shows 72% of recoveries.	[18, 19]
SiO_2	• Less oil recovery at room temperature. • Considered as suitable EOR agent in all different wettability conditions. • Forming stable foam, more stable bubbles penetrate further inside the pore which can displace more oil.	[4, 9, 20–22]

Nanoparticles	Findings	References
Hydrophobic silicon oxide (SiO_2)	• The small size of these particles in the range of several to tens of nanometres reduces the risk of blocking the pore in an EOR process.	[3, 4]
Inorganic silica core or polymer-shell nanocomposite	• Can reduce IFT. • Increasing the viscosity at critical concentrations. • Can be an excellent EOR agent for sandstone reservoirs, especially when dispersed in ethanol.	[23, 24]

Table 1.
Summary of nanotechnology method in EOR process.

are more stable than foam when they meet the residual oil due to bigger bubbles being flushed and squeezed into smaller ones towards the dead-end. The more stable bubbles penetrate further inside the pore which can displace more oil. When the stable bubbles invade the dead-end pore, the microforces acting on the oil droplet also help to recover more oil. The attached nanoparticles on the bubbles reduce the surface area available for interbubble gas diffusion, which stabilize foam against Ostwald ripening [22]. Other than that, the use of SiO_2 nanoparticles during core-floods conducted at room temperature result in less recovery, and it is still considered as a suitable EOR agent in all different wettability conditions from water-wet to intermediate and oil-wet. Researchers explain that alumina coated silica nanoparticles on the SiO_2 nanoparticles entirely alter their properties.

The coating creates a positive charge on the surface of a nanoparticle. The study proves that alumina coated with SiO_2 possesses higher surface area compared to those without coating, and when disposed into the environment, it shows lower toxicity. The study also comes out with fascinating results in which alumina coated silica nanoparticles with modified surface form a more stable foam and result in good oil recovery from sandstone cores compared to the bare nanoparticle or any surfactant flooding. For hydrophobic silicon oxide nanoparticles, it is demonstrated with addition of silanol (Si-OH) group into the silica nanoparticles surface, and completed with a satisfying result, showing that it is a better EOR agent in sandstone reservoir compared to the metal oxide nanoparticles [4].

Other than that, most researchers use the spherical fumed silica nanoparticles as a stabilizing agent for oil/water emulsion [23]. In EOR studies, it has limited direct use in flooding experiment and has not been studied yet. However, the size of the nanoparticles is suitable for EOR activities which can reduce the risk of blocking the pores. After that, as for inorganic silica core or polymer-shell nanocomposite, its build is illustrated as SiO_2 nanoparticles in the core covered with a shell of synthetic polyacrylamide polymer. The composite nature of the nanoparticles is suitable to be applied for higher temperature and salinity with the presence of hard ion that can be found in offshore reservoir [24]. **Table 1** show the summary of nanotechnology method in EOR process.

3. Carbon dioxide injection

Another effort in increasing the production of oil from the reservoir is the type of method in the EOR process. In secondary production including water flooding and gas injection, it is employed to increase production by boosting depleted pressure in formation. After the oil and natural gas in a formation is produced, the remaining trapped hydrocarbon in the reservoir due to the pressure in the formation is reduced. Therefore, the production is either slow dramatically or stop altogether [25]. In the secondary phase, gas injection is used on a reservoir in enhancing waning pressure within the formation. It will systematically spread throughout the field, and the gas-injection reservoir is used to inject gas and effectively sweeps the formation for remaining petroleum and boosts production [26].

The gas injection, also known as miscible flooding, maintains the reservoir pressure and improves the oil displacement due to the reducing interfacial tension between water and oil. The techniques remove the interface between the two interacting fluids, and this allows for total displacement efficiency [27]. The gasses used are carbon dioxide (CO_2) and natural gas or nitrogen, but CO_2 is commonly used for miscible displacement because it reduces the oil viscosity and cheaper [28]. Oil displacement by CO_2 injection depends on the mixtures of the gas and the crude phase behavior, which are strongly dependent on reservoir temperature, pressure and crude oil composition.

The use of CO_2 as an injection fluid in oil reservoirs has been widely investigated [29]. The characteristics of CO_2 include ease to dissolve oil, can reduce the viscosity of the oil, use moderate pressure to extract the oil's light component, and can form heterogeneously with oil at relatively small pressure [30]. Besides, CO_2 has minimum miscibility pressure (MMP) compared to other gasses. From the study, oil recovery has been improved by manipulating injection pressure (3000, 3144 and 3400 psi) for both horizontal and 450 down dip displacement processes. It can also be said that oil recovery and displacement efficiency increase with the increase of injection pressure. Thus, high pressure can produce maximum displacement efficiency and oil recovery.

The solvent-based process such as cyclic solvent injection has shown a significant contribution in enhancing heavy oil recovery. The examples of solvent are CO_2, flue gas, and light hydrocarbon gases such as natural gas, methane, ethane, propane, and butane. The cyclic solvent injection is an initiative for cyclic steam injection for heavy oil. This is done by injecting the gas into a well (huff cycle), followed by a short shut-in time and then the well is returned to production after a soaking time to allow solvent interaction with oil formation (puff cycle) [31]. Firouz and Torabi [31] investigate the efficiency of the huff-and-puff recovery technique through eight sets of cyclic injection experiments at different operating pressure, utilizing pure CO_2 and pure methane to enhance heavy oil recovery. 71% of oil recovery is obtained by injecting pure CO_2 at the near supercritical condition of 7239 kPa and 28°C, while 50% of oil recovery is obtained by using pure methane at the highest operating pressure of 6895 kPa. The production is contributed by several governing mechanisms such as solution gas drive, viscosity reduction, extraction of lighter components, the formation of foamy oil and to a lesser degree of diffusion process.

Consequently, a study is also done via a long core in the CO_2 huff "n" puff process. When the CO_2 is injected into the core, diffusion occurs to prove viscosity reduction and oil swelling. The IFT between the CO_2 and heavy oil declines [32, 33]. The mobility then increases, and oil recovery enhances significantly. The ultimate heavy oil recovery factors are as high as 32.75%, which is a very good production performance for a cold heavy oil production process [34]. Another study is done in which the CO_2 injection is compared with the injection of nitrogen gas. Both gases show a positive result in oil recoveries and the recoveries are led with the injection of CO_2 gas with 15.8% based on OOIP.

When CO_2 is in the soaking stage, it can also lower the viscosity, and alter the relative permeability hysteresis of higher oil recovery. Then, by using this gas injection, the recovery obtained is ultimately higher at 85.9% based on OOIP compared to the injection of nitrogen gas at 64.7%. In 2017, further study was done and the team investigated the probability of improving oil recoveries for 21 samples at reservoir conditions. The oil transportability in the small pores improves, then the CO_2 can extract oil from the unconventional core samples by diffusion. The result proves that by injecting this gas, it is able to recover up to 99% of oil samples in 24 hours under reservoir condition exposure, and it summarizes that the CO_2 could be injected to highly fractured tight reservoirs through fractures to recover oil [35, 36].

Particularly, injecting the CO_2 in a supercritical state is effective in reservoirs with depth of about 2000 feet. It can be applied in high pressure with lighter oil, as a result of oil swelling, in reducing the viscosity and possible in reducing IFT with the reservoir rock. For low pressure or heavy oils case, CO_2 will form an immiscible fluid, or it can only partially mix with the oil. Some oil may be swelling and significantly the oil viscosity still can be reduced [37]. In this application, there is about one-half and two-third of injected CO_2 return with the produced oil. Usually, it is reinjected into the reservoir to minimize the cost. Thus, the use of CO_2 as a solvent is beneficial for being more economical than other similar miscible fluids such as propane and butane [38].

Besides that, water-alternating-gas (WAG) is another method implemented in the EOR process. As water mixing with CO_2 is used, the saline solution is used to not disturb the carbonate formation in the reservoir. Water mixed with CO_2 is injected into the reservoir for a larger recovery as the mixture has lower miscibility with oil. The use of both water and CO_2 also lowers the mobility of CO_2 gas, for instance, making the gas more effective at displacing the oil in the reservoir [39]. The researcher states that using a small slug of both CO_2 and water allows for a quick recovery of the oil. Additionally, in a study done by Saxena K. [40], using water

Findings	References
• CO_2 can easily dissolve in oil. • Injecting CO_2 can reduce the oil viscosity in the reservoir. • Only requires moderate pressure to be applied for oil recovery. • CO_2 can form heterogenous when mixing with oil in the reservoir.	[30]
• The diffusion occurs when CO_2 injected into the reservoir and this leads reduction of viscosity and oil swelling.	[32, 33]
• The mobility of oil increases after the injection of CO_2. • Heavy oil recovery achieved as high as 32.75% when CO_2 is injected.	[34]
• The injection of CO_2 gas able to recover up to 99% of oil samples. • The CO_2 could be injected on to highly fractured tight reservoirs through fractures to recover oil.	[35, 36]
• CO_2 gas will form an immiscible fluid for low pressure or heavy oils case. • The gas only partially mixes with the oil then some oil may be swelling, and the oil viscosity still can be reduced.	[37]
• CO_2 gas is more economical than other miscible gas such as propane and butane.	[38]
• Water-alternating-gas (WAG) making the gas more effective at displacing the oil in the reservoir.	[39]
• Using water allows for greater oil removal and greater geochemical interactions in WAG.	[40]

Table 2.
Summary of carbon dioxide injection method in EOR process.

with lower salinity allows for greater oil removal, and greater geochemical interactions [40]. **Table 2** show the summary of CO_2 injection method in EOR process.

4. Thermal recovery

Thermal EOR is another technique used for more than 50% in the EOR process. Steam injection is the most common method in thermal EOR including the in-situ combustion [41]. This process involved by heating the reservoir and then the injected high-oxygen gas mixture was burnt to create the combustion front. Basically, for the steam injection, it is applied in a shallow reservoir containing high viscosity oil usually for heavy crude oil, for example, the reservoir in the San Joaquin Valley of California or the oil sands of Alberta, Canada [42]. In the 1960s, steam injection is commercially used and well understood by the researchers in EOR. Steam injecting heats the crude oil in the formation whilst lowering the oil viscosity and at the same time the steam will vaporize some of the oil to increase its mobility.

Besides, when the crude oil viscosity decreases the surface tension also reduces. It increases the permeability of oil and improves the reservoir seepage condition. When oil vaporizes, it allows the oil flowing freely through the reservoir and forms a better quality of oil once it has been condensed. The steam injection EOR varies with two distinct categories: cyclic steam stimulation and steam flooding [43]. For cyclic steam stimulation, the same reservoir is used for steam injection and oil production. Firstly, the steam is injected for a period from a couple of weeks to a couple of months. The introduction of the steam into the reservoir immediately allows the oil to heat up through convective heating, and at the same time it is lowering the oil viscosity. After the targeted oil viscosity is achieved, the steam injection stops to allow the heat to redistribute evenly in the formation. By doing that, it can contribute by increasing the amount of oil recovered after this stage. These steps are repeated when the reservoir temperature drops and the viscosity of oil increases again.

Other than that, steam flooding injection and production wells vary from each other. Steam is introduced through the injection wells and move towards the oil by physically displace while heating the oil to reduce its viscosity. The steam flooding is in the continuous form which allows the steam to drive the oil towards the production wells. Compared to cyclic steaming, this steam flooding is more costly due to this method which requires more steam during the process. Nonetheless, this method usually recovers a big portion of oil. In some cases, both methods can be implemented together for cyclic stimulation followed by steam flooding [44].

Thermal Oil Recovery is by far the most popular method used in the world during the tertiary stage of oil recovery. Steam injection is the most common method

Findings	References
• Using steam injection to heat the crude oil in lowering the oil viscosity.	[41]
• The steam injection was applied in a shallow reservoir containing high viscosity oil which usually suitable for heavy crude oil.	[42]
• Thermal recovery method used to recover a big portion of oil. • Both methods can be implemented in cyclic stimulation followed by steam flooding in EOR.	[44]

Table 3.
Summary of thermal recovery method in EOR process.

used in thermal EOR. It helps produce up to 30% of OOIP. Steam injection does not pose as many environmental risks as other EOR methods might have. This helps implement this technology in different countries, even with strict regulations. The economy is the main factor that determines if this technology should be implemented in one field or the other. **Table 3** shows the summary of thermal recovery method in EOR process.

5. Chemical injection

Usually, the injection of any type of chemical as a dilute solution is used in mobility aid and the reduction of IFT. The injection of an alkaline or caustic solution into the reservoir with oil which consists of organic acids that occur in the oil naturally will produce a soap that may reduce the IFT, and sufficient to increase the production of oil [45]. Other than that, water-soluble polymer diluted solution is injected to increase the viscosity of injected water in the reservoir which can improve the amount of oil recovered in some formations. For example, the use of petroleum sulfonates as surfactant or biosurfactant such as rhamnolipids in dilute solutions can lower the IFT or capillary pressure that impedes oil droplets from moving through the reservoir. So then, this is analyzed in terms of the number of the bond, relating the capillary forces to gravitational ones.

Special formulations of oil, water, surfactant and microemulsions can be particularly effective in reducing interfacial tension. Concerning this application of these methods, they are usually limited by the cost of the chemicals and their adsorption and loss onto the rock of the oil-containing formation. All the chemicals are injected into several reservoirs and the production occurs in other nearby wells. These methods include the polymer flooding, microbial injection, and plasma pulse.

For polymer flooding, it consists of long-chain polymer molecules mixed with the injected water to improve the water viscosity. It also implements the vertical and areal sweep efficiency to improve the water/oil mobility ratio [46]. The surfactant may be used in conjunction with polymer, it decreases the IFT between oil and water. This reduces the residual oil saturation and improves the macroscopic efficiency of the process [47]. Primary surfactants usually need the addition of co-surfactant, activity booster and co-solvent in fixing the stability of the formulation. As caustic flooding is the addition of sodium hydroxide into the injection of water, therefore, it lowers the IFT, reversing the rock wettability, oil emulsification, the oil mobility and aids in drawing the oil out of the rock.

Other than that, the microbial injection which is part of microbial EOR is a method rarely used due to the higher cost and not preferable. These microbes help in EOR by generating biosurfactant in partially digesting long hydrocarbon molecules or by emitting CO_2 gas. There are a few studies in achieving this microbial injection. Bacterial cultures are mixed with food source such as molasses injected into the reservoir. Then, the second approach is where the nutrients are injected into the ground to nurture the existing microbial bodies. The bacteria tend to help in increasing the production of natural surfactants which they normally used to metabolize the underground crude oil [48]. After the injected nutrients are utilized, the microbes will be terminated where their exterior surface will become more hydrophilic. At the same time, the microbes will migrate to the oil-water interface area where it will cause the oil droplets to form a larger oil mass. Thus, making the droplets to be more likely to migrate to the wellhead. **Table 4** shows the chemical injection method in the EOR process.

Findings	References
• The injection of an alkaline or caustic solution into the reservoir with oil produces a soap that can reduce the IFT. • Water-soluble polymer diluted solution also can be used in reducing the IFT. • This may lead in increasing the production of oil.	[45]
• Polymer flooding consists of long-chain polymer molecules mixed with the injected water can improve the water viscosity. • It implements the vertical and areal sweep efficiency due to improving the water/oil mobility ratio in the reservoir.	[46]
• The surfactants may be used in conjunction with polymer, it decreases the IFT between oil and water. • This will reduce the residual oil saturation and improves the macroscopic efficiency of the process.	[47]
• Using microbial injection in generating the biosurfactant or emitting CO_2 in decreasing the IFT of oil. • Nutrients are injected into the ground to nurture the existing microbial bodies and these nutrients cause the bacteria to help increase the production of natural surfactant which they normally used to metabolize crude oil underground. • Causing the oil droplets to form from the larger oil mass and making the oil move to the reservoir surface.	[48]

Table 4.
Summary of chemical injection method in EOR process.

6. Conclusion

Most studies show that nanoparticles can be used in increasing the oil recovery from an oil reservoir in which this nanotechnology in displacement fluid can lower the interfacial tension, change the wettability of rock to a more water-water state. It also lowers the adsorption of surfactant on reservoir rock, but the stability of nanoparticles is the most challenging problem as well as the aggregation of nanoparticles. Besides, there is a nanoparticle technology that has not been discovered and needs further investigation in EOR. The most concerning problem in nanoparticle technology is costly in the application and its effects on the environment. Nevertheless, using the recommended nanofluid to flush a depleted reservoir or using ethanol itself may boost or improve the oil recovery.

After that, the CO_2 injection has the potential in the application for enhancing heavy oil recovery with CO_2 injection, and it is increasing in reservoir pressure and driving higher forces to produce more oil. Sometimes injected gas mixes and dissolves in oil. Thus, the oil viscosity decreases because oil moves easier than before and oil production improves. Therefore, the prominent benefits using CO_2 gas is the miscibility of gas in crude oil, less expensive and is an excellent method in EOR where the injection may improve oil recovery at the same time, the greenhouse gas profile is improved as well.

Besides, thermal EOR is a commonly used method in the world during the tertiary stage which helps improve the production of oil about 30% of OOIP. It also does not contribute to any environmental risk or pollution as another method in EOR might have. For the chemical injection, this method uses many types of chemicals including polymers and surfactants. It can reduce the IFT and increase the flooded water viscosity. This method is used followed by the waterflood where it captures residual oil then the production of oil is up to 15% incremental. In a nutshell, both nanoparticles, CO_2 injection, thermal recovery and chemical injection in enhancing oil recovery are reviewed. The results are promising and there is still a chance for these methods to become better.

Acknowledgements

The authors would like to acknowledge the Centre for Biofuel and Biochemical Research (CBBR) and Chemical Engineering Department, Universiti Teknologi PETRONAS for the support.

Conflict of interest

The authors declare no conflict of interest.

Author details

Mohd Afnan Ahmad[1,2], Shafirah Samsuri[1,2]* and Nurul Aini Amran[1,2]

1 Chemical Engineering Department, Universiti Teknologi PETRONAS, Seri Iskandar, Perak, Malaysia

2 Centre for Biofuel and Biochemical Research, Universiti Teknologi PETRONAS, Seri Iskandar, Perak, Malaysia

*Address all correspondence to: shafirah.samsuri@utp.edu.my

References

[1] Planckaert M. Oil reservoirs and oil production. In: Petroleum microbiology. Washington, DC: American Society of Microbiology ASM Press; 2005. pp. 3-19. DOI: 10.1128/9781555817589.ch1

[2] Guler B, Wang P, Delshad M, Pope GA, Sepehrnoori K. Three-and four-phase flow compositional simulations of CO_2/NGL EOR. In: SPE Annual Technical Conference and Exhibition. New Orleans, Louisiana: Society of Petroleum Engineers; 2001. DOI: 10.2118/71485-MS

[3] Negin C, Ali S, Xie Q. Application of nanotechnology for enhancing oil recovery–A review. Petroleum. 2016;**2**:324-333. DOI: 10.1016/j.petlm.2016.10.002

[4] Ogolo NA, Olafuyi OA, Onyekonwu MO. Enhanced oil recovery using nanoparticles. In: SPE Saudi Arabia Section Technical Symposium and Exhibition. Al-Khobar, Saudi Arabia: Society of Petroleum Engineers; 2012. p. 9

[5] Shahrabadi A, Bagherzadeh H, Roostai A, Golghanddashti H. Experimental investigation of HLP nanofluid potential to enhance oil recovery: A mechanistic approach. In: SPE International Oilfield Nanotechnology Conference and Exhibition, Noordwijk, Netherlands. 2012. DOI: 10.2118/156642-MS

[6] Onyekonwu MO, Ogolo NA. Investigating the use of nanoparticles in enhanced oil recovery. In: 34th Annual SPE International Conference and Exhibition, Tinapa, Calabar, Nigeria. 2010. DOI: 10.2118/140744-MS

[7] Zaid HM, Yahya N, Latiff NRA. The effect of nanoparticles crystallite size on the recovery efficiency in dielectric nanofluid flooding. Journal of Nano Research. 2013;**21**:103-108

[8] Roustaei A, Saffarzadeh S, Mohammadi M. An evaluation of modified silica nanoparticles' efficiency in enhancing oil recovery of light and intermediate oil reservoirs. Egyptian Journal of Petroleum. 2013;**22**:427-433

[9] Morrow NR. Wettability and its effect on oil recovery. Journal of Petroleum Technology. 1990;**42**(12):1-476

[10] Li S, Genys M, Wang K, Torsæter O. Experimental study of wettability alteration during nanofluid enhanced oil recovery process and its effect on oil recovery. In: SPE Reservoir Characterisation and Simulation Conference and Exhibition. Abu Dhabi, UAE: Society of Petroleum Engineers; 2015. DOI: 10.2118/175610-MS

[11] Hendraningrat L, Torsaeter O. Unlocking the potential of metal oxides nanoparticles to enhance the oil recovery. In: Offshore Technology Conference-Asia. 2014. DOI: 10.4043/24696-MS

[12] Alomair OA, Matar KM, Alsaeed YH. Nanofluids application for heavy oil recovery. In: SPE Asia Pacific Oil & Gas Conference and Exhibition. 2014. DOI: 10.2118/171539-MS

[13] Hendraningrat L, Li S, Torsaeter O. Enhancing oil recovery of low-permeability Berea sandstone through optimised nanofluids concentration. In: SPE Enhanced Oil Recovery Conference. Kuala Lumpur, Malaysia: Society of Petroleum Engineers; 2013. DOI: 10.2118/165283-MS

[14] Lee J-H, Jang J-T, Choi J-S, Moon SH, Noh S-H, Kim J-W, et al. Exchange-coupled magnetic nanoparticles for efficient heat 356 induction. Nature Nanotechnology. 2011;**6**(7):418-422. DOI: 10.1038/nnano.2011.95

[15] Palchoudhury S, Lead JR. A facile and cost-effective method for separation of oil–water mixtures using polymer-coated Iron oxide nanoparticles. Environmental Science & Technology. 2014;**48**(24):14558-14563. DOI: 10.1021/es5037755

[16] Naje AN, Norry AS, Suhail AM. Preparation and characterization of SnO_2 nanoparticles. International Journal of Innovative Research in Science, Engineering and Technology. 2013;**2**:7068-7072

[17] Ehtesabi H, Ahadian MM, Taghikhani V. Enhanced heavy oil recovery using TiO_2 nanoparticles: Investigation of deposition during transport in core plug. Energy & Fuels. 2014;**29**(1):1-8

[18] Chandran K. Multiwall Carbon Nanotubes (MWNT) Fluid in EOR Using Core Flooding Method under the Presence of Electromagnetic Waves. Malaysia: Petronas University of Technology; 2013

[19] Friedheim JE, Young S, Stefano G, Lee J, Guo Q. Nanotechnology for oilfield applications-hype or reality? In: SPE International Oilfield Nanotechnology Conference and Exhibition. Engineers. Noordwijk, The Netherlands: Society of Petroleum Engineers; 2012. DOI: 10.2118/157032-MS

[20] Wang D, Han P, Shao Z, Hou W, Seright RS. Sweep-improvement options for the Daqing oil field. SPE Reservoir Evaluation & Engineering. 2008;**11**(1):18-26

[21] Sun Q, Li Z, Li S, Jiang L, Wang J, Wang P. Utilization of surfactant-stabilized foam for enhanced oil recovery by adding nanoparticles. Energy & Fuels. 2014;**28**(4):2384-2394. DOI: 10.1021/ef402453b

[22] Stocco A, Garcia-Moreno F, Manke I, Banhart J, Langevin D. Particle-stabilised foams: Structure and aging. Soft Matter. 2011;**7**(2):631-637

[23] Zhang T, Davidson D, Bryant SL, Huh C. Nanoparticle-stabilized emulsions for applications in enhanced oil recovery. In: SPE Improved Oil Recovery Symposium. Tulsa, Oklahoma, USA: Society of Petroleum Engineers; 2010. DOI: 10.2118/129885-MS

[24] Nguyen PT, Do BPH, Pham DK, Nguyen QT, Dao DQP, Nguyen HA. Evaluation on the EOR potential capacity of the synthesized composite silica-core/polymer-shell nanoparticles blended with surfactant systems for the HPHT offshore reservoir conditions. In: SPE International Oilfield Nanotechnology Conference and Exhibition. Noordwijk, The Netherlands: Society of Petroleum Engineers; 2012. DOI: 10.2118/157127-MS

[25] Mohsenzadeh A, Escrochi M, Afraz MV, Karimi G, Al-Wahaibi Y, Ayatollahi S. Non-hydrocarbon gas injection followed by steam–gas co-injection for heavy oil recovery enhancement from fractured carbonate reservoirs. Journal of Petroleum Science and Engineering. 2016;**144**:121-130

[26] Ren B, Ren S, Zhang L, Chen G, Zhang H. Monitoring on CO_2 migration in a tight oil reservoir during CCS-EOR in Jilin oilfield China. Energy. 2016;**98**:108-121

[27] Guo K, Li H, Yu Z. In-situ heavy and extra-heavy oil recovery: A review. Fuel. 2016;**185**:886-902

[28] Alagorni AH, Yaacob ZB, Nour AH. An overview of oil production stages: Enhanced oil recovery techniques and nitrogen injection. International Journal of Environmental Science and Development. 2015;**6**(9):693-701

[29] AlOtaibi FM, Zhou X, Kokal SL, Senthilmurugan B, Alhashboul AA,

Alabdulwahab AM. A novel technique for enhanced oil recovery: In-situ CO_2-emulsion generation. In: SPE Asia Pacific Enhanced Oil Recovery Conference. Kuala Lumpur, Malaysia: Society of Petroleum Engineers; 2015. p. 13

[30] Abdassah D, Siregar S, Kristanto D. The potential of carbon dioxide gas injection application in improving oil recovery. In: International Oil and Gas Conference and Exhibition in China. Beijing, China: Society of Petroleum Engineers; 2000. DOI: 10.2118/64730-MS

[31] Firouz AQ, Torabi F. Utilization of carbon dioxide and methane in huff-and-puff injection scheme to improve heavy oil recovery. Fuel. 2014;**117**:966973. DOI: 10.1016/j.fuel.2013.10.040

[32] Seyyedsar SM, Sohrabi M. Visualization observation of formation of a new oil phase during immiscible dense CO_2 injection in porous media. Journal of Molecular Liquids. 2017;**241**:199-210

[33] Zhou X, Yuan Q, Rui Z, Wang H, Feng J, Zhang L, et al. Feasibility study of CO_2 huff "n" puff process to enhance heavy oil recovery via long core experiments. Applied Energy. 2019;**236**:526-539. DOI: 10.1016/j.apenergy.2018.12.007

[34] Li S, Li B, Zhang Q, Li Z, Yang D. Effect of CO_2 on heavy oil recovery and physical properties in huff-n-puff processes under reservoir conditions. Journal of Energy Resources Technology. 2018;**140**(7):072907

[35] Jin L, Sorensen JA, Hawthorne SB, Smith SA, Pekot LJ, Bosshart NW, et al. Improving oil recovery by use of carbon dioxide in the Bakken unconventional system: A laboratory investigation. SPE Reservoir Evaluation & Engineering. 2017;**20**(3):602-612. DOI: 10.2118/178948-pa

[36] Eide Ø, Fernø MA, Alcorn Z, Graue A. Visualization of carbon dioxide enhanced oil recovery by diffusion in fractured chalk. SPE Journal. 2016;**21**(1):112-120. DOI: 10.2118/170920-pa

[37] Lashkarbolooki M, Vaezian A, Hezave AZ, Ayatollahi S, Riazi M. Experimental investigation of the influence of supercritical carbon dioxide and supercritical nitrogen injection on tertiary live-oil recovery. The Journal of Supercritical Fluids. 2016;**117**:260-269. DOI: 10.1016/j.supflu.2016.07.004

[38] Dhuwe A, Lee J, Cummings S, Beckman E, Enick R. Small associative molecule thickeners for ethane, propane and butane. The Journal of Supercritical Fluids. 2016;**114**:9-17

[39] Mokhtari R, Ayatollahi S, Hamid K, Zonnouri A. Co-optimization of enhanced oil recovery and carbon dioxide sequestration in a compositionally grading Iranian oil reservoir; technical and economic approach. In: Abu Dhabi International Petroleum Exhibition and Conference. Abu Dhabi, UAE: Society of Petroleum Engineers; November 2016. DOI: 10.2118/183560-MS

[40] Saxena K. Low Salinity Water Alternate Gas Injection Process for Alaskan Viscous Oil EOR. Fairbanks, AK, US: University of Alaska Fairbanks; 2017

[41] Al-Nakhli AR, Sukkar LA, Arukhe J, Mulhem A, Mohannad A, Ayub M, et al. In-situ steam generation a new technology application for heavy oil production. In: SPE Heavy Oil Conference and Exhibition. Kuwait City, Kuwait: Society of Petroleum Engineers; 2016. DOI: 10.2118/184118-MS

[42] Kovscek AR. Emerging challenges and potential futures for thermally enhanced oil recovery. Journal of

Petroleum Science and Engineering. 2012;**98**:130-143

[43] Alvarez J, Han S. Current overview of cyclic steam injection process. Journal of Petroleum Science Research. 2013;**2**(3):16-27

[44] Ageeb AA, Al-siddig MH, Nor-aldeen MR, Soliman MS, Ibrahim IH. The influence of steam injection volume on sand and oil production in cyclic steam stimulation (CSS) wells [Doctoral dissertation]. Khartoum, Sudan: Sudan University of Science and Technology; 2017

[45] Pogaku R, Fuat NHM, Sakar S, Cha ZW, Musa N, Tajudin DNAA, et al. Polymer flooding and its combinations with other chemical injection methods in enhanced oil recovery. Polymer Bulletin. 2018;**75**(4):1753-1774

[46] Leon JM, Izadi M, Castillo A, Zapata JF, Chaparro C, Jimenez J, et al. Use of cross-linked polymer systems to improve volumetric sweep efficiency and alternative full field development strategy for a mature Waterflooding optimization processes-dina cretaceous field case. In: SPE Improved Oil Recovery Conference. Tulsa, Oklahoma, USA: Society of Petroleum Engineers; 2018. DOI: 10.2118/190313-MS

[47] Mandal A. Chemical flood enhanced oil recovery: A review. International Journal of Oil, Gas and Coal Technology. 2015;**9**(3):241-264

[48] Chellappan SK, Al Enezi F, Marafie HA, Bibi AH, Eremenko VB. First application of plasma technology in KOC to improve well's productivity. In: SPE Kuwait Oil and Gas Show and Conference. Mishref, Kuwait: Society of Petroleum Engineers; 2015

Chapter 3

PVT Properties of Black Crude Oil

Abdelaziz El-Hoshoudy and Saad Desouky

Abstract

Precise PVT studies and behavior of phase-equilibrium of petroleum reservoir fluids are essential for describing these fluids and appraising their volumetric behavior at several pressure stages. There are numerous laboratory studies that can be performed on a reservoir sample. The amount of data desired determines the number of tests to be performed in the laboratory. Generally, there are three laboratory tests which characterize hydrocarbon fluids, namely primary study, constant mass depletion, and differential vaporization test. Generally, PVT properties are determined either experimentally or calculated theoretically through published correlations. This chapter presents different PVT laboratory tests that are required to understand the phase behavior of black oils.

Keywords: PVT experiments, black oil properties, petroleum reservoirs

1. Introduction

Petroleum (an equivalent term is a crude oil) is a complex mixture consisting predominantly of hydrocarbons and containing sulfur, nitrogen, oxygen, and helium as minor constituents. The physical and chemical properties of crude oils vary considerably and depend on the concentration of the various types of hydrocarbons and minor constituents present. Crude oil reservoirs are classified according to initial reservoir pressure into the following categories: (1) undersaturated oil reservoir; in which initial reservoir pressure is greater than the bubble point pressure of reservoir fluid; (2) saturated oil reservoir; in which initial reservoir pressure is equal to the bubble point pressure of reservoir fluid; (3) gas-cap reservoir; in which initial reservoir pressure is below the bubble point pressure of reservoir fluid, so reservoir is termed as gas-cap or two-phase reservoir, since the gas or vapor phase is underlain by an oil phase. Petroleum hydrocarbons exist as gaseous or liquid phase depending on reservoir temperature. If it is higher than the critical temperature of the fluid, the reservoir fluid is gas. Otherwise, the reservoir fluid is oil [1]. Black oil reservoir is considered as one of the most precious reservoir fluids. It constitutes the majority of oil reservoirs and exists in every basin. An accurate description of physical properties of crude oils is of considerable importance in the fields of both applied and theoretical science and, especially, in the solution of petroleum reservoir engineering problems. Some of these physical properties are of primary interest in petroleum engineering studies and detected through PVT tests which aim to determine reservoir fluid behavior at simulated reservoir conditions. Data on these fluid properties is usually determined by laboratory experiments performed on samples of actual reservoir fluids. In the absence of experimentally measured properties of crude oils, it is necessary for the petroleum engineer to determine the properties from empirically derived correlations.

However; correlations are approximations and may be useful only in regional geological provinces.

Crude oils cover a wide range of physical properties and chemical compositions and are classified into black oils and near-critical or volatile-oils [2] depending on their phase behavior. The phase diagrams of two reservoir types are illustrated in **Figures 1** and **2**, while their physical properties are indicated in **Table 1** according to classifications established by Moses [1], McCain [3], Whitson and Brule [4]. From the phase diagram, following the pressure reduction path as indicated by the vertical line EF, the iso-lines are scattered uniformly around the reservoir temperature which can be located at any region between points C&F. It is apparent that the

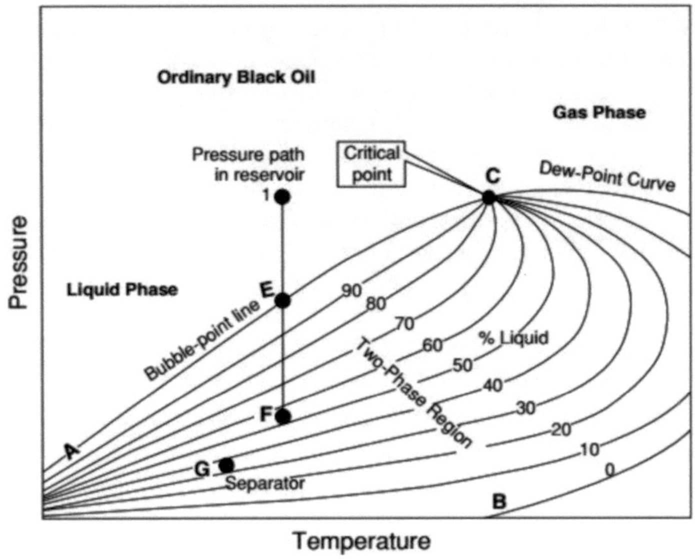

Figure 1.
Typical phase diagram for black oil [8].

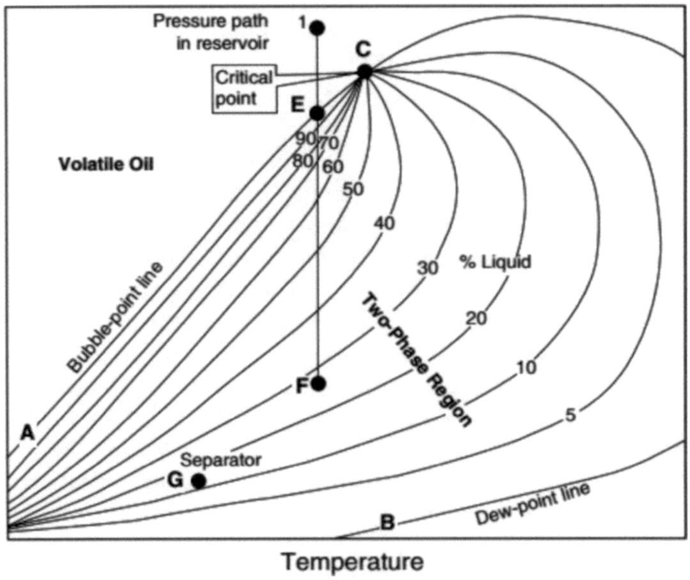

Figure 2.
Typical phase diagram for volatile oil [8].

Property	Black oil [7, 9]	Volatile oil [1]
Fluid color	Deep black	Light colored turn to slightly reddish
Mole % of heptane plus (C_7^+)	>17.5% up to 26.5%	12.5–17.5%
(GOR)	~1500–1750 & up to 2000 scf/STB	2000–3000 scf/STB
API	<45°	~40° or higher
C_o	3–150 × 10^{-6} psi^{-1} (under-saturated to highly saturated GOR oils)	......
β_o	<2 rbbl/STB	>2 rbbl/STB

Table 1.
Criteria of black and volatile oil reservoirs.

reservoir temperature (T_{res}) in black oil reservoirs is less than the critical temperature (T_c). Moreover, the liberated gas below the bubble point pressure is considered as lean gas for industrial applications [5]. While in volatile oils, T_{res} is closest to the critical temperature. Another important differentiation lies in the volume of evolved gas below bubble point which is much greater in volatile oil than that observed for black oils [6]. Another important factor comprising gas oil ratio (GOR), and mole % of C_7^+, where GOR is obtained from field data and the mole % of C_7^+ is obtained from composition analysis. PVT properties for black oils in almost all reservoirs are required for production and surface facilities calculations to manage, develop, and forecast oil field behavior [7], as well as maximizing economic profit.

PVT analysis is utilized by reservoir engineers to distinguish physical properties of reservoir fluids and variations in the volume and phase state that occurs during oil production [10, 11]. The phase envelope is characterized by some physical criteria such as oil formation volume factor, saturation pressure, gas oil ratio, oil density, oil viscosity, and oil isothermal compressibility. This phase envelope may be drawn through experimental data points or via software packages like PVTp, PVTsim, Eclipse and so on, based on the differential equation of states. The importance of PVT properties for reservoir performance analysis makes a dire need to develop a convenient way to predict these properties mathematically either by empirical correlations or equations of state. These empirical relations comprise two types. The first one belongs to the black oil type which forecast PVT behavior from the available PVT data, including saturation pressure, reservoir temperature, oil API index, gas-oil ratio, and reservoir pressure. The second one belongs to compositional models which rely on the equation of states and its hybrids. In these models besides PVT parameters, other measurements including fluid composition, critical temperature, molar masses and components acentric factor are utilized [12, 13]. Although accuracy of the empirical PVT correlations is often limited due to variations and complexity of multicomponent reservoir fluid systems, these correlations still used on field scale [14–17].

2. PVT physical parameters and black oil modeling

In a black oil reservoir, the oil and gas formation volume factors, gas densities, the solution gas-oil ratio, and the viscosities of oil and gas are measured at reservoir temperature as a function of pressure. Once these measurements are, they can be applied in empirical correlations to obtain the relative in situ amounts of oil and gas during the production life of the reservoir.

2.1 Crude oil API gravity

Crude oil density is defined as the mass of a unit volume of the crude at a specified pressure and temperature and expressed in g/cc. In typical oil reservoirs, the oil density decreases with pressure depletion until it reaches the lowest value at the bubble point. The higher oil density at higher pressures resorts to the high oil molecules compressing per the unit volume. By reducing pressure beneath the bubble-point pressure, the gas is ejected from the solution which forms a free gas. The released gas is saturated with some of oil intermediate components, while the heavier one remains in the reservoir. These intermediates release results in the density reversal trend versus pressure beneath the saturation pressure. It is calculated either by compositional models or by correlating parameters. The API gravity is the preferred gravity scale, and related to the specific gravity which is defined as the ratio of the oil density to water density at 60°F and atmospheric pressure by the following relation:

$$API = \frac{141.5}{\gamma_o} - 131.5 \tag{1}$$

2.2 Solution gas specific gravity

It is a dimensionless property that represents the ratio between hydrocarbon gas density and the air density at standard conditions. It is ascribed practically by the weighted average of the specific gravities of the separated gas from each separator, and expressed by the following relation:

$$\gamma_g = \frac{\Sigma_{i=1}^{n}\left(R_{sep}\right)_i\left(\gamma_{sep}\right)_i + R_{st}\gamma_{st}}{\Sigma_{i=1}^{n}\left(R_{sep}\right)_i + R_{st}} \tag{2}$$

2.3 Gas solubility

It is defined as the number of standard cubic feet of gas that dissolve in one stock-tank barrel of crude oil at certain pressure and temperature and defined in SCF/STB [18].

$$R_s = \frac{(V_g)_{P,T}}{(V_o)_{sc}} \tag{3}$$

At standard conditions, dissolved gas is completely released from the oil, and therefore the oil contains no gas. Gas solubility increase with pressure increasing until it reaches the maximum value at saturation pressure. Below the bubble point pressure, gas is evolved with reduction in pressure and the gas solubility also decreases. Generally, the lighter the oil the higher the gas solubility, therefore volatile oils often has a higher solution GOR than black oils.

2.4 Bubble point pressure

It is defined as the highest pressure at which a bubble of gas is first liberated from the oil, and is determined experimentally by conducting a constant mass depletion test or estimated from empirical correlations. By reservoir pressure depletion beneath the saturation pressure, a released gas is formed in the reservoir. Since gas mobility is greater than the oil mobility, the produced GOR increase with continual pressure depletion beneath the saturation pressure. Other PVT properties are also greatly affected by reservoir pressure depletion.

2.5 Isothermal oil compressibility

It is the change in the fluid volume with respect to the change in pressure at isothermal conditions, expressed in psi^{-1} and calculated above the bubble point pressure from Eqs. (4 and 5).

$$C_o = -\frac{1}{d_v}\frac{d_v}{d_p} \tag{4}$$

$$C_o = -\frac{1}{B_o}\left[\left(\frac{\partial B_o}{\partial P}\right)_T - B_g\left(\frac{\partial R_s}{\partial P}\right)_T\right] \tag{5}$$

Oil compressibility is determined during constant mass depletion test. In that procedure, the oil in the PVT cell is allowed to expand by reducing the pressure, and both the increase in oil volume and the decrease in oil pressure are measured. Oil compressibility above the saturation pressure greatly affects the material balance calculations for estimating volumetric reserve in oil reservoirs, as well as pressure transient analysis for under-saturated oils. Moreover, it is used in the estimation of vertical lift performance.

2.6 Gas compressibility factor

It is a dimensionless quantity used for gas correction at higher pressures and temperature, calculated from the basic equation of state as

$$Z = \frac{PV}{nRT} \tag{6}$$

2.7 Oil formation volume factor, B_o

It is defined as the ratio of oil volume either at the reservoir or separator condition to the volume of oil at standard conditions.

$$(B_o)_{res} = \frac{(V_o)_{(p,T)res}}{(V_o)_{sc}} \tag{7}$$

$$(B_o)_{sep} = \frac{(V_o)_{(p,T)sep}}{(V_o)_{sc}} \tag{8}$$

The oil formation volume factor is expressed in units of reservoir volume over standard volume (bbl/STB). At the surface, the oil loses the dissolved gas it contained under high pressure and temperature in the reservoir. By pressure decline from reservoir to surface conditions, the following scenarios are generated:
(1) The oil weight decrease owing to the loss of dissolved gas by pressure decrease;
(2) The oil volume reduced slightly owing to the temperature reduction, however, the oil slightly expands owing to the pressure decrease. The volume reduction and expansion due to temperature and pressure successively usually cancel each other and no significant changes occur. Practically, (B_o) is determined from residual oil

volume at 60°F at the end of differential liberation. The increase in B_o above bubble point pressure represents oil expansion by pressure decline so the oil volume at P & T (the numerator in Eq. 8) increase and consequently B_o increase. Below bubble point, the oil shrinkage due to gas evolving tends to overcome its expansion by pressure reduction, consequently B_o decline steeply.

2.8 Gas formation volume factor, B_g

Defined as the ratio of the gas volume at the temperature and pressure of any stage below the bubble point and the volume of the same gas at standard conditions

#	Property		Input parameters
1	Bubble point pressure (P_b)		Reservoir temperature (T_{res}); API; gas solubility (R_s); the specific gravity of the solution gas (γ_g); oil specific gravity (γ_o); separator gas gravity ($\gamma_{g\ sep}$); reservoir pressure (P_{res}); separator pressure (P_{sep}); separator temperature (T_{sep})
2	Crude oil gravity (API)		The specific gravity of the solution gas (γ_g); oil specific gravity (γ_o)
3	The specific gravity of the solution gas (γ_g)		Separator GOR (R_{sp}); separator gas gravity (γ_{sep}); GOR from the stock tank (R_{st}); gas gravity from the stock tank (γ_{st})
4	Oil density (ρ_o)		The molecular weight of the stock tank oil (M_{STO}); API; gas solubility (R_s); the specific gravity of the solution gas (γ_g); reservoir temperature (T_{res}); reservoir pressure (P_{res})
5	Gas solubility (R_s)		Reservoir temperature (T_{res}); reservoir pressure (P_{res}); oil density (ρ_o); API; the specific gravity of the solution gas (γ_g); oil specific gravity (γ_o); separator pressure (P_{sep}); separator temperature (T_{sep}); oil formation volume factor (B_o)
6	Oil formation volume factor (B_o)		Reservoir temperature (T_{res}); reservoir pressure (P_{res}); oil density (ρ_o); API; specific gravity of the solution gas (γ_g); oil specific gravity (γ_o); separator pressure (P_{sep}); separator temperature (T_{sep}); isothermal compressibility (C_o); bubble point pressure (P_b); gas solubility (R_s)
7	Isothermal compressibility (C_o)	Above bubble point (P_b)	Reservoir pressure ($P_{res} > P_b$); oil density (ρ_o); API; specific gravity of the solution gas (γ_g); separator pressure (P_{sep}); separator temperature (T_{sep}); bubble point pressure (P_b); gas solubility at the bubble point pressure (R_s)
		Below bubble point (P_b)	Reservoir temperature (T_{res}); reservoir pressure ($P_{res} < P_b$); API; specific gravity of the solution gas (γ_g); oil specific gravity (γ_o); oil formation volume factor (B_o); gas formation volume factor (B_g); bubble point pressure (P_b); gas solubility (R_s)
8	Oil viscosity (μ_o)	Dead oil viscosity (μ_{od})	Reservoir temperature (T_{res}); API
		Saturated oil viscosity (μ_{ob})	Reservoir temperature (T_{res}); reservoir pressure (P_{res}); API; specific gravity of the solution gas (γ_g); oil specific gravity (γ_o); bubble point pressure (P_b); gas solubility (R_s); oil density at bubble point (ρ_{ob}); dead oil viscosity (μ_{od})
		Under-saturated oil viscosity (μ_{ou})	Reservoir pressure (P_{res}); API; bubble point pressure (P_b); gas solubility (R_s); dead oil viscosity (μ_{od}); saturated oil viscosity (μ_{ob})

Table 2.
Major input parameters for PVT equations.

through differential liberation (Eq. 9). B_g can also be calculated from composition analysis as a function of Z-factor as stated in Eq. 10.

$$B_g = \frac{(V_g)_{P,T}}{(V_g)_{sc}} \tag{9}$$

$$B_g = \frac{0.028269Z(T)_R}{(P)_{psia}} \tag{10}$$

2.9 Crude oil viscosity

It is defined as the resistance of the fluid to flow, expressed in centipoise (cp). It is affected by the oil composition, GOR, and reservoir temperature. Oil viscosity represents internal dynamic fluid friction relevant to fluid layers. Higher and lower viscosity depends on the gas/oil ratio. The behavior of viscosity is similar to that of density since oil viscosity shrinks with pressure decrease, owing to reduced friction between fluid layers, and reaches its minimal at the saturation pressure. With further pressure depletion more and more gas escaped from the oil, so the oil becomes denser and more viscous owing to loss of lighter components. Practically oil viscosity is measured by falling ball viscometer.

Table 2 summarizes some of the input physical parameters used to characterize black oil reservoirs.

3. Experimental PVT analysis

Crude oil samples can be obtained from downhole using a wireline sampling chamber, or at the separator, using separation facilities [19]. To carry a successful PVT analysis, the samples should be generated through the following procedure:

3.1 Sample handling

The pressurized liquid sample is heated to 200°F and stabilized in single phase prior to removal of any sample in order to ensure sample integrity and avoid any wax precipitation.

3.2 Sample validation

Carried out for both surface and bottom hole samples to ensure that no leakage occurs for the samples and the fluid virtually represents the reservoir.

3.2.1 Surface samples obtained from separator

They are subjected to assuring cylinders integrity during transportation through measurement of opening pressures of the separator oil cylinder and separator gas cylinder at ambient temperature and separator temperature. If the measured pressure equals the separator pressure, the sample is valid, otherwise, the sample is invalid. Moreover, quality control of separator samples carried out through measurement of the saturation pressure (bubble point) of the separator oil sample at the separator temperature and compared with the separator pressure. The sample was then transferred to the Lab cylinder, and the amount of water if present in the separator oil was determined, and discharged.

3.2.2 Bottom hole sample

Also, check cylinders' integrity during transportation and sampling carried out by measurement of sample opening pressures at transferring temperature and reservoir temperature. The sample cylinder was then agitated and transferring to Lab cylinder at reservoir temperature and pressure of 1000 psi above reservoir pressure.

3.3 Primary study

3.3.1 Surface samples obtained from separator

A primary study carried only for surface samples is as follows:

3.3.1.1 Atmospheric flashing of separator oil

A portion of the separator oil sample was charged to a high-pressure visual PVT cell and thermally expanded to the separator temperature till equilibrium. The sample was subjected to flash separation from separator conditions to standard conditions of 14.73 psia and 60°F. Dissolved gas-oil ratio and oil formation volume factor were calculated using the volumetric data obtained so that the supplied stock-tank oil rate could be adjusted to separator oil rate for the well-stream calculations. The density and API gravity of the stock-tank oil were also measured using density meter. The dissolved gas composition was determined using extended gas chromatography and the flashed liquid by high-resolution capillary gas chromatography. The two analyses were then mathematically recombined according to the dissolved gas-oil ratio to get the composition of separator oil. **Figure 3** illustrates the primary study procedure.

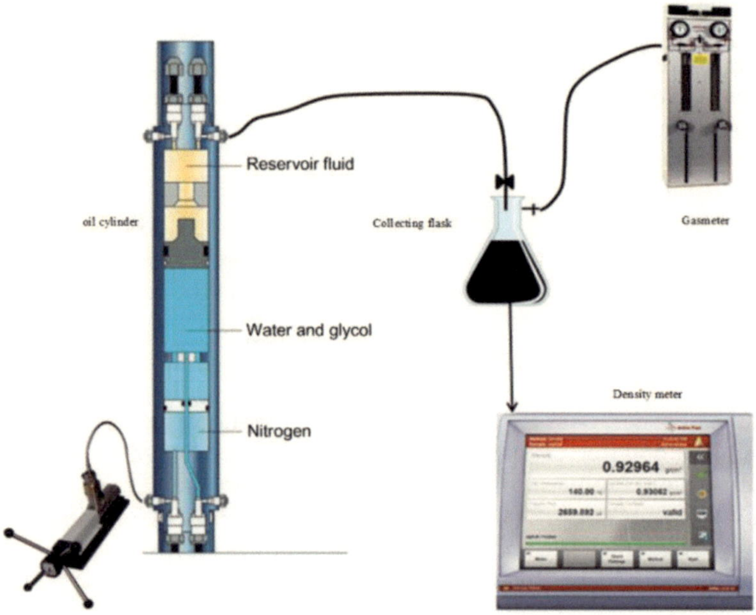

Figure 3.
Diagram of the primary study steps.

3.3.1.2 Composition analysis of stock-tank oil

The compositions of paraffins and aromatics in liquid samples were determined using Clarus 500 Perkin Elmer Gas Chromatograph connected with flame ionization detector (FID) using selective PIONA capillary column of 100 m in length and 0.25 mm internal diameter. Suitable sample capacity was injected into split / splitless injector through a microsyringe according to both response and linear range of FID detector. Nitrogen gas was used as carrier gas at its optimum flow rate. The system enables to detect the composition up to C_{36}^{+}. Further oil analysis includes density, viscosity, pour point, and distillation analysis, in addition to water, basic sediment (BS) and sulfur contents, in order to determine the oil quality.

3.3.1.3 Gas composition analysis

The analysis of both separator gas and dissolved gas were determined according to ASTM 1945, using Clarus 500 PerkinElmer Gas Chromatograph online with the condensate cell utilizing four packed columns connected in series with thermal conductivity detector and DP-1 capillary column matched with flame ionization detector. Helium was used as carrier gas and the oven isothermal temperature 100°C in order to provide highly efficient detection of aromatics and hydrocarbons paraffin (normal and *iso*) from methane (C_1) up to undecane (C_{11}) components and non-hydrocarbons including H_2S, CO_2, and N_2. Once repeatability is verified, the validated bottle generated to proceed the complete PVT analysis, which, in the case of the reservoir fluid be a black oil, is composed by three main assays; constant mass depletion (CMD), differential liberation (DL), and separator test.

3.3.1.4 Measurement of density

The density of oil is measured by Anton Paar Density Meter. The density meter is calibrated with air and distilled water. The procedure reference is ASTM D-4052.

3.3.1.5 Measurement of molecular weight

The average molecular weight number is determined by using cryoscopy, (freezing point depression of benzene).

3.3.1.6 Corrected and adjustment of field separator GOR

Adjustment of field GOR carried out in the lab through the following equations, owing to higher deviations of the measured field gas deviation factor (Z_f)

$$\text{corrected separator GOR} = \left(\text{field GOR} \times \frac{\rho_f \times Z_f}{\rho_l \times Z_l}\right)^{0.5} \tag{11}$$

$$Zf = \left(\frac{1}{fpv}\right)^{0.5} \tag{12}$$

$$\text{adjusted separator GOR} = \frac{\text{corrected separator GOR}}{B_o} \tag{13}$$

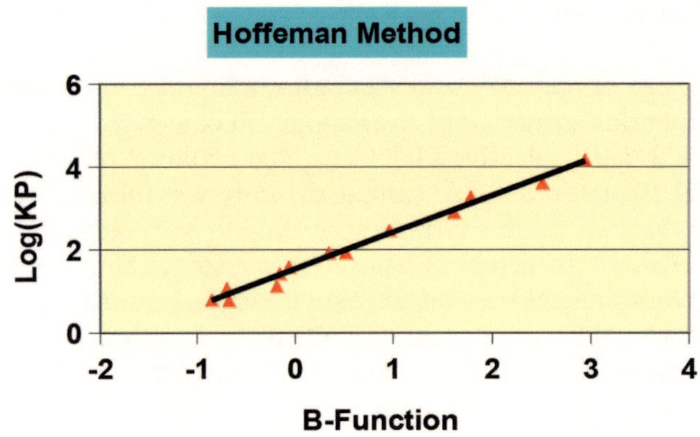

Figure 4.
Hofmann QC-plot.

3.3.1.7 Quality control of primary study

Carried out by applying Hoffmann method to data of separator oil & gas through plotting of ($K_i\,P_{sep}$) against $[1/B_i(1/T_{bi} - 1/T_{sep})]$ on a Cartesian scale which results in a straight line, as shown in **Figure 4**.

where,

$$Ki = \frac{yi}{xi} \tag{14}$$

$$B_i = \log\left(P_{ci}/14.7\right)/\left(1/T_{bi} - 1/T_{ci}\right) \tag{15}$$

3.3.1.8 Physical recombination of well stream

The samples were physically recombined at separator conditions (P_{sep} & T_{sep}) using the corrected separator (GOR), (B_o) and (B_g) to form the wellstream. The wellstream composition can mathematically be determined by applying material balance equation to the separator product compositional analysis and their physical properties such as corrected separator GOR, dissolved GOR, B_o, the density of stock tank oil, separator oil and gas densities as well as B_g.

3.3.2 Bottom hole sample (BHS)

A known portion of the reservoir fluid was charged from (BHS) cylinder to a high-pressure visual cell and thermally expanded to the reported reservoir temperature and pressure. After establishing thermal equilibrium, the fluid sample was subjected to a flash separation at standard conditions (P = 14.7 psia & T = 60°F). The flashed data were then checked using Hoffmann method.

3.4 Constant mass depletion (CMD)

An integrated and automated mercury free system (PVT-cell) is used for PVT analysis of black-oil and volatile oil. A portion of the sample is pumped into the cell, and the volume and, therefore, pressure of the sample chamber are controlled with a piston. Most sample chambers are equipped with a magnetic stirrer to ensure that the system is well mixed. Constant mass depletion also known as flash liberation, in which a set of volume measurements as a function of pressure at fixed reservoir temperature are performed within the PVT cell. The recombined well stream was pressurized and heated to reservoir temperature and pressure. A portion of the sample (~100 cc of live oil) was charged to a high-pressure visual PVT cell and thermally expanded to the reservoir temperature and the pressure was adjusted to a pressure of 1000 psi higher than the reservoir pressure. The measurements of pressures and volumes were recorded and then used to determine the bubble point pressure of the under-saturated reservoirs at reservoir temperature. Above bubble point pressure, the relative volume, oil formation volume factor, oil compressibility, density, and viscosity were determined, while below bubble point pressure the Y-Function is calculated at each pressure step.

3.5 Differential liberation (DL)

After the constant composition mass depletion, the sample was pressurized and equilibrated to a single phase condition. A differential vaporization procedure is then performed for the purpose of measuring two-phase properties as a function of reservoir pressure depletion. As with the flash expansion test, the volume of the sample chamber is increased incrementally; however, in this case, the evolved solution gas is withdrawn from the sample chamber after each incremental expansion. At each pressure depletion, the oil formation volume factor, gas formation volume factor, two-phase volume factor, gas gravity, gas solubility, density, and viscosity were determined. The viscosity of the evolved gas is usually calculated through Lee et al. correlations rather than measured. While the viscosity

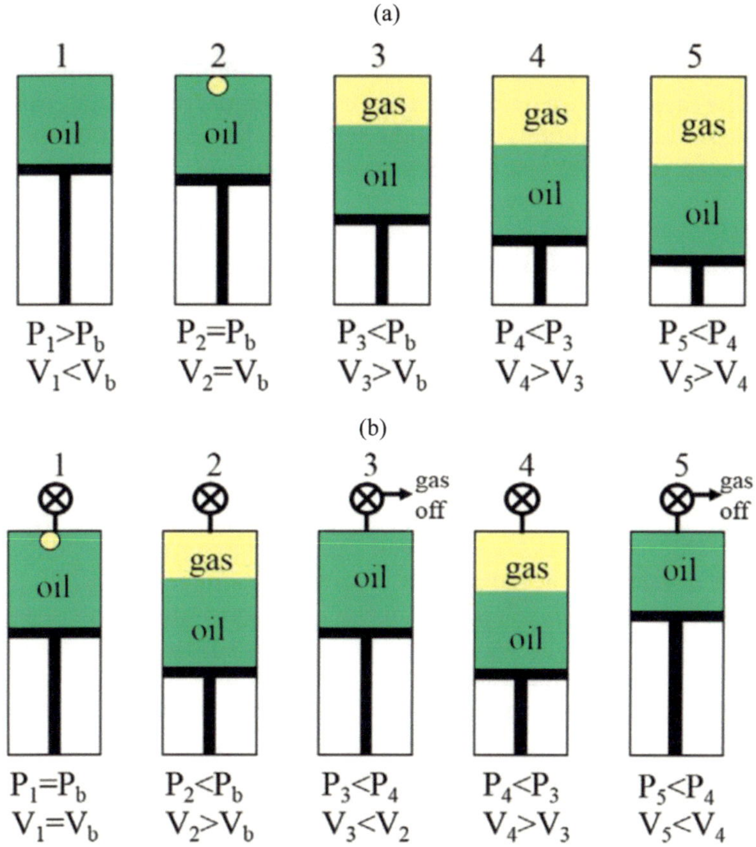

Figure 5.
Schematic illustration of (a) constant mass depletion and (b) differential liberation [6].

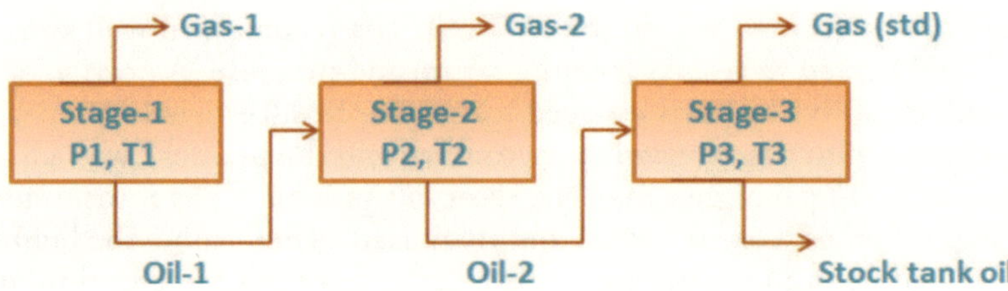

Figure 6.
Multistage separator test.

measurement for both live and dead oil is carried out over a wide range of high pressure and temperature by a rolling ball viscometer (RUSKA). Schematic representation of CMD and differential liberation are shown in **Figure 5**.

3.6 Separator test

Separator test is used to determine the effect of surface separation parameters on the gas-oil ratio, stock tank oil gravity and formation volume factor at separator conditions. About 100 cc of the recombined sample is raised to the saturation pressure at the reservoir temperature. Then the sample condition is lowered and adjusted to the separator conditions, where the oil and gas volumes are measured. In presence of the second separator, the sample is adjusted to the second separator temperature and pressure, where the oil and gas volumes are also measured. After testing all separator stages, the sample is flashed to stock tank conditions, where the volumes of stock tank oil and gas are monitored. The separator test volumetric data and the compositional analysis of the evolved gases from each pressure stage are reported. The separator test data were then used to adjust differential vaporization result to the surface conditions used in the separator test. Schematic representation of the multistage separator test is represented in **Figure 6**.

4. PVT correlations and software

Numerous correlations to calculate PVT properties for oils have been developed. The correlations are usually simple expressions (equations) that allow engineers to calculate the PVT properties from readily available inputs in case of absence of experimentally measured data. Correlations are generated from PVT Datasets associated with the particular geographical region. The most commonly known geographical correlations are Al-Marhoun [20] correlations for Saudi Arabian crudes; Standing [21] for California crudes; Petrosky [22] for Gulf of Mexico crudes; Hanafy et al. [23] for Egyptian crudes; and Omar and Todd [24] for Malaysian crudes. Some developed correlations use common oil samples of worldwide origin. The correlation is usually developed by fitting a predetermined form of an equation (or more) with the measured data and calculating the equation(s) parameters by use of multiple regression analysis. It is essential to note that the accuracy of correlation performance and applicability is associated with the range of data used in correlation development. PVT correlations developed as a function of PVT properties, or as groups of correlations developed from the same dataset for several PVT properties. **Table 3** gives a detailed survey on the published PVT correlations. Numerous software were developed for prediction of PVT phase behavior and physical properties such as PVTp, PVTsim, Eclipse E100, Eclipse E300, Petrel, and Ecrin. A detailed description of this software will be described elsewhere.

Correlation name	P_b	R_s	C_{ob}	B_{ob}	ρ_{ob}	μ_{ob}	C_{ou}	$(GOR)_{STO}$	B_{ou}	μ_{ou}	μ_{od}
Abdul-Majeed and Salman [25]				✓							
Abdul-Majeed et al. [26]						✓				✓	
Abu-Khamsim and Al-Marhoun [27]						✓					
Ahmed [28]			✓	✓	✓		✓				
Ahmed [29]									✓		
Al-Khafaji et al. [30]						✓				✓	✓
Al-Marhoun [31]	✓										
Al-Marhoun [20]		✓		✓							
Al-Marhoun [32]			✓	✓			✓		✓		
Al-Marhoun [33]				✓			✓				
Al-Marhoun [34]							✓				
Al-Mehaideb [35]	✓		✓	✓		✓				✓	
Al-Najjar et al. [36]	✓			✓							
Al-Shammasi [37]	✓	✓		✓							
Andrade [38]											✓
Frick [39]				✓							
Asgarpour et al. [40]		✓		✓							
Asgarpour et al. [40]	✓										
Beal [41]										✓	✓
Beggs and Robinson [42]						✓				✓	✓
Bennison [43]											✓
Bergman [44]						✓				✓	✓
Bergman and Sutton [45]						✓					
Sutton and Bergman [46]										✓	✓
Bolondarzadeh et al. [47]	✓			✓							
Calhoun [48]			✓								
Casey and Cronquist [49]		✓		✓							
Chew and Connally [50]			✓			✓					
De Ghetto et al. [51]	✓	✓	✓			✓				✓	✓
Dindoruk and Christman [52]	✓	✓	✓	✓		✓	✓			✓	✓
Doklah and Osman [53]	✓	✓		✓							
Elam [54]	✓			✓							
El-Banbi et al. [55]	✓			✓							
Elmabrouk et al. [56]	✓			✓				✓			
Elsharkawy and Alikhan [57]	✓	✓	✓	✓		✓					
Elsharkawy and Alikhan [58]										✓	✓
Elsharkawy and Gharbi [59]											✓
Farshad et al. [60]	✓	✓	✓	✓							
Fitzgerlad [61]											✓
Glaso [62]	✓	✓		✓		✓				✓	✓

Correlation name	P_b	R_s	C_{ob}	B_{ob}	ρ_{ob}	μ_{ob}	C_{ou}	$(GOR)_{STO}$	B_{ou}	μ_{ou}	μ_{od}
Hanafy et al. [23]	✓	✓		✓	✓	✓					
Hassan [63]	✓	✓									
Hemmati and Kharrat [64]	✓	✓		✓							
Hossain et al. [65]						✓				✓	✓
Ikiensikimama et al. [66]			✓			✓	✓			✓	✓
Ikiensikimama and Azubuike [67]	✓										
Ikiensikimama and Ogboja [68]	✓	✓		✓							
Karimnezhad et al. [69]	✓			✓							
Kartoatmdjo and Schmidt [70]	✓	✓	✓	✓		✓				✓	✓
Kartoatmdjo and Schmidt [71]	✓	✓	✓	✓		✓				✓	✓
Kaye [72]											✓
Khairy et al. [73]	✓			✓							
Khamehchi et al. [74]	✓	✓				✓					
Khan et al. [75]						✓				✓	✓
Khazam et al. [76]	✓	✓	✓	✓		✓				✓	✓
Knopp and Ramsey [77]		✓		✓							
Kouzel [78]										✓	
Labedi [79]		✓	✓	✓		✓	✓			✓	✓
Labedi [80]	✓	✓									
Labedi [81]	✓					✓				✓	✓
Lasater [82]	✓	✓									
Levitan and Murtha [83]	✓			✓							
Macary and El Batanony [84]	✓	✓		✓							
Mazandarani and Asghari [85]	✓	✓		✓							
McCain [14]	✓	✓	✓	✓	✓	✓				✓	✓
McCain et al. [86]			✓				✓				
Mehran and Movagharnejad [87]	✓			✓							
Moradi et al. [88]				✓							
Moradi et al. [89]	✓										
Mehran and Movagharnejad [87]	✓										
Naseri et al. [90]						✓				✓	✓
Naseri et al. [91]											✓
Ng and Egbogah [92]											✓
Obomanu and Okpobiri [93]	✓			✓							
Okeke and Sylvester [94]								✓			
Okoduwa and Ikiensikimama [95]	✓										
Omar and Todd [24]	✓			✓							
Osorio [96]						✓					
Ostermann and Owolabi [97]	✓			✓							
Owolabi [98]	✓			✓							

Correlation name	P_b	R_s	C_{ob}	B_{ob}	ρ_{ob}	μ_{ob}	C_{ou}	$(GOR)_{STO}$	B_{ou}	μ_{ou}	μ_{od}
Oyedeko and Ulaeto [99]											✓
Petrosky [22]	✓	✓	✓	✓		✓				✓	✓
Petrosky and Farshad [100]	✓	✓	✓	✓		✓				✓	✓
Petrosky and Farshad [101]						✓				✓	✓
Petrosky and Farshad [102]	✓	✓	✓	✓		✓				✓	✓
Rollins et al. [103]								✓			
Sattarina et al. [104]											✓
Spivey et al. [105]							✓				
Standing [21]	✓	✓	✓	✓	✓	✓				✓	✓
Standing [106]	✓	✓									
Standing [107]			✓	✓	✓	✓	✓			✓	✓
Sim [108]	✓	✓	✓			✓					
Sulaimon et al. [109]				✓							
Twu [110]										✓	
Oyedeko and Ulaeto [99]											✓
Valko and McCain [111]								✓			
Vasquez and Beggs [112]	✓	✓	✓	✓						✓	
Velarde et al. [113]	✓	✓									
Whitson and Brule [4]											✓

Table 3.
Summary of the published PVT correlations.

5. Conclusion

PVT studies are very critical and precious for reservoir engineers to monitor the phase behavior of petroleum fluids and generating material balance calculations. Traditionally, PVT experiments are carried out through high pressurized PVT-cell, but in case of unavailability of experimental facilities, we resort to empirically derived correlations. Several empirical correlations were published depending on geographical provinces. In this chapter, a brief description of the phase diagram for black and volatile oil was introduced as well as the importance of PVT properties in predicting petroleum fluids behavior. Then PVT physical parameters and its meanings were briefly illustrated and a summary of the input parameters in their predictions as well as a survey of the published correlations were provided. Moreover, a detailed description of the experimental PVT procedures and tests were illustrated to help engineers and analysts in conducting such tests.

Nomenclature

$(GOR)_{STO}$	stock-tank gas/oil ratio
μ_{ob}	saturated oil viscosity (at or below bubble point)
μ_{od}	dead oil viscosity correlations
μ_{ou}	under-saturated oil viscosity
bbl/STB	barrel/stock tank barrel

B_g	gas formation volume factor
B_o	oil formation volume factor, bbL/STB
B_{ob}	oil formation volume factor at the bubble point
B_{ou}	under-saturated oil formation volume factor
Co	oil compressibility
C_{ob}	saturated oil compressibility (at or below bubble point)
C_{ou}	under-saturated oil compressibility (above bubble point)
GOR	producing gas-oil ratio
K_i	the equilibrium constant of component i in the recombined sample
N	number of separators
P_b	bubble point pressure
P_{ci}	critical pressure of component i in the recombined sample
PVT	pressure-volume-temperature
R_s	gas solubility
R_{sp}	separator GOR, scf/STB
R_{st}	GOR from the stock-tank, SCF/STB
Scf/STB	standard cubic foot/stock tank barrel
T_{bi}	true boiling point temperature of component i in the recombined sample
T_{ci}	the critical temperature of component i in the recombined sample
x_i	the mole fraction of component i in the separator oil
y_i	the mole fraction of component i in the separator gas
Z	gas compressibility factor
Z_f	gas compressibility factor measured at the field
Z_l	gas compressibility factor measured at PVT lab
γ_g	solution gas specific gravity
γ_{sep}	separator gas gravity
γ_{st}	gas gravity from the stock-tank
ρ_f	gas gravity measured at the field
ρ_l	gas gravity measured at PVT lab
ρ_{ob}	oil density at the bubble point

Author details

Abdelaziz El-Hoshoudy[1,2]* and Saad Desouky[1]

1 Production Department, Egyptian Petroleum Research Institute, Naser City, Cairo, Egypt

2 Faculty of Energy and Environmental Engineering, British University in Egypt, Elshorouk City, Cairo, Egypt

*Address all correspondence to: azizchemist@yahoo.com

References

[1] Moses PL. Engineering applications of phase behavior of crude oil and condensate systems (includes associated papers 16046, 16177, 16390, 16440, 19214 and 19893). Journal of Petroleum Technology. 1986;**38**(07):715-723

[2] Mukerji T, Jørstad A, Avseth P, Mavko G, Granli J. Mapping lithofacies and pore-fluid probabilities in a North Sea reservoir: Seismic inversions and statistical rock physics. Geophysics. 2001;**66**(4):988-1001

[3] McCain WD Jr. Revised gas-oil ratio criteria key indicators of reservoir fluid type; part 5. Petroleum Engineer International; (United States). 1994;**66**:4

[4] Whitson C, Brule M. Phase behavior. In: Monograph, Vol. 20 SPE Henry L. Doherty Series. Richardson, Texas: SPE Monograph series; 2000

[5] Holditch S, Hill AD, Zhu D. Advanced Hydraulic Fracturing Technology for Unconventional Tight Gas Reservoirs. US department of energy; Texas Engineering Experiment Station; 2007

[6] Danesh A. PVT and Phase Behaviour of Petroleum Reservoir Fluids. Scothland: Elsevier; 1998

[7] Jarrahian A, Moghadasi J, Heidaryan E. Empirical estimating of black oils bubblepoint (saturation) pressure. Journal of Petroleum Science and Engineering. 2015;**126**:69-77

[8] Ahmed T. Reservoir Engineering Handbook. Amestrdam: Elsevier; 2006

[9] McCain WD, Spivey JP, Lenn CP. Petroleum Reservoir Fluid Property Correlations. Houston, Texas: PennWell Books; 2011

[10] Di Primio R, Dieckmann V, Mills N. PVT and phase behaviour analysis in petroleum exploration. Organic Geochemistry. 1998;**29**(1–3):207-222

[11] Mansour E, Farag A, El-Dars F, Desouky S, Batanoni M, Mahmoud M. Predicting PVT properties of Egyptian crude oils by a modified soave–Redlich–Kowng equation of state. Egyptian Journal of Petroleum. 2013;**22**(1): 137-148

[12] El-Hoshoudy A, Farag A, Ali O, El-Batanoney M, Desouky S, Ramzi M. New correlations for prediction of viscosity and density of Egyptian oil reservoirs. Fuel. 2013;**112**:277-282

[13] El-Hoshoudy A, Desouky S, Betiha M, Abdelraouf S, Elaily M. Mathematical correlation predicting stock tank oil density for Egyptian crude oils at standard conditions. International Journal of Oil, Gas and Coal Engineering. 2014;**2**(4):55-58

[14] McCain JW. Reservoir-fluid property correlations-state of the art (includes associated papers 23583 and 23594). SPE Reservoir Engineering. 1991;**6**(02):266-272

[15] El-Hoshoudy AN, Desouky SM. An empirical correlation for estimation of formation volume factor of gas condensate reservoirs at separator conditions. Petroleum & Petrochemical Engineering Journal. 2018;**2**(2):9

[16] El-Hoshoudy AN. Numerical prediction of oil formation volume factor at bubble point for black and volatile oil reservoirs using nonLinear regression models. Petroleum & Petrochemical Engineering Journal. 2018;**2**(2):11

[17] El-Hoshoudy AN, Desouky SM, Gomaa S. Prediction of dew point pressure in gas condensate reservoirs based on a combination of gene expression programming (GEP) and

multiple regression analysis. Petroleum & Petrochemical Engineering Journal. 2018;**2**(5)

[18] Vazquez M, Beggs HD. Correlations for fluid physical property prediction. In: SPE Annual Fall Technical Conference and Exhibition. Society of Petroleum Engineers; 1977

[19] Starling KE. Fluid Thermodynamic Properties for Light Petroleum Systems. Houston, Texas: Gulf Publishing Company; 1973

[20] Al-Marhoun MA. PVT correlations for Middle East crude oils. Journal of Petroleum Technology. 1988;**40**(05): 650-666

[21] Standing M. A pressure-volume-temperature correlation for mixtures of California oils and gases. In: Drilling and Production Practice. Washington, United states: American Petroleum Institute; 1947

[22] Petrosky GE. PVT Correlations for Gulf of Mexico Crude Oils. Lafayette, USA: University of Southwestern Louisiana; 1990

[23] Hanafy H, Macary S, ElNady Y, Bayomi A, El Batanony M. Empirical PVT correlations applied to Egyptian crude oils exemplify significance of using regional correlations. In: International Symposium on Oilfield Chemistry; Society of Petroleum Engineers. 1997

[24] Omar M, Todd A. Development of new modified black oil correlations for Malaysian crudes. In: SPE Asia Pacific Oil and Gas Conference; Society of Petroleum Engineers. 1993

[25] Abdul-Majeed GH, Salman NH. An empirical correlation for oil FVF prediction. Journal of Canadian Petroleum Technology. 1988;**27**(06): 118-122

[26] Abdul-Majeed GH, Kattan RR, Salman NH. New correlation for estimating the viscosity of undersaturated crude oils. Journal of Canadian Petroleum Technology. 1990;**29**(03):80-85

[27] Abu-Khamsin SA, Al-Marhoun MA. Development of a new correlation for bubble-point oil viscosity. Arabian Journal of Science and Engineering. 1991;**16**:99-106

[28] Ahmed T. Hydrocarbon Phase Behavior (Contributions in Petroleum Geology and Engineering). Houston, Texas: Gulf Publishing Company; 1989

[29] Ahmed T. Oil Recovery Mechanisms and the Material Balance Equation Reservoir Engineering Handbook. Houston, Texas: Gulf Professional Publishing; 2010. pp. 733-809

[30] Al-Khafaji AH, Abdul-Majeed GH, Hassoon SF. Viscosity correlation for dead, live and undersaturated crude oils. Journal of Petroleum Research. 1987; **6**(2):1-16

[31] Al-Marhoun M. Pressure-Volume-Temperature Correlations for Saudi Crude Oils. Saudi Arabia: University of Petroleum and Minerals; 1985

[32] Al-Marhoun MA. New correlation for formation volume factor of oil and gas mixtures. Journal of Canadian Petroleum Technology. 1992;**31**(3): 22-26

[33] Al-Marhoun MA. The coefficient of isothermal compressibility of black oils. In: Middle East Oil Show. Bahrain: Society of Petroleum Engineers; 2003

[34] Al-Marhoun MA. A New Correlation for Undersaturated Isothermal Oil Compressibility. 2006

[35] Al-Mehaideb R. Improved PVT correlations for UAE offshore crudes. Journal of the Japan Petroleum Institute. 1997;**40**(3):232-235

[36] Al-Najjar H, Al-Soof N, Al-Khalisy K. Correlations for bubble-point pressures, gas oil ratios and formation volume factors for Iraqi crude oils. Journal of Petroleum Research. June 1988;**1988**:13

[37] Al-Shammasi A. Bubble point pressure and oil formation volume factor correlations. In: Middle East Oil Show and Conference; Society of Petroleum Engineers. 1999

[38] Andrade EC. The Viscosity of Liquids. London, United Kingdom: Nature Publishing Group. 1930

[39] Frick TC. Petroleum Production Handbook. Newyork, United states: McGraw-Hill; 1962

[40] Asgarpour S, McLauchlin L, Wong D, Cheung V. Pressure-volume-temperature correlations for western Canadian gases and oils. Journal of Canadian Petroleum Technology. 1989; **28**(04):103-112

[41] Beal C. The viscosity of air, water, natural gas, crude oil and its associated gases at oil field temperatures and pressures. Transactions of the AIME. 1946;**165**(01):94-115

[42] Beggs HD, Robinson J. Estimating the viscosity of crude oil systems. Journal of Petroleum Technology. 1975; **27**(09):1140-1141

[43] Bennison T. Prediction of heavy oil viscosity. Presented at the IBC Heavy Oil Field Development Conference. 2-4 December; 1998

[44] Bergman D. Don't forget viscosity. Petroleum Technology Transfer Council 2nd Annual Reservoir Engineering Symposium; Lafayette, Louisiana. 28 July; 2004

[45] Bergman DF, Sutton RP. An update to viscosity correlations for gas-saturated crude oils. In: SPE Annual Technical Conference and Exhibition; Society of Petroleum Engineers. 2007

[46] Sutton RP, Bergman DF. Undersaturated oil viscosity correlation for adverse conditions. In: SPE Annual Technical Conference and Exhibition; Society of Petroleum Engineers. 2006

[47] Bolondarzadeh A, Hashemi S, Solgani B. The new PVT generated correlations of Iranian oil properties. In: 4th Iranian Petroleum Engineering Student Conference. 2006

[48] Calhoun J Jr. New methods for determining permeability. The Petroleum Engineer. February 1947: 103-108

[49] Casey J. Estimate GOR and FVF using dimensionless PVT analysis. World Oil (United States). 1992;**213** (11):12-16

[50] Chew J-N, Connally CA Jr. A Viscosity Correlation for Gas-Saturated Crude Oils. 1959

[51] De Ghetto G, Paone F, Villa M. Pressure-volume-temperature correlations for heavy and extra heavy oils. In: SPE International Heavy Oil Symposium; Society of Petroleum Engineers. 1995

[52] Dindoruk B, Christman PG. PVT properties and viscosity correlations for Gulf of Mexico oils. SPE Annual Technical Conference and Exhibition; Society of Petroleum Engineers; 2001

[53] Dokla M, Osman M. Correlation of PVT properties for UAE crudes (includes associated papers 26135 and 26316). SPE Formation Evaluation. 1992;**7**(01):41-46

[54] Elam FM. Prediction of Bubble Point Pressures and Formation Volume Factors from Field Data. University of Texas; 1957:53-79

[55] El-Banbi AH, Fattah KA, Sayyouh H. New modified black-oil PVT correlations for gas condensate and volatile oil fluids. In: SPE Annual Technical Conference and Exhibition; Society of Petroleum Engineers. 2006

[56] Elmabrouk SK, Zekri AY, Shirif EE. Prediction of bubblepoint pressure and bubblepoint oil formation volume factor in the absence of PVT analysis. In: SPE Latin American and Caribbean Petroleum Engineering Conference; Society of Petroleum Engineers. 2010

[57] Elsharkawy AM, Alikhan AA. Correlations for predicting solution gas/oil ratio, oil formation volume factor, and undersaturated oil compressibility. Journal of Petroleum Science and Engineering. 1997;**17**(3–4): 291-302

[58] Elsharkawy A, Alikhan A. Models for predicting the viscosity of Middle East crude oils. Fuel. 1999;**78**(8): 891-903

[59] Elsharkwy A, Gharbi R. Comparing classical and neural regression techniques in modeling crude oil viscosity. Advances in Engineering Software. 2001;**32**(3):215-224

[60] Frashad F, LeBlanc J, Garber J, Osorio J. Empirical PVT correlations for Colombian crude oils. In: SPE Latin America/Caribbean Petroleum Engineering Conference; Society of Petroleum Engineers. 1996

[61] Fitzgerald D. A Predictive Method for Estimating the Viscosity of Undefined Hydrocarbon Liquid Mixtures. PA, United States: Pennsylvania State University, State College; 1994

[62] Glaso O. Generalized pressure-volume-temperature correlations. Journal of Petroleum Technology. 1980; **32**(05):785-795

[63] Hassan OF. Correlation for solution gas-oil ratio of Iraqi oils at pressures below the bubble point pressure. Iraqi Journal of Chemical and Petroleum Engineering. 2011;**12**(2):1-8

[64] Hemmati MN, Kharrat R. A correlation approach for prediction of crude oil PVT properties. In: SPE Middle East Oil and Gas Show and Conference; Society of Petroleum Engineers. 2007

[65] Hossain MS, Sarica C, Zhang H-Q, Rhyne L, Greenhill K. Assessment and development of heavy oil viscosity correlations. In: SPE International Thermal Operations and Heavy Oil Symposium; Society of Petroleum Engineers. 2005

[66] Ikiensikimama S, Egbe T, Effiong E, Ogboja O. New screening technique for PVT correlations as applied to the Niger Delta crude. In: Refereed Proceedings. 2008

[67] Ikiensikimama S, Azubuike I. Modeling approach for Niger-delta oil formation volume factor prediction using artificial neural network. In: Nigeria Annual International Conference and Exhibition; Society of Petroleum Engineers. 2012

[68] Ikiensikimama SS, Ogboja O. New bubblepoint pressure empirical PVT correlation. In: Nigeria Annual International Conference and Exhibition; Society of Petroleum Engineers. 2009

[69] Karimnezhad M, Heidarian M, Kamari M, Jalalifar H. A new empirical correlation for estimating bubble point oil formation volume factor. Journal of Natural Gas Science and Engineering. 2014;**18**:329-335

[70] Kartoatmodjo TR, Schmidt Z. New Correlations for Crude Oil Physical Properties. 1991

[71] Kartoatmodjo T, Schmidt Z. Large data bank improves crude physical property correlations. Oil and Gas Journal;(United States). 1994;**92**(27)

[72] Kaye S. Offshore California viscosity correlations. COFRC, TS85000940; 1985

[73] Khairy M, El-Tayeb S, Hamdallah M. PVT correlations developed for Egyptian crudes. Oil and Gas Journal. 1998;**96**(18):114-116

[74] Khamehchi E, Rashidi F, Rasouli H, Ebrahimian A. Novel empirical correlations for estimation of bubble point pressure, saturated viscosity and gas solubility of crude oils. Petroleum Science. 2009;**6**(1):86-90

[75] Khan S, Al-Marhoun M, Duffuaa S, Abu-Khamsin S. Viscosity correlations for Saudi Arabian crude oils. In: Middle East Oil Show. Society of Petroleum Engineers; 1987

[76] Khazam M, Alkhaboli M, Shlak M. New PVT Correlations for Libyan Crudes

[77] Knopp C, Ramsey L. Correlation of oil formation volume factor and solution gas-oil ratio. Journal of Petroleum Technology. 1960;**12**(08):27-29

[78] Kouzel B. How pressure affects liquid viscosity. Hydrocarbon Processing. March 1965;**1965**:120

[79] Labedi RM. PVT Correlations of the African Crudes. Colorado School of Mines; 1982

[80] Labedi RM. Use of production data to estimate the saturation pressure, solution GOR, and chemical composition of reservoir fluids. In: SPE Latin America Petroleum Engineering Conference; Society of Petroleum Engineers. 1990

[81] Labedi R. Improved correlations for predicting the viscosity of light crudes. Journal of Petroleum Science and Engineering. 1992;**8**(3):221-234

[82] Lasater J. Bubble point pressure correlation. Journal of Petroleum Technology. 1958;**10**(05):65-67

[83] Levitan LL, Murtha M. New correlations estimate PH, FVF. Oil & Gas Journal. 1999;**97**(10):10

[84] Macary S, El-Batanoney M. Derivation of PVT correlations for the Gulf of Suez crude oils. Journal of the Japan Petroleum Institute. 1993;**36**(6): 472-478

[85] Mazandarani MT, Asghari SM. Correlations for predicting solution gas-oil ratio, bubblepoint pressure and oil formation volume factor at bubblepoint of Iran crude oils. In: European Congress of Chemical Engineering (ECCE-6). 2007. pp. 16-20

[86] McCain WD Jr, Rollins JB, Lanzi AJV. The coefficient of isothermal compressibility of black oils at pressures below the bubblepoint. SPE Formation Evaluation. 1988;**3**(03):659-662

[87] Mehran F, Movagharnejad K, Didanloo A. New correlation for estimation of formation volume factor and bubblepoint pressure for Iranian oil fields. In: 1st Iranian Petroleum Engineering Conference. 2006

[88] Moradi B, Malekzadeh E, Mohammad A, Awang M, Moradie P. New oil formation volume factor empirical correlation for Middle East crude oils. International Journal of Petroleum and Geoscience Engineering (IJPGE);**2013**:12-23

[89] Moradi B, Malekzadeh E, Amani M, Boukadi FH, Kharrat R. Bubble point pressure empirical correlation. In: Trinidad and Tobago Energy Resources

Conference; Society of Petroleum Engineers. 2010

[90] Naseri A, Nikazar M, Dehghani SM. A correlation approach for prediction of crude oil viscosities. Journal of Petroleum Science and Engineering. 2005;**47**(3–4):163-174

[91] Naseri A, Yousefi S, Sanaei A, Gharesheikhlou A. A neural network model and an updated correlation for estimation of dead crude oil viscosity. Brazilian Journal of Petroleum and Gas. 2012;**6**(1)

[92] Ng J, Egbogah E. An improved temperature-viscosity correlation for crude oil systems. In: Annual Technical Meeting; Petroleum Society of Canada. 1983

[93] Obomanu D, Okpobiri G. Correlating the PVT properties of Nigerian crudes. Journal of Energy Resources Technology. 1987;**109**(4): 214-217

[94] Okeke H, Sylvester O. Improved correlation for predicting stock tank gas-oil ratio in Niger Delta. In: SPE Nigeria Annual International Conference and Exhibition; Society of Petroleum Engineers. 2016

[95] Okoduwa I, Ikiensikimama S. Bubble point pressure correlations for Niger Delta crude oils. In: Nigeria Annual International Conference and Exhibition; Society of Petroleum Engineers. 2010

[96] Osorio JG. Empirical PVT Correlations for Colombian Crude Oil. University of Southwestern Louisiana; 1990

[97] Ostermann R, Owolabi O. Correlations for the reservoir fluid properties of Alaskan crudes. In: SPE California Regional Meeting; Society of Petroleum Engineers. 1983

[98] Owolabi OO. Reservoir Fluid Properties of Alaskan Crudes. Fairbanks: University of Alaska; 1984

[99] Oyedeko KF, Ulaeto UW. Predicting the dead oil viscosity of reservoir fluids: A case study of the Niger Delta. Journal of Energy Technology and Policy. 2011; **3**:1-7

[100] Petrosky JG, Farshad F. Pressure-volume-temperature correlations for Gulf of Mexico crude oils. In: SPE Annual Technical Conference and Exhibition; Society of Petroleum Engineers. 1993

[101] Petrosky G Jr, Farshad F. Viscosity correlations for Gulf of Mexico crude oils. In: SPE Production Operations Symposium; Society of Petroleum Engineers. 1995

[102] Petrosky JG, Farshad F. Pressure-volume-temperature correlations for Gulf of Mexico crude oils. SPEREE. 1998;**1**(5):416-420. SPE-51395-PA. DOI: 10.2118/51395-PA

[103] Rollins JB, McCain WD Jr, Creeger TJ. Estimation of solution GOR of black oils. Journal of Petroleum Technology. 1990;**42**(01):92-94

[104] Sattarin M, Modarresi H, Bayat M, Teymori M. New viscosity correlations for dead crude oils. Petroleum & Coal. 2007;**49**(2):33-39

[105] Spivey JP, Valkó PP, McCain WD. Applications of the coefficient of isothermal compressibility to various reservoir situations with new correlations for each situation. SPE Reservoir Evaluation & Engineering. 2007;**10**(01):43-49

[106] Standing MB. Volumetric and Phase Behavior of Oil Field Hydrocarbon Systems: PVT for Engineers. California Research Corporation; 1951

[107] Standing M. Volumetric and Phase Behavior of Oil Field Hidrocarbon Systems. Richardson, Texas: SPE; 1981

[108] Sim SSK. Pressure-volume-temperature correlations for crude oils from the Illinois Basin. In: Illinois Petroleum. University of Illinois Urbana-Champaign. Vol. 140; 1993

[109] Sulaimon A, Ramli N, Adeyemi B, Saaid I. New correlation for oil formation volume factor. In: SPE Nigeria Annual International Conference and Exhibition; Society of Petroleum Engineers. 2014

[110] Twu CH. Internally consistent correlation for predicting liquid viscosities of petroleum fractions. Industrial & Engineering Chemistry Process Design and Development. 1985; **24**(4):1287-1293

[111] Valko P, McCain W Jr. Reservoir oil bubblepoint pressures revisited; solution gas–oil ratios and surface gas specific gravities. Journal of Petroleum Science and Engineering. 2003;**37**(3–4): 153-169

[112] Vazquez M, Beggs H. Correlations for fluid physical property prediction. JPT. 1980;**32**(6):968-970. SPE-6719-PA. DOI: 10.2118/6719-PA

[113] Velarde J, Blasingame T, McCain W Jr. Correlation of black oil properties at pressures below bubble point pressure—A new approach. In: Annual Technical Meeting; Petroleum Society of Canada. 1997

Chapter 4

Pumping System of Heavy Oil Production

Tarek Ganat

Abstract

The geological features of the hydrocarbon reservoir and the inconsequential mobility of the heavy oil make the recovery process challenging. Recently, commercial interest has been developed in heavy oil production systems with the advent of improved drainage area by drilling multilateral and horizontal wells and highly deviated wells at shallow reservoirs. Moreover, other new recovery methods were developed such as downhole technologies that include cold or thermal production. Commonly, artificial lift techniques are utilized when the well cannot offload naturally at its economical rate. This is applicable for heavy oil reservoirs, where high viscosity along with the reservoir pressure drop will avoid the wells to produce naturally. Producing heavy oil together with associated water from the reservoir can create emulsions, which may cause high loads on artificial lift methods, along with high power consumption and requirements of expensive chemicals. The optimization and the selection of handling viscous oils; had a fundamental impact on the development of pimps. This chapter reviews the applications and types of pumping systems as an artificial lift in the heavy oil production process and reviews the pumping system performance, and its future development, as well as the expected technical challenges.

Keywords: artificial lift, pump systems, multiphase flow, heavy oil, oil recovery

1. Introduction

Globally, the heavy oil reserves have become more important as a future energy source. An excess of 50% of the world's hydrocarbons have an oil gravity of <20° API. These hydrocarbon assets are normally bitumen and heavy oil, where most of these deposits are located in China, Canada, Venezuela, and Russia. Typically, the heavy oil viscosity is varied within the range of 500 and 15,000 cP, and for bitumen, it's about 100,000 cP. Such crudes are usually found in shallow reservoirs (300–600 m depths), and normally the average flow rate of an individual well can be from 1 to 70 m^3/day. Therefore, to reach economic production rates, all oil wells need to be pumped at low bottom-hole pressures. Besides, hydrocarbons are typically produced from unconsolidated reservoirs which are susceptible to sand production that can exceed above 30% by volume.

Typically, 60% of producing oil wells need some additional lift systems to pump the reservoir oil. Conventionally, heavy oil wells are using beam pumping as an artificial lift system. However, beam pump is used for low flow rate wells; besides this pump has many operating problems. Alternatively, there are many types of pumps which are recently employed as a primary option in heavy oil wells, such as progressing cavity pump (PCP), jet pumps, and electric submersible pump (ESP). These wells are normally producing at low bottom-hole pressure, low gas-oil ratio, and low bubble-point pressure, high water-cut, or low °API oil gravity. This chapter reviews the application of different pumping systems as an artificial lift in the heavy oil production process. The main focus of the chapter is on types of pump and their applications and reviews the pumping system performance, and its future development, as well as the expected technical challenges.

2. Production of heavy oils

Heavy oil production is a developing skill for producing heavy oil in economical amounts. There are several ways to produce heavy oil and bring to the surface such as primary, secondary, and tertiary recovery. Since oil mobility is a function of effective permeability and oil viscosity, the efficiency of a well production is related to the delivery of reservoir zone thickness and mobility [2]. **Figure 1** displays the oil recovery mechanism as identified by Pinczewski [3] and Ershagi (1994) [c], and **Figure 2** shows the expected recovery factor from every method. The enhanced oil recovery (EOR) processes can be categorized into three main groups as stated below. The approaches have their own characteristics and mostly linked to the kind of oil remaining in the reservoir and reservoir characteristics.

1. Chemical

    a. Surfactant flooding

    b. Micellar polymer flooding

    c. Polymer flooding

    d. Alkaline atau caustic flooding

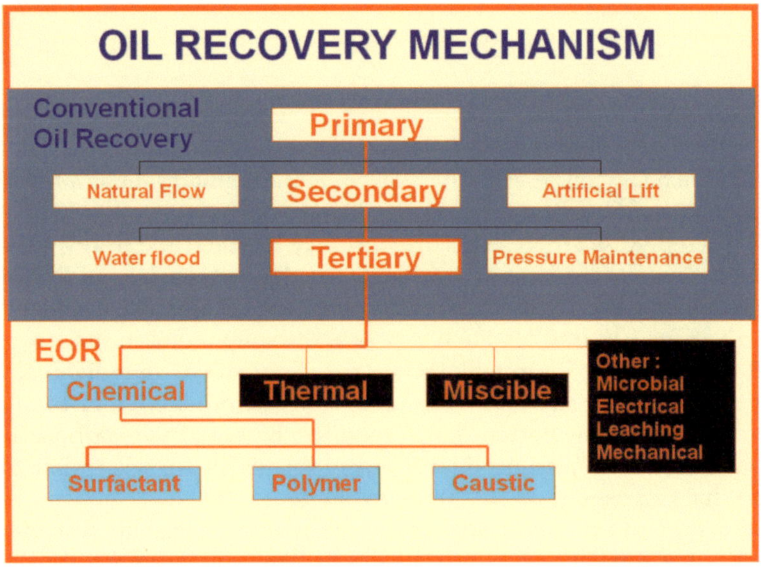

Figure 1.
Oil recovery mechanism (source: Pinczewski (1993)) [3].

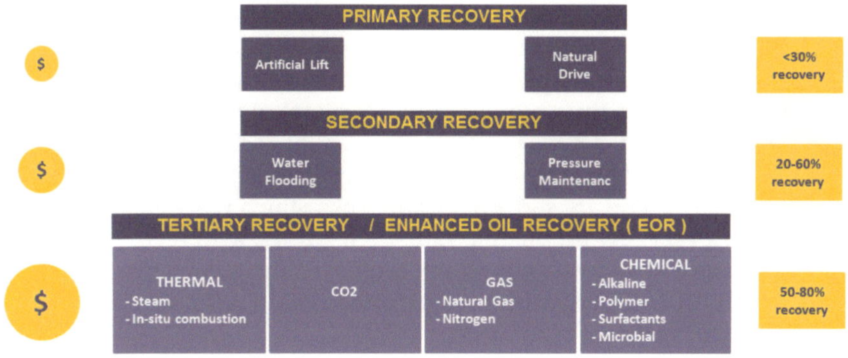

Figure 2.
Oil recovery from each stage (source: DALEEL) [4].

2. Thermal

    a. Steam flooding

    b. Fire flooding

3. Miscible

    a. Carbon dioxide flooding

    b. Nitrogen and flue gas flooding

    c. Enriched hydrocarbon gas flooding

Things that are essential to be considered in the EOR method

1. Physical properties

2. Reservoir type

3. Structure and physical properties of porous media

4. Fluid condition in porous media

5. Mobilization of oil remaining

6. Adsorption process

2.1 Primary recovery

2.1.1 Natural flow

The first oil extraction process from the reservoir rock is known as primary recovery (utilizing the natural energy). It's the first production phase of hydrocarbons, where the well depends on the natural flow of the oil because of pressure differences between reservoir pressure and the well bottom-hole pressure. Besides, using pumping lift systems such as an electrical submersible pump is also known as a primary recovery technique. These approaches are normally named as natural drive mechanism. The recovery factor from conventional oil production is above 30%, but for heavy oil, it is within 5–10% [5]. When natural lift pressure is not enough to move the oil to the wellbore or to lift the oil to the surface, once the reservoir depleted, then the primary recovery stage has reached its maximum extraction limit. Normally, heavy oils cannot be produced via natural flow from the reservoir to surface. There are some heavy oil wells which that can be produced naturally, but at a very low production rate (±20 stb/d), with recovery factor within the low range of 6–9% of the oil in place; specific to densities ranging between 9 to 20° API, and viscosities range from 1000-13,000 cP or more, together with low reservoir pressures.

2.2 Secondary recovery

Once primary recovery methods seize to produce the oil naturally, secondary recovery methods will kick off which is the next phase of producing the oil from the reservoir and to bring the oil to the surface. Basically, these approaches include injecting additional energy sources (supplementary energy) into the reservoir to maintain and increase the reservoir pressure. These artificial approaches contain natural gas reinjection, water injection, and CO_2 injection as show in **Figures 3** and **4**. With time the artificial pressure loses efficiency as the residual heavy oil is extremely viscous to flow and is detained by sandstone in the reservoirs [6]. The total recovery factor of the heavy oil including the primary recovery approaches will be within the range of 10–25% [7].

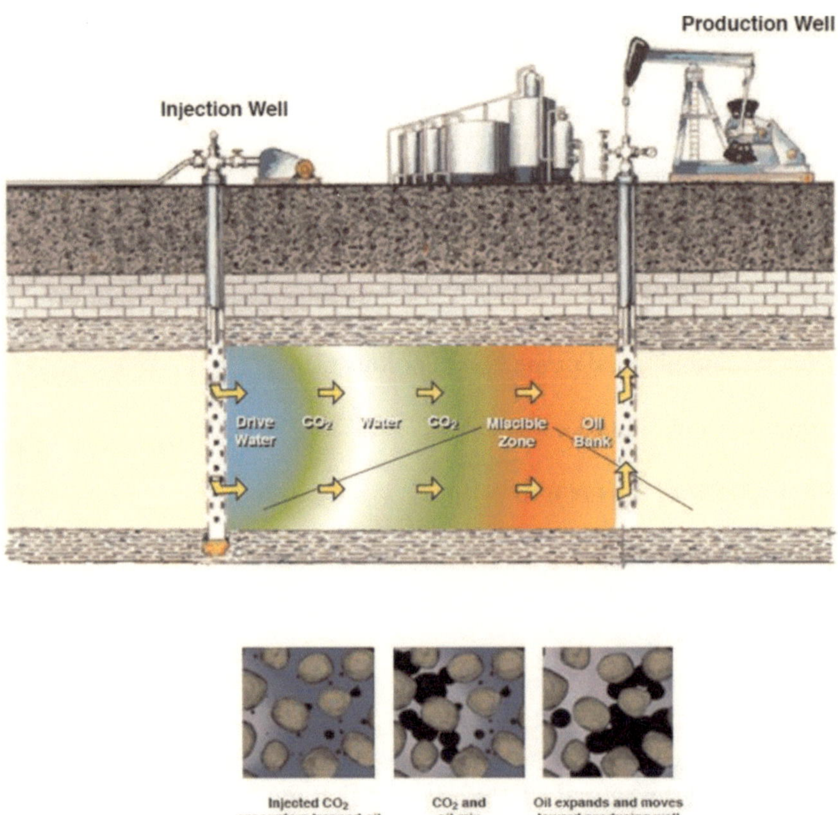

Figure 3.
Enhanced oil recovery using CO_2 injection (source: NETL (2010)) [8].

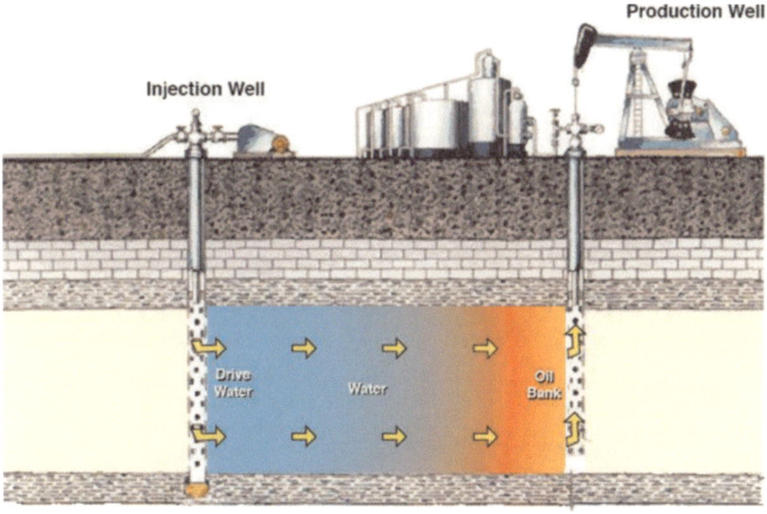

Figure 4.
Illustrating waterflooding technique of secondary recovery (source: NETL (2010)) [8].

2.3 Tertiary recovery

Tertiary recovery is generally denoted as enhanced oil recovery. It is an approach of extracting oil through thermal and nonthermal processes after most of the oil has been extracted by primary and secondary recovery methods [9]. Mainly, EOR is used to extract the heavy oil trapped in porous media of reservoir rock which is too viscous to flow. The most common approaches for tertiary recovery are thermal, chemical, and miscible enhanced recovery.

For nonthermal approaches, chemicals and microbes are used to release trapped heavy oil and carbon dioxide under pressure. However, thermal approaches are generally steam injection which is the most effective means of decreasing viscosity and mobilizing heavy oil [7].

2.4 Thermal methods of recovery steam-based processes

2.4.1 Steam injection

Steam injection is commonly used for high viscous oil. The main objective of the steam injection is to heat up and force the oil to the wellbore by the pressurized steam depicted in **Figure 5**. Generally, the EOR methods are costly because of the required external energy resources and materials. Consequently, the volume of heavy oil to be extracted from a reservoir rock is a function of economics [10]. As a result of this, engineers must start to study in more details the reservoir rock permeability, pore media, and oil viscosity, together with the reservoir heterogeneity, where all these issues affect the success of any recovery technique. Overall steam injection efficacy is the product of the sweeping capability and displacement competence.

2.4.2 Cyclic steam stimulation

Cyclic steam stimulation (CSS) is one of the main EOR approaches for heavy oil production. The notion of the CSS is that the steam is injected into the reservoir via a production well for a period of time. Then the well is closed and permitted to soak by steam for some period of time before it returns to production. CSS was applied in

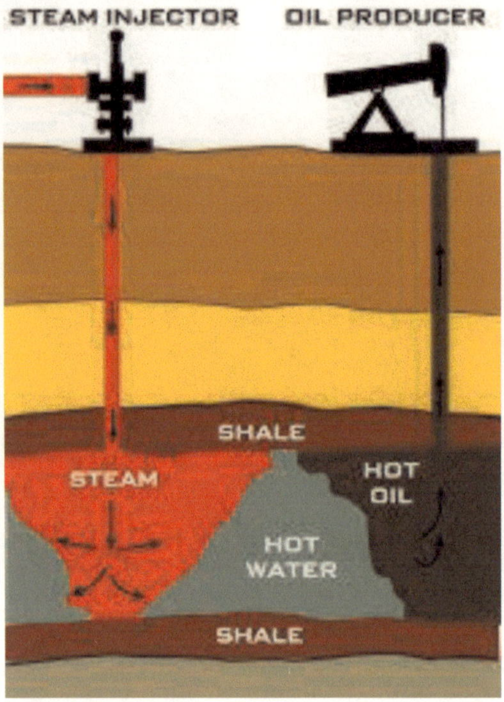

Figure 5.
Illustrating the thermal flooding technique (source: Steam EOR 1.Jpg (2008)) [8].

many heavy oil wells in the Middle East. Various cycles were done in these wells. However, the total amount of produced water for each cycle was considerably less than estimated.

2.4.3 Steam-assisted gravity drainage

Steam-assisted gravity drainage (SAGD) comprises pairs of a high-angle injection wells with an adjacent production well drilled along a parallel trajectory depicted in **Figure 6**. Normally, steam is injected via the upper well. Once the steam rises and spreads, it will heat up the heavy oil trapped in the porous media, decreasing its viscosity. Then assisted by the gravity forces, the oil will be drained into the lower well where it is produced [11]. Generally, the steam injection involves two core approaches, cyclic steam injection, and steam flooding.

2.4.4 Cyclic steam injection

The fundamental idea of the cyclic steam injection (CSC) is to inject hot steam through a single well for a period of time. A CSC method contains three phases (see **Figure 7**). The first phase is injection, through which a slug of steam is injected into the reservoir. The second phase requires that the well is shut for some days to permit equal heat spreading to thin the oil. Finally, throughout the third phase, the thinned oil is produced over the same well. The same cycle process is repeated many times as far as oil production is still profitable. Commonly, the cyclic steam injection process is used widely in heavy oil reservoirs and tar sands and also can be used to enhance injectivity prior to steam flood process and in combustion processes. Steam injection

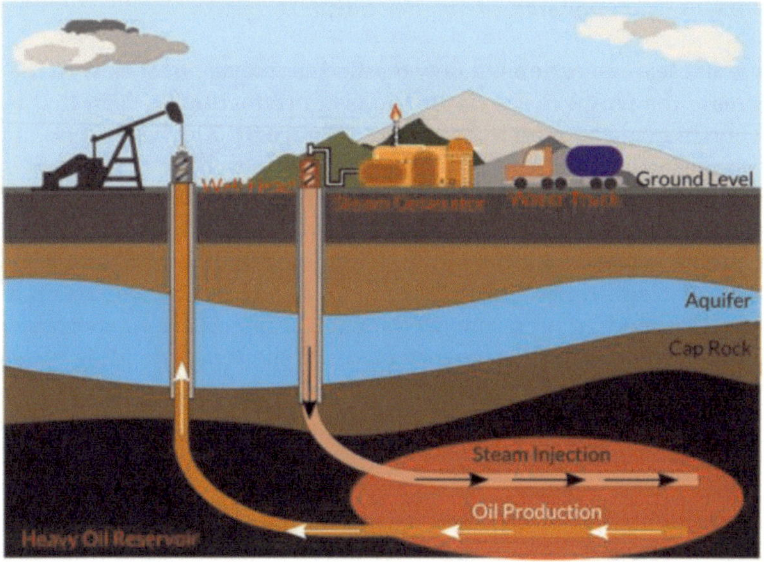

Figure 6.
Illustrating steam-assisted gravity drainage (source: Markham Hislop (2017)) [12].

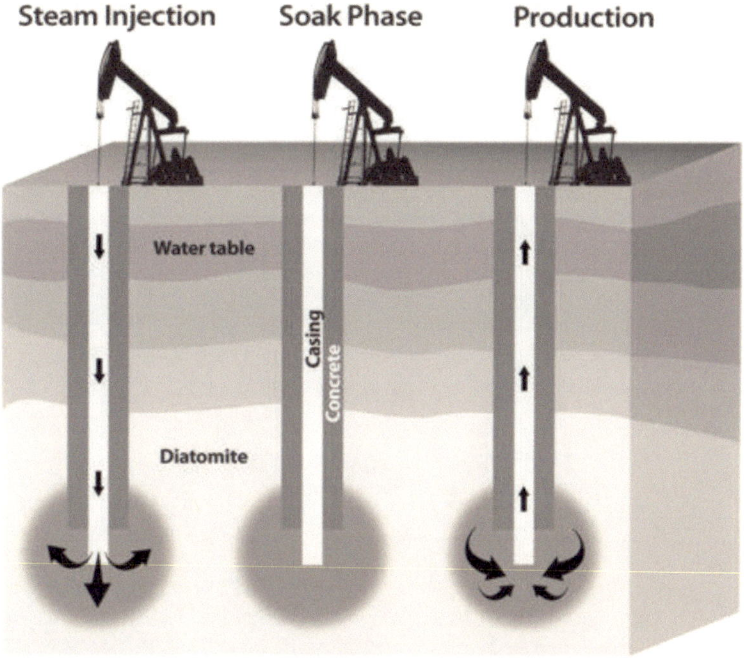

Figure 7.
Cyclic steaming processes. First, steam is injected at high pressure; second, the steam thins the viscous oil; and third, the oil is produced up to the surface (source: Lyz Hoffman (2014)) [13].

is typically used in both horizontal and vertical wells producing high viscosity as high as 100,000 cP. Normally in CSC wells, oil can be both viscous and solid. Usually, CSC is termed "steam soak" or the "huff 'n puff" (slang) technique. Normally, CSC recovery factor is within the range of 10–20% of the initial oil in place [13].

2.4.5 Continuous steam injection (steam flooding)

Continuous steam injection is a new method to manage heat flow in the reservoir and to decrease the temperature of the fluids at production wells in that way avoiding shut-in made by high temperature (**Figure 8**). The best injection plan with different pressure and fixed steam quality has shown to have the biggest oil recovery given the same quantity of energy injected under various reservoir sets. Steam is injected at high temperature and high pressure via an injector well. The oil recovery of this technique is more than the cyclic steam injection method, and it's more practical and efficient. It has lower thermal efficacy than CSC and needs a larger surface area. This method needs one well for steam injection and another well for oil production [15]. Typically, the recovery factor of the steam flooding method is about 50% of the initial oil in place [9].

2.4.6 Combustion processes

The combustion process, also referred as fire flooding, challenges the recovery of more heavy oil by a flare-up of a part of the oil in place by injecting either oxygen, air, or chemical or by electrical shock (see **Figure 9**). This decreases the heavy oil viscosity and heats the oil in place, and the oil is moved out by a combination of steam, hot water, and gas drive. This method appeared always to be very suitable, mainly when we need to recover bitumen from tar sand deposits. Normally, the temperature of the combustion process can reach up to 700°C. This temperature can be observed at the combustion front [17–23]. Typically, the combustion technique is applied to hydrocarbon reservoirs of low API gravity oil. By using the combustion method, the heavy oil gravity can be upgraded from 2 to 6° API (Ramey et al., 1992). Forward combustion includes a drive of the burning front in a similar direction as the injected air; where the reverse combustion includes a drive of the burning front opposite to the direction of the injected air.

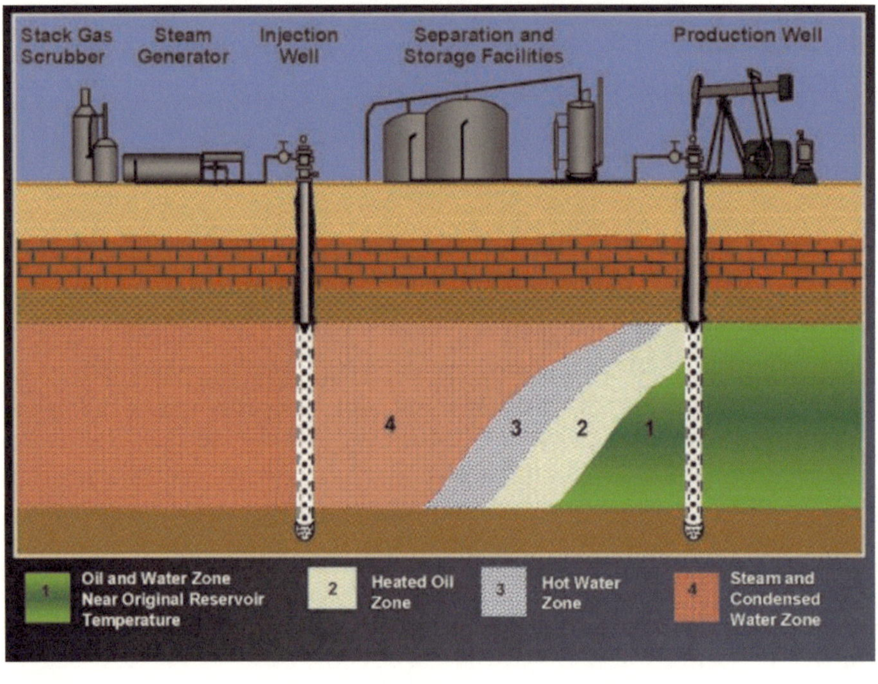

Figure 8.
Diagram shows steam flooding (source: Alhakiki (2012)) [16].

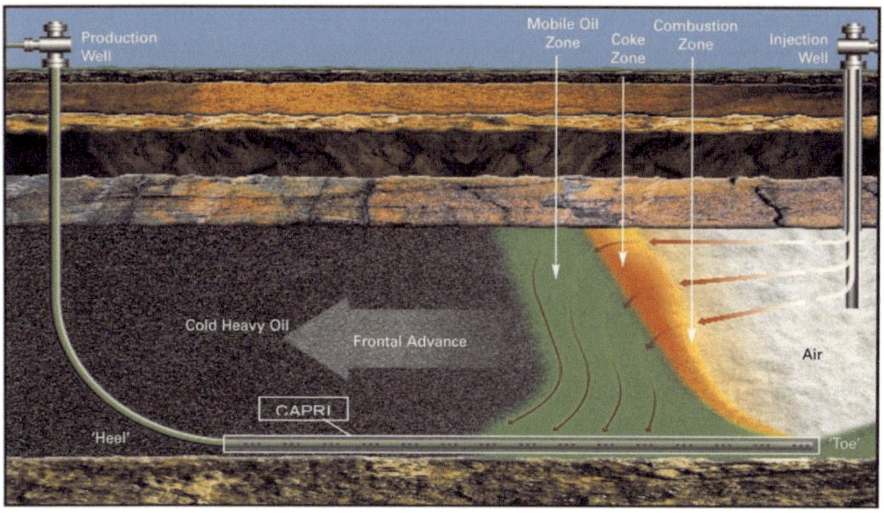

Figure 9.
Schematic of combustion processes (source: Rob Kendall (2009)) [23].

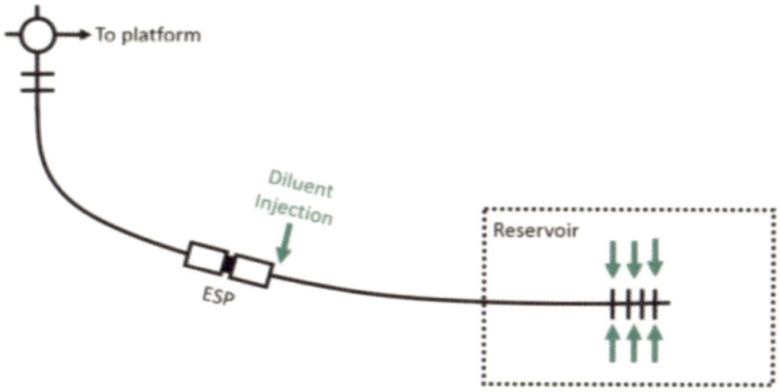

Figure 10.
Schematic of downhole diluent injection method (source: Arnaud Hoffmann (2016)) [25].

2.5 Downhole diluent injections for a heavy oil field

Many previous studies proved the diluent injection with ESP can be an efficient artificial lift method for heavy oil reservoirs. It consists of injecting a light hydrocarbon liquid to reduce the oil density and viscosity depicted in **Figure 10**. This recovery method makes use of several kinds of diluents injected downhole to decrease heavy oil viscosity in the reservoir. Normally the diluents used such as kerosene, naphtha, and light oil or may be injected into the drain and blended with heavy oil. Numerous field experiences show that the use of diluents allows the viscosity of the heavy oil to decrease from 100,000 to <1000 cP. The diluted fluid is then returned back to the surface by a downhole pump such as progressive cavity pumps (PCP). There are many diluent injection operations that extracted heavy oil of 7.5–9° API. The API quality of heavy oil has been upgraded to high commercial quality (32° API). For instance, a case study for extra heavy oil is produced from the 1300 meter horizontal section through downhole ESP pumps. The diluent fluid used has 46° API gravity of naphtha. In the diluent fluid injected at the bottom of the slotted liner, the diluent fluid transfers gradually in the horizontal part of the drain through the influence of the pressure differential created by pump process. The heavy oil gradually moves from the reservoir to the liner and reaches the pump intake section with decreased heavy oil viscosity that is suitable for the pump efficiency. The heavy oil viscosity at reservoir conditions was 10,000 cP at 50°C, but once mixed with 20% naphtha, the viscosity value decreased down to 200 cP. The mixed pumped fluid viscosity can be simply attuned by the rate of diluent injected downhole through the injection line.

3. Artificial lift systems

Well artificial lift plan is a strategic aspect in the production of heavy oils. Obviously, some types of artificial lift was required in order for the oil to flow and return the flow rate of the oil to their normal rates to maximize the ultimate oil recovery. The choice of which artificial lift technique is to be used is very significant for the long-term profitability of the oil field. An inappropriate selection of artificial lift can decrease production and raise the operating cost significantly. After a decision has been taken, it can be hardly changed whether or not the technique selected was suitable for the existing conditions.

The selection procedure of the lifting method to be used, which are confined to the operating life of surface and downhole equipment, maintenance, environmental concerns, and cost. Therefore, there are several configurations of downhole oil pump systems including pumps and drivers as described below.

3.1 Hydraulic pumping systems

Hydraulic pumping is one of the artificial lift methods used since the early 1930s. Hydraulic pumping systems can be used at different oil well production conditions (**Figure 11**). This type of pump was installed at different setting depths ranging from 400 to 20,000 ft. with varying production flow rates from 80 to more than 20,000 STB/D. The pump has surface speed drive box ranging from 15 to 625 hp. which makes the downhole pumping rate to be controlled on the surface.

The basic operating process of the pump is that the power fluid is pumped from the surface facilities to activate the downhole piston together with a reciprocating piston pump (refer to **Figure 12**). The power fluid acts on a piston like a steam engine, and the power fluid could be oil or water. The power fluid transfers to the piston and returns back to surface over another pipe if a closed loop power fluid is used. In the open power fluid design, the power fluid is combined with the production once flowing to the surface. To control the corrosion, chemicals can be injected downhole along with the power fluid. The advantage of using this pump is the power fluid which can be heated for handling heavy oil. The pump is appropriate for deviated wells which might be difficult for other artificial lift methods. The pump surface facilities have a small footprint and can be assembled into one main battery to service many wells. Commonly, hydraulic pumps are applied primarily in very deep oil wells that are producing at great volumes which cannot be handled by using beam pump systems.

3.1.1 Types of hydraulic pump systems

Generally, it can be used in low API oil gravity wells and in wells with high paraffin contents. Also used in the wells that failed to use any other artificial lift techniques or, because of well conditions. It's used in deep and deviated wells including sandy and corrosive wells. Hydraulic pumping systems are quite

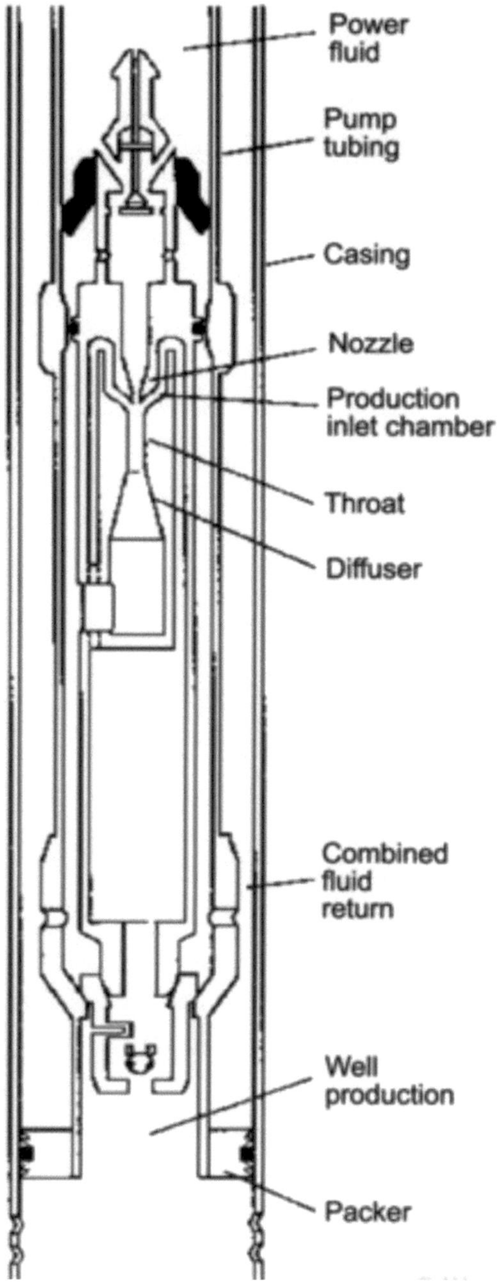

Figure 11.
A schematic of the downhole assembly piston pump (source: Cholet H, 2004) [26].

expensive, but they may have a good application where other artificial lift techniques may not be possible.

3.1.1.1 Piston pump

A piston pump includes a motor at one end and a plunger pump at the other end (**Figure 13**). Hydraulic fluid is forced down the completion string at very high pressure and goes in a reciprocating motor. The motor mechanism is piston-like pump which is forces the produced hydrocarbons to the surface throughout the

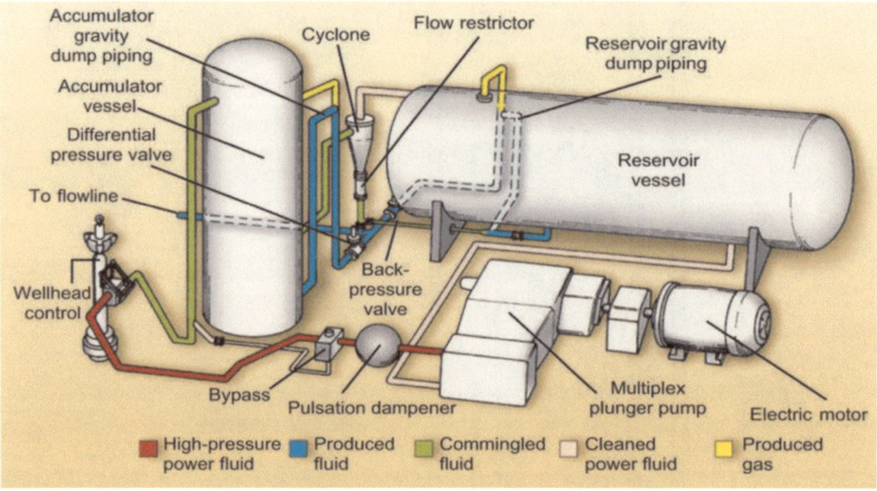

Figure 12.
Diagram of hydraulic pumping surface facility system (source: SPE) [27].

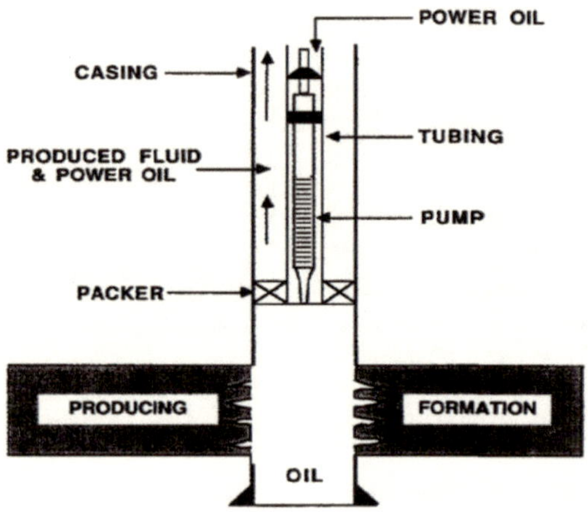

Figure 13.
Hydraulic piston pumps (source: Conoco Inc., 1990 [29]).

casing annulus. Normally, the hydraulic fluid is used as a power fluid. Once both hydraulic fluid and produced fluids reach the surface, the hydraulic fluid is separated and reused again as the power fluid.

3.1.1.2 Jet pump

Jet pumps have been applied in the oil industry for more than 75 years. A jet pump is one of artificial lift methods, and it can be applied when depth and deviation of producing wells increase and reservoir pressure depleted (**Figure 14**). A jet pump is used in thermal production as it's not directly affected by the high temperature of the fluids. This pump can considerably reduce the risk of equipment failure in the wellbores because it has no moving parts. These pumps are in the family of thermo-compressors, and they are categorized as "eductors" as they are considered for a liquid to pump a liquid (incompressible liquid). The pump is set downhole and pulled up the well for retrieval by using pressurized fluid.

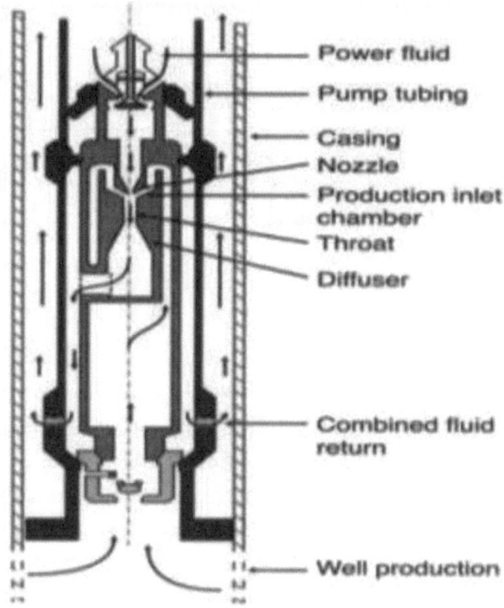

Figure 14.
Typical jet pump system (source: Cholet H, 2004) [26].

Normally, this process is using a pump at the surface to generate high-pressure power fluid to be transmitted downhole via tubing or through an annular casing. The power fluid moves into the pump and goes via a nozzle, where almost the whole pressure of the power fluid is transformed into a velocity head if no loss happens. The velocity of the power fluid can reach up to 70 m/s from the nozzle outlet section into the production inlet chamber linked to the pump intake section. Both production fluid and power fluid returned back to the surface over the production pipes. Comparing with other pump systems, hydraulic efficacies of jet pumps are lower in the range of 20–30%. A jet pump can produce high rates and can handle free gas as well, though it's not as efficient as a positive displacement pump, therefore needing higher horsepower requirements at the surface [28].

A jet pump is very suitable in specific conditions, for instance, when high production rates are preferred. The pump is suitable for handling viscous, corrosive, and heavy crude oils. Besides, locations where beam pumping units cannot be installed, such as inhabited regions, offshore oil fields, and gas lift, are not accessible. The pump is attractive in horizontal, deviated, deep, and high-temperature wells. Also, it has excellent solid-handling capabilities and has long operational life (average of 4 years). Moreover, it has tolerance for gas and solids production, and it has low installation and workover costs.

3.2 Beam pump systems

Beam pump is an artificial lift pumping system applying power source at a surface to transmit the energy to a downhole pump assembly. Producing heavy oil from shallow reservoirs using beam pump systems needs accurate design mainly for downhole assembly to get maximum production performance and maximize the run life.

A beam and crank assembly creates reciprocating motion in a sucker rod string that attaches to the downhole pump assembly. The pump comprises a plunger and valve assembly to transform the reciprocating motion to vertical fluid movement. **Figure 15** shows typical beam pumps.

3.2.1 Classification

The hollow sucker rod electric heating device can be divided into a pumping unit device, subsurface pump and screw pump hollow sucker rod device. **Tables 1–3** show the sucker rod electric technical data.

3.3 Progressive cavity pump systems

PCP was developed in 1930. Currently, this pump is used to produce heavy oil in any kind of wells: vertical, deviated, or horizontal. A PCP is essentially made of two helical gears, one inside the other (**Figure 16**). **Table 4** displays the lift selection guidelines. The suitable use of each lift type is reliant to the type of the reservoir fluid, reservoir pressure, and production rate as estimated by inflow and outflow system.

The metallic rotor is a single helical "rotating" inside the stator based on a double helical elastomer-lined nitrile in most cases. The external gear or stator has a double helical shape, one more than the internal single helical gear rotor. When the rotor is rotating, the fluid transfers together with the pump axis inside the cavities

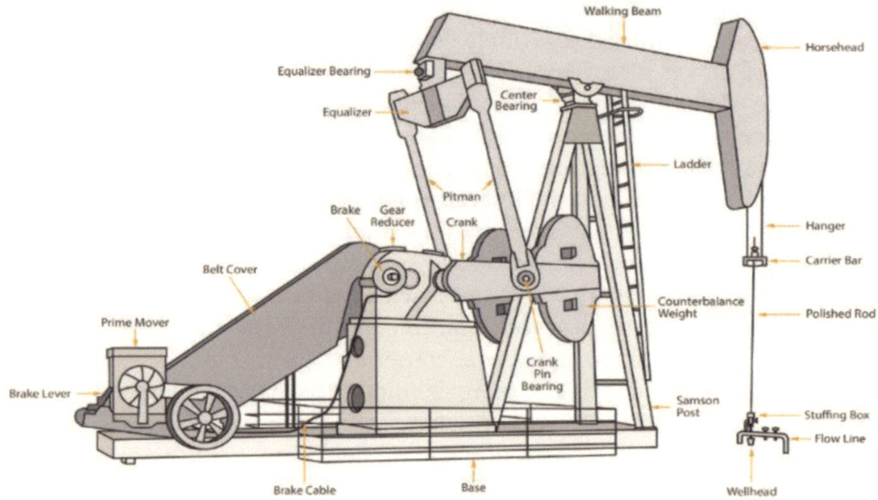

Figure 15.
Typical beam pumps (source Conoco, 1990) [29].

Rod outer diameter	34 mm	36 mm	42 mm
Rod thickness	5.0 and 5.5 mm	5.5, 6.0 and 6.5 mm	6.0 mm
Coupling outer diameter	50 mm	50 mm	60 mm
Thread size	1–9/16"	1–9/16"	1–7/8"
Grade	D		

Table 1.
Hollow sucker rod.

Heating core sectional area mm^2	3×8.4 mm^2
Outer diameter of cable	18–20 mm
Outer protect layer	Stainless steel wire armor weave
Withstand voltage	≥2500 V
Length	Underground length + ground length
Temperature-resist grade	Long-time working temperature: 200°C (C grade)
Electrical performance	Insulation resistance >50 MΩ, AC withstand voltage test 2500 V
Tensile strength	≥50 Mpa

Table 2.
Heating cable.

Rated power KVA	35, 50, 75, 100, 135
Input voltage	Three phase 380 V ± 5% 50HZ three phase 440 V ± 5%, 60 HZ
Output voltage	Single phase 160–900 is available
Power factor	cos θ ≥ 0.95
Three phase imbalance percent	≤10%

Table 3.
Electric control cabinet.

present between the rotor and stator. The flow rate is a function of many parameters, for example, pumps eccentricity, rotor diameter, length of the stator pitch, and rotation speed. Manufacturers can provide a catalog of pumps with a wide range of well conditions as a function of reservoir fluid types, flow rates, pressure heads, and for any type of fluid viscosity.

PCP is normally driven at the surface, but it can be driven by a downhole electrical submersible motor. When PCPs are driven from the surface (refer to **Figure 17**), the stator is screwed at the tubing extremity, and the rotor is fixed to the drive string of sucker rods. On the surface, the drive head, absorbing the force of the sucker rods, is operated by an electric motor and a speed reducer. Most of the downhole assembly is driven in this way.

The PCP pumps can operate at high efficiency for high viscosity cruds, high sand, low productivity wells, and in horizontal and deviated wells. Besides, it has a small footprint on the surface. Besides, the pump has some disadvantages, where the pump has restricted production rate, lift depth, and temperature tolerance. In the case of horizontal and deviated well profile, where malfunction can cause tubing leaks made by wear or failure of the sucker rod drive shaft, the PCP is not allowed to pump dry, and the completion string must be pulled out of the well to change the pump.

3.4 Electrical submersible pump

An ESP is a centrifugal pump driven by a downhole electrical motor (see **Figure 18**). ESP surface facilities contain power system and transformers and connectors to the wellhead. Normally, ESP pumps are installed for high flow rates (from 150 to 150,000 bopd) dependent on size and pressure gain and variable speed controllers. These pumps are not suitable for very viscous untreated oil, but ESPs can be applied to lift oil production after injection of diluents fluids (reduced the

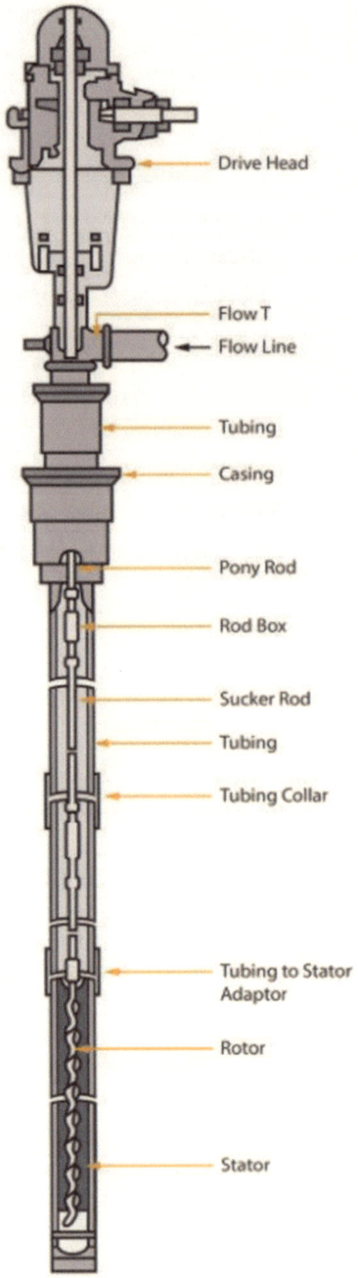

Figure 16.
Progressive cavity pump (source: John Martinez (2017)) [28].

viscosity). An example of the electric submersible pump/motor assembly is given in **Figure 18**. The efficiency of the pump is very dependent on the design of the flow rate which must carefully be optimized based on the reservoir deliverability. Moreover, the wellhead of the ESP system needs to have an electric cable entering. The downhole electrical submersible assembly contains an electrical power cable, motor, motor protector, and centrifugal pump.

Thermal recovery offers several limitations for ESP pumps because of the high temperature applied to the electric motor, electric parts, power cable, and pump assemblies. ESP manufacturers present different designs for high-temperature

	Rod lift	Progressing cavity	Gas lift	Plunger lift	Hydraulic piston	Hydraulic jet	Electric submersible
Operating depth (ft)	To 16,000 TVD	To 6000 TVD	To 15,000 TVD	To 19,000 TVD	To 17,000 TVD	To 15,000 TVD	To 15,000 TVD
Operating volume	To 5000 BPD	To 4500 BPD	To 30,000 BPD	To 50 BPD	50–4000 BPD	300–>15,000 BPD	200–30,000 BPD
Operating temperature	100/500°F	75/250°F	100/400°F	120/500°F	100/500°F	100/500°F	100/400°F
Corrosion handling	Good to excellent	Fair	Good to excellent	Excellent	Good	Excellent	Good
Gas handling	Fair to good	Fair	Excellent	Excellent	Fair	Good	Poor to fair
Solid handling	Fair to good	Excellent	Good	Poor to fair	Poor	Good	Poor to fair
Fluid gravity	>8° API	<35° API	>35° API	GLR required 300scf/bbl./1000′ depth	>8° API	>8° API	>10° API
Servicing	Workover or pulling rig	Workover or pulling rig	Wireline or workover	Wellhead catcher or wireline	Hydraulic or wireline	Hydraulic or wireline	Workover or pulling rig
Prime mover	Gas engine or electric	Gas engine or electric	Compressor	Wells' natural energy	Gas engine or electric	Gas engine or electric	Electric motor
Offshore application	Limited	Good	Excellent	N/A	Good	Excellent	Excellent
Overall system efficiency	40–60%	40–70%	10–30%	N/A	45–55%	10–30%	35–60%

Table 4.
Lift selection guidelines (John Martinez (2017)) [28].

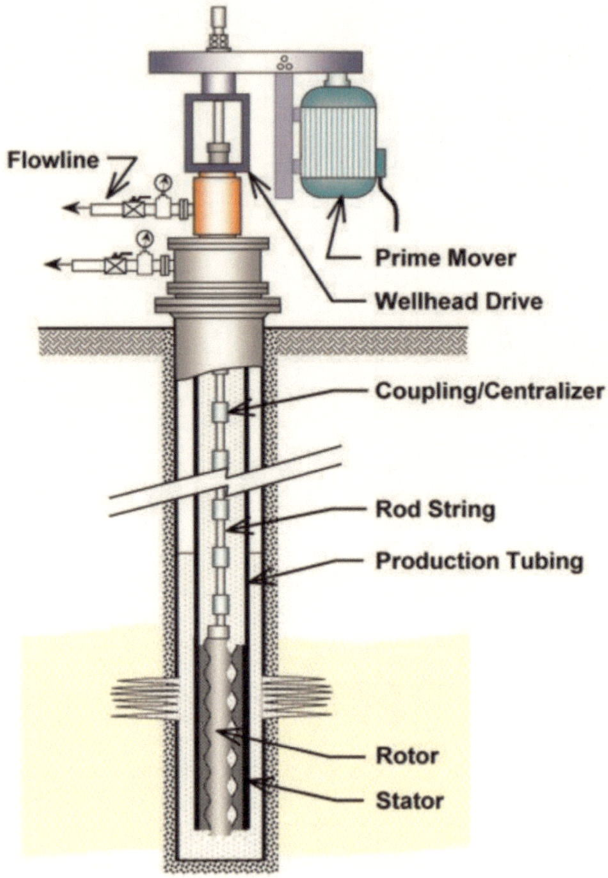

Figure 17.
Typical progressing cavity pumping system (source: SPE) [27].

reservoir fluids, for example, "hotline production" equipment with operating temperature ratings up to 550°F for the power cable and motor.

3.5 Gas lift systems

Gas lift is a type of artificial lift techniques used to lower the producing bottom-hole pressure to achieve a higher oil production rate. The principle of gas lift method is that gas injected into the tubing string decreases the density of the fluids in the pipe and lets the two-phase mixture to flow up to the surface. There are two main kinds of gas lift techniques being applied today which are continuous and intermittent flow. Typically, natural gas is continuously injector under high pressure through tubing or through the annular between casing and production pipe into the pocket mandrels along the production tubing. At high-temperature the multiphase flow will be produced at the surface (refer to **Figure 19**). Normally, gas compressor pressure and rate parameters are modified based on the gas lift constraint depicted in **Figure 20**. Gas lift is commonly used with SAGD heavy oil production in Canada.

3.5.1 Applications of gas lift

Gas lift is mainly appropriate for lifting fluids in wells that have a low amount of gas produced with the oil. Gas compressors are almost mounted to collect the

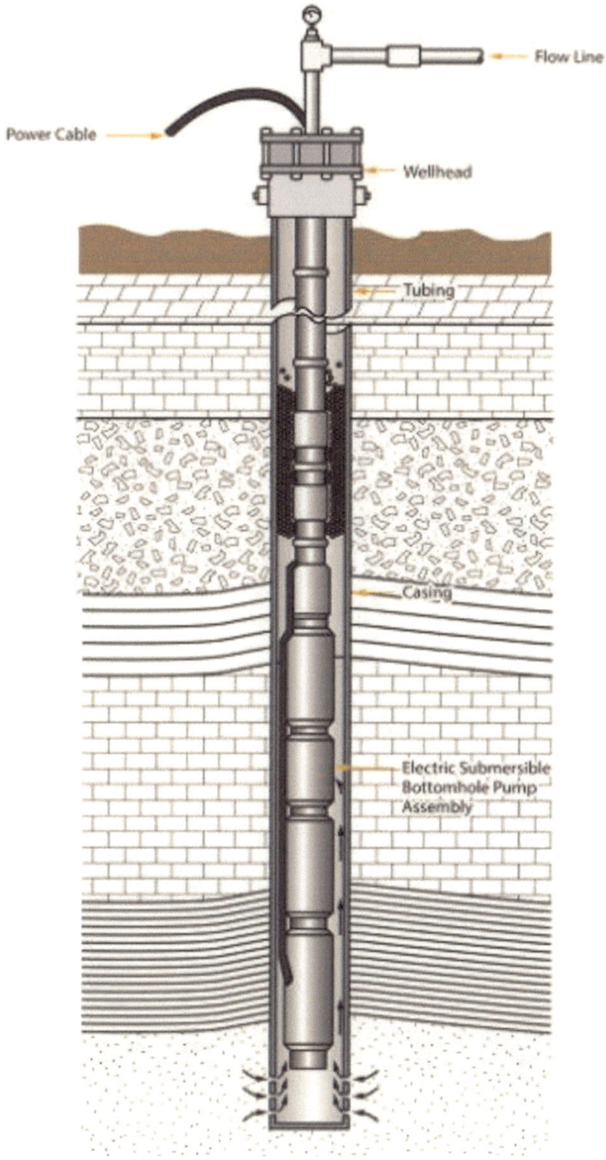

Figure 18.
Typical electric submersible pump (source: John Martinez (2017)) [28].

produced gas and designed to be used for the gas lift system. The flexibility of gas lift, in terms of production rates and depth of lift, can seldom be matched by other methods of artificial lift if adequate injection gas pressure and volume are available. Gas lift is very suitable for highly deviated wells which produce sand and high gas-liquid ratios. There is no other method that suitable for through-flowline ocean-floor completions as a gas lift system. Besides, wireline-retrievable gas lift valves can be replaced without killing a well or pulling the tubing. Individual well downhole tools are low-cost. The surface gas lift facilities for injection gas control are simple and need low maintenance and nearly no space for installation. Usually, the reported high overall reliability and lower effective costs for a gas lift method are more to other techniques of lift.

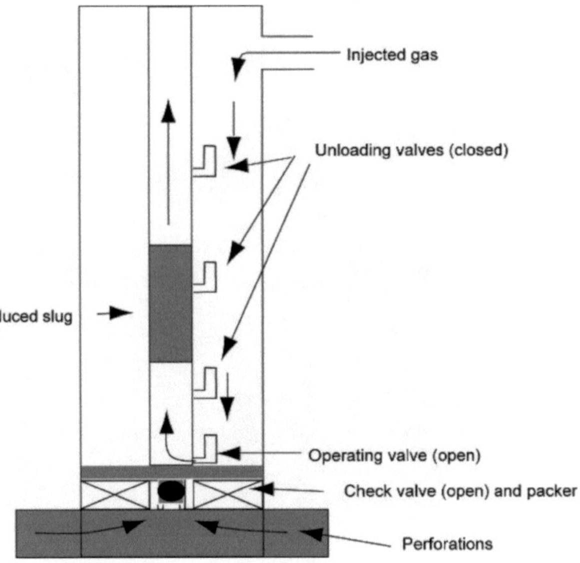

Figure 19.
Diagram of injection gas cycle for gas lifting well (courtesy of Schlumberger).

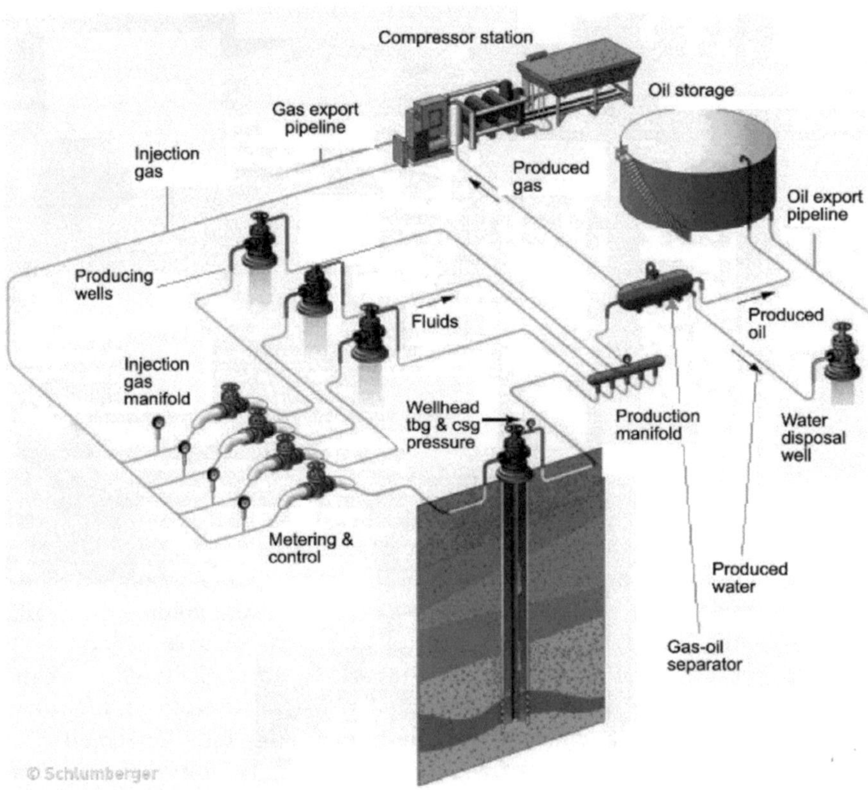

Figure 20.
Diagram of a gas lift system (courtesy of Schlumberger).

4. Future development of pumping system performance

Pump monitoring is a crucial factor in prolonging the working life of all kinds of downhole pump systems. Currently, it is feasible to measure the downhole and surface pressure and temperatures using sensors that are connected to data controllers to decrease the risk of "pump off": a lack of fluid to lubricate the pump, allowing heat to build up which would damage the elastomer stator of any type of pump. Advanced controller alarms can set parameters to reduce risk, extend pump working life and improve the total oil produced.

The pumping system needs more development to handle the operating cost, mainly for operation and maintenance either for single equipment or the whole system. Regularly, oil producers are just focused on the instant demands of the equipment, and they do not realize how the system parameters are affecting this equipment. A system method studies for both supply and demand sides of the system and how they can act together, shifting the attention from a single component to the total pumping system performance. Normally, most types of pump failure are leakage, fouling, valve failure, and cracks in pipe chains. Assessing pumping system performance is required to solving such pump failure and finding enhancement opportunities. In endeavoring to find out solutions or search for techniques to develop pump performance, assessing only the components instead of the entire pumping system can cause analysts to manage potential cost savings. For instance, although a pump may be functioning efficiently, it could be producing more flow than the system needs. Therefore, it is essential to evaluate system efficacy based on how the end uses are worked by the pumps.

In the future, pumps need to be more reliable and proficient in functioning for a long time before requiring maintenance. Pumps must be safer to work, use less space, use less power, less noise and temperature. Additionally, the need to improve corrosion resistance and as well as the reliability of working in the high-temperature environment. Consequently, pumps will be friendlier to the environment and running with less power to decrease their carbon footprint. As well, the use of more recycled materials with fewer consumables, in that way helping to decrease whole pumps costs. Pumps are required to be easier to clean, overhaul, and reconstructed. Generally, there are several chances to increase the reliability, performance, and efficacy of pumping systems in the oil industry. Definitely, the next generation of important savings for pump operators belongs to a broader pumping system optimization method. This certainly needs a middle way, taking the proper mix of the best suitable available technology combinations for certain applications.

5. Technical challenges

Engineers and facilities and pump designers will encounter enormous challenges in developing heavy oil reservoirs, such as crude oil properties and composition, flow assurance, lifting process, and operations.

5.1 Fluid properties and composition

Proper experimental methods are required to properly characterize heavy oil emulsions at conditions that will come across in the actual production process. Correctly describing the apparent viscosity of an emulsion phase is an even bigger challenge. The viscosity of the crude could be very important description tool than the API gravity. Transport of high viscosity crudes can be a major flow assurance challenge for future developments. Skills at many fields in the design phase are important trends in heavy crude properties and description needs. Therefore, high oil viscosity, low reservoir energy, and cold ambient temperatures make recovery and transport of heavy oils a challenge. Heavy crudes have a lower market value due to low oil gravity, high sulfur content, and higher TAN numbers. Crude oil upgrading processes have CAPEX and OPEX intensives. With the vapor extraction process, a vaporized hydrocarbon solvent is injected into the reservoir to dilute the heavy crude and extract the lighter components while leaving the heavier ends behind. This technology essentially performs in situ upgrading of the heavy crude. The possibility of organizing any upgrading process will be technically and economically challenging.

Typically, some of the reservoirs are producing heavy oil and water. This can create emulsions, which generate high loads on lifting systems with more chemical and energy consumption. The capability to drive sand together with high viscosity fluids has made the pump systems the best alternative option for managing heavy oil production. Many challenges will meet pump designers to develop heavy oil reservoirs that have high viscosity and low initial pressure and temperature. Besides, heavy oil producers may face possible gas and water inflow on low-producing mature fields.

5.2 Reservoir characterization

The main challenge related to the characterization of deepwater and heavy oil reservoirs is that this must be accomplished with quite fewer reservoir penetrations. This means having to address and manage more uncertainties and risks with less information. This is a function of the greater drilling costs in deep water and the more marginal economics of these types of developments. Fewer well penetrations mean fewer database available such as logs, cores, tests, and fluid samples that are very significant in characterizing, measuring, and managing heavy oil reservoir uncertainty and risk. Consequently, deepwater heavy oil reservoirs are integrally characterized not only by greater unit development and production costs and lower product cost but also by greater reservoir uncertainty and well performance risk.

5.3 Impact of heavy oil on flow assurance mitigation strategies

The reservoir describes the main flow assurance challenges based upon reservoir fluid properties, phase behavior, composition, and initial reservoir conditions. Other ecological elements such as water depth, offset distance, ambient conditions, and development model also influence the approaches and processes employed to control flow assurance risks. Hydrate formation is a probability in essentially all offshore production systems if water is existing and ambient temperatures are cool. Reservoir fluid composition impacts the potential for wax and asphaltene problems. For steady-state conditions, the heavy oil viscosities should be controllable with proper protection. Appropriate modeling of transient operations could pose a bigger challenge.

5.4 Operations

There are several operational concerns that are essential to be considered when designing lift systems for viscous crudes. These contain start-up and gravity segregation. Starting up a system full of viscous heavy crude may be very challenging, if not impossible, for most lift systems. This is particularly true in cold or deep water. For ESPs and PCP pump systems, this has to be considered during the design phase to certify the required horsepower is installed for start-up. Equipment failure can simply happen when a downhole pump is made to turn from 0 to 3500 rpm in <1 second in a high viscous condition. Fluid's resistance and gravity segregation are the main problems with start-up in heavy oils in gas lift process. Even if a system may have sufficient gas lift injection pressure to flow gas over the operating valve, it does not offer greatly in terms of reducing the mixture density if the gas segregates to the high side of the tubulars. The gravity segregation can cause severe slugging. Using transient multiphase simulation programs during the design phase could predict slug volumes. Besides, the programs can propose solutions for slug mitigation. The injection of diluents may help to decrease the mixture viscosity, but some completion components, if elastomeric, might react by swelling and losing mechanical strength. Both gas lift equipment, ESPs, and PCPs contain elastomers, and exposure to diluents has to be cautiously assessed during the lift selection and design process.

6. Conclusions

Internationally, the heavy oil reserves have become more important as a future energy source. There are three techniques to produce heavy oil and bring to the surface which are primary, secondary, and tertiary recovery. The EOR processes can be categorized into three main groups, chemical, thermal, and miscible. Commonly, artificial lift techniques are utilized when the well cannot produce naturally at its economical rate. This is applicable for heavy oil reservoirs, where high viscosity along with the reservoir pressure drop will avoid the wells to produce naturally. Conventionally, heavy oil wells are using beam pump as primary artificial lift system. However, beam pumps are used for low flow rate wells; besides this pump has many operating problems. Alternatively, there are several pump systems currently employed as the first option in heavy oil wells, such as PCP, hydraulic pumps, and ESP.

The pumping system needs more development to handle the operating cost, mostly for operation and maintenance either for single equipment or the whole system. In the future, pumps need to be more reliable and capable of running for a long time before requiring maintenance. Pumps must be safer to work, use less space, use less power, and have less noise and high temperature. Accordingly, pumps will be friendlier to the environment and running with less power to decrease their carbon footprint. Pump designers and the technology are faced with enormous challenges in developing heavy oil reservoirs that have high viscosity and low initial pressure and temperature. Besides, heavy oil producers may also face possible gas and water inflow on low-producing mature fields.

Acknowledgements

The authors wish to thank the Universiti Teknologi PETRONAS, Malaysia, for supporting this work. A special thanks to the production technology team of PETRONAS Carigali. Last but not least, a special thanks to Mr. Taha S. Abouargub for his generous assistance and for providing technical support, collaboration, and words of encouragement on the success of this chapter.

Conflict of interest

The corresponding author confirms that there have been no involvements that might raise the question of bias in the work reported or in the conclusions or implications.

Author details

Tarek Ganat
Universiti Teknologi PETRONAS, Seri Iskandar, Perak, Malaysia

*Address all correspondence to: tarekarbi.ganat@utp.edu.my

References

[1] The Canadian Heavy Oil Association: Reservoir Handbook. Calgary, Alberta; 1991

[2] Jayasekera AJ, Goodyear SG. Improved hydrocarbon recovery in the United Kingdom Continental Shelf: Past, present and future. SPE/DOE Improved Oil Recovery Symposium. Society of Petroleum Engineers. 2002:13

[3] Pinczewski W. Diffusion-controlled swelling of reservoir oil by indirect contact with injection gas. 1993;**48**(18). DOI: 10.1016/0009-2509(93)80208-8

[4] DALEEL. Enhanced oil recovery (EOR) summary [Internet]. Available from: https://www.scmdaleel.com/category/enhanced-oil-recovery-eor-summary/164

[5] Ambastha A. Heavy oil recovery. SPE Reprint Series no. 61. Tulsa, OK: Society of Petroleum Engineers. [Accessed: 26 March 2008]

[6] Moulds TP, Trussell P, Haseldonck SA, Carruthers RA. Magnus field: Reservoir management in a mature field combining waterflood, EOR and new area developments. In: SPE Offshore Europe Conf.; 6-9 September 2005; Society of Petroleum Engineers; DOI: 10.2118/96292-MS

[7] Istchenko CM, Ian G. Well/wormhole model of cold heavy-oil production with sand. SPE Journal;**19**(2):260-269. DOI: 10.2118/150633-PA

[8] NETL. Carbon sequestration FAQ information portal: Permanence and safety of CCS [Internet]. 2012. Available from: 10.2118/150633-PA

[9] Muggeridge Ann CA, Webb K, Frampton H, Collins I, Moulds T, Salino P. Recovery rates, enhanced oil recovery and technological limits. Philosophical Transactions of the Royal Society A: Mathematical, Physical and Engineering Sciences. 2006;**372**:20120320. DOI: 10.1098/rsta.2012.0320. PMC 3866386. PMID 24298076

[10] Alusta G, Mackay E, Fennema J, Collins I. EOR vs. infill well drilling: How to make the choice? In: SPE Enhanced Oil Recovery Conference 2011, EORC 2011; Society of Petroleum Engineers. DOI: 10.2118/143300-MS

[11] Ren Z, Al-Haqqan H, Ahmad F, Tinnin J, Hamilton A. 3D VSP surveillance of thermal heavy oil recovery in Kuwait. SPE. search.spe.org. DOI: 10.2118/184127-MS [Accessed: 26 March 2018]

[12] Markham H. SAGD, big data help cenovus gain steam [Internet]. 2017. Available from: https://www.epmag.com/sagd-big-data-help-cenovus-gain-steam-1664056

[13] Hoffman L. The anatomy of cyclic steaming. Santa Barbara. [Internet]. 2014. Available from: https://www.independent.com/news/2014/jun/26/anatomy-cyclic-steaming/

[14] Shabelansky A, Malcolm A, Fehler M. Monitoring viscosity changes from time-lapse seismic attenuation: Case study from a heavy oil reservoir. Geophysical Prospecting;**63**(5):1070-1085. DOI: 10.1111/1365-2478.12229. ISSN 1365-2478

[15] Al-Mutairi A, Ren Z, Tinnin J, Randazzo S. Case study of monitoring steam flood projects in thin vertically stacked reservoirs using 3DVSP technology. In: 79th EAGE Conference and Exhibition 2017; DOI: 10.3997/2214-4609.201700510. ISBN 978-94-6282-217-7

[16] Alhakiki. Stemflood [Internet]. 2012. Available from: https://alhakiki.

wordpress.com/2012/12/30/mature-field/steamflooding/

[17] Dabbous M, Fulton P. Low temperature oxidation kinetics and effects on the in-situ combustion process. Paper No. SPE 4143. In: Society of Petroleum Engineers American Institute of Mechanical Engineers 47th Annual Fall Meeting; 8-11 October; San Antonio, Texas; DOI: 10.2118/4143-PA

[18] Fassihi M, Brigham W. Reaction kinetics of in-situ combustion. Society of Petroleum Engineers Journal. 1984; **24**(4):408-416. DOI: 10.2118/9454-PA

[19] He B, Chen Q, Castanier M, Kovscek A. The effect of metallic salt additives on in-situ combustion performance. M.S. Report. California: Stanford University, Stanford; SPE-93901-MS. DOI: 10.2118/93901-2004

[20] He B, Chen Q, Castanier M, Kovscek A. Improved in-situ combustion performance with metallic salt additives. Paper No. SPE 93901. In: Proceedings of the Society of Petroleum Engineers Western Regional Meeting; March 30-April 1; Irvine, California; DOI: 10.2118/93901-MS

[21] Shallcross D, Rios F, De Los Castanier M. Modifying in-situ combustion performance by the use of water-soluble additives. SPE Reservoir Engineering;**6**(3):287-294. DOI: 10.2118/19485-PA

[22] Strycker A, Sarathi P, Wang S. Evaluation of in situ combustion for schrader bluff [Internet]. Topical Report; Washington, DC: National Petroleum Technology Office, United States Department of Energy; March 1999. Available from: http://www.osti.gov/bridge/

[23] Castanier L, Kovscek A. Heavy oil upgrading in-situ via solvent injection and combustion: A new method. In: Proceedings of the EAGE 67th Conference and Exhibition; 13-16 June 2005; Madrid Spain

[24] Rob Kendal. Using time lapse seismic to monitor the THAI heavy oil production process. In: Paper presented at the 2009 CSEG CWLS Convention; Calgary, Alberta, Canada; Vol. 34 No. 07. doi.org/10.1190/1.3255694

[25] Hoffmann A. Diluent injection optimization for a heavy oil field. In: Conference: SPE Heavy Oil Conference and Exhibition; DOI: 10.2118/184119-MS

[26] Cholet H. Well Production Practical Handbook. Paris, France: Institut français du pétrole publications, Éditions Technip; 2000

[27] Christ FC, Zublin JA. The application of high volume jet pumps in north slope water source wells. In: Presented at the SPE California Regional Meeting; 23-25 March 1983; Ventura, California; SPE-11748-MS. http://dx.doi.org/10.2118/11748-MS

[28] John Martinez. Introduction to artificial lift. 2017. Available from: https://www.petroskills.com/blog/entry/00_totm/sept17-sub-totm-artificial-lift?page=4#.XKMfrCIzaCg

[29] Conoco Inc. Beam pumping engineering and operating manual. Available from: https://wiki.aapg.org/Artificial_lift#cite_note-pt09r5-1 [Accessed: July 1990]

Chapter 5

Modeling Friction Losses in the Water-Assisted Pipeline Transportation of Heavy Oil

Sayeed Rushd, Rasel A. Sultan and Shahriar Mahmud

Abstract

In the lubricated pipe flow (LPF) of heavy oils, a water annulus acts as a lubricant and separates the viscous oil from the pipe wall. The steady state position of the annular water layer is in the high shear region. Significantly, lower pumping energy input is required than if the viscous oil was transported alone. An important challenge to the general application of LPF technology is the lack of a reliable model to predict frictional pressure losses. Although a number of models have been proposed to date, most of these models are highly system specific. Developing a reliable model to predict pressure losses in LPF is an open challenge to the research community. The current chapter introduces the concept of water lubrication in transporting heavy oils and discusses the methodologies available for modeling the pressure drops. It also includes brief descriptions of most important pressure loss models, their limitations, and the scope of future works.

Keywords: lubricated pipe flow, continuous water-assisted flow, core annular flow, CFD, friction factor, data analysis

1. Introduction

1.1 Background

The reserve of nonconventional heavy oils is one of the most important petroleum resources in the current world [1, 2]. These oils are highly asphaltic, dense, and viscous compared to conventional oils, such as Brent and West Texas Intermediate [3, 4]. The density is comparable to that of water, and the viscosity can be greater than that of water by more than five orders of magnitude at room temperature [5, 6]. This type of highly viscous oils is produced using a variety of mining and *in situ* techniques [7]. After extraction, the oil is delivered from the production site to a central processing/upgrading facility. A number of pipeline transportation methods are available for the transportation. The conventional transporting technologies involve viscosity reduction through heating or dilution [1, 3, 4, 8].

The focus of the current chapter is the lubricated pipe flow (LPF) of heavy oils, where a water annulus separates the viscous oil-core from the pipe wall. It is an alternative flow technology, which is more economic and environmentally friendly than conventional heavy oil transportation technologies [9, 10]. The benefit of LPF is that it is a specific flow regime in which a continuous layer of water can be found near the pipe wall. As wall shear stresses are balanced by pressure losses in any kind of pipeline transportation, this flow system requires significantly less pumping energy than would be required to transport the viscous oil alone at comparable process conditions [10, 11–17].

A number of industrial scale applications of LPF are reported in the literature. For example, a 38.9-km long lubricated pipeline having 6 inch diameter was successfully operated by Shell for more than 12 years in California [18]. The frictional pressure loss for this pipeline was not only orders of magnitude less than that for transporting heavy oil but also quite comparable to the loss for transporting water [6]. The pipeline was operated by adding up to 30 vol% water. A number of water lubricated pipelines were used to transport heavy oil at Lake Maracaibo in Venezuela [1]. One of the challenges the operators faced to run these pipelines was cumulative wall fouling. Different operational measures, such as increasing water fraction or water flow rate and changing the water composition were taken to control the fouling. However, these measures were never sufficient to stop wall fouling. Water lubricated pipe flow technology was also used in Spain for the purpose of transporting heavy fuel oil [6]. Syncrude Canada Ltd. transported bitumen froth (a mixture of 60% bitumen, 30% water, and 10% solids) from a remote extraction plant to upgrading facility; they used a 35-km long and 36-inch diameter lubricated pipeline [12, 19, 20]. The lubrication process in the Syncrude pipeline produced a fouling layer of oil on the pipe wall. The thickness of the fouling layer was approximately 5% of the pipe's internal diameter [12, 19]. At present, Brazilian oil producers are in the process of producing viscous oils from off-shore reservoirs with the application of water lubricated flow in vertical pipelines [21, 22].

A concerning phenomenon during the lubricated pipe flow of viscous heavy oil or bitumen is wall fouling [1, 3]. The probable LPF regime is presented in **Figure 1**. A wall fouling layer of oil is shown to surround a water annulus lubricating the viscous oil core. Although a number of experimental studies demonstrated the fouling layer to be a natural and inevitable consequence of the lubrication process, the mechanism of wall fouling in LPF has not been studied in detail [7, 12, 17, 19]. The application of LPF where the phenomenon of wall fouling must be accepted under regular operating conditions is sometimes referred to as "continuous water assisted flow (CWAF)" [13].

1.2 Lubricated pipe flow

Successful operation of a water lubricated pipeline is dependent on a few critical flow conditions. The preliminary requirement for establishing LPF is the simultaneous pumping of heavy oil and water in the pipeline. This kind of pumping into a horizontal pipeline can result in different flow regimes, depending upon the superficial velocities and the properties of oil [18, 24, 25]. The prominent flow regimes

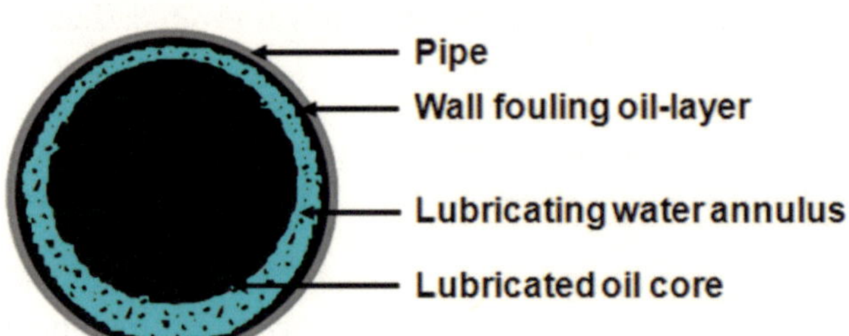

Figure 1.
Hypothetical presentation of the flow regime in a water lubricated pipeline [23].

are dispersed, stratified flow, bubbles, slugs, and lubricated flows. The boundaries between the flow regimes are not well defined [7, 18]. It is possible to describe qualitatively the transition from a flow regime to the other one on the basis of similar regime transitions in gas-liquid flow systems [13]. At lower flow rates of the fluids, stratified flow can be expected [26, 27]. The relative positions of the oil and water in this kind of flow regime are controlled by the effect of gravity, that is, the difference between the liquid densities. If the density of water is higher than that of oil, oil is to float on water and vice versa. By increasing the water flow rate, the stratified flow regime may be transformed into bubble or slug flow. The increased flow rate is likely to increase the kinetic energy and turbulence of the water, which results in waves at the oil-water interface and, ultimately, transforms the stratified oil into bubbles or slugs. Further increase in the water flow rate may split bubbles or slugs into smaller droplets of oil. Contrariwise, increasing the oil flow rate at a constant water flow can promote coalescence of bubbles or slugs, which may produce the water lubricated flow regime [24, 25].

The minimum velocity for the mixture of heavy oil and water required to obtain the water lubricated flow regime in a horizontal pipeline has been reported as 0.1–0.5 m/s for different applications [10, 12, 13, 15, 28]. In addition to the minimum velocity criterion, sustainable lubricated pipe flow also requires a minimum water fraction, typically between 10 and 30% [1]. A greater percentage of lubricating water does not cause a significant reduction in the pressure loss; even if it reduces the pressure loss to some extent, it also reduces the amount of oil transported per unit of energy consumed [10, 13, 20]. Water lubrication is usually identified from pressure loss measurements [13]. The establishment of lubricated pipe flow is typically associated with a significant and nearly instantaneous reduction in frictional pressure losses [20].

As mentioned earlier, a significant concern during the application of lubricated pipe flow is that a minor fraction of the transported oil tends to adhere to the pipe wall, which eventually leads to the formation of an oil layer on the pipe wall [1, 3, 12, 13, 15, 18, 19, 29]. Frictional pressure losses in a "fouled" pipe, that is, with an oil coating on the wall, are higher compared to those for transportation of the same mixture in an unfouled pipe [15, 30]. Nevertheless, the frictional losses with wall fouling are substantially lower than that would be expected for transporting only heavy oil [10, 20, 29].

Wall fouling is practically unavoidable in the water lubricated pipeline transportation of viscous oils [10, 12, 13, 15, 20]. Varying degrees of wall fouling are experienced in the applications of this pipe-flow technology. Different descriptions have been used in the literature to classify these applications, for example:

a. Core annular flow [11, 30]

b. Self-lubricated flow [12]

c. Continuous water assisted flow [10, 13]

Lubricated pipe flow has been used in this chapter to refer to any of these flow types, despite the fact that they exhibit quite different characteristics.

Core annular flow (CAF) primarily denotes an idealized version of lubricated pipe flow. It involves a core of viscous oil lubricated by a water annulus through a pipe with a clean (unfouled) wall [11, 29, 31]. Many research studies published in the 1980s and 1990s focused exclusively on CAF, for example, [11, 31, 32]. In most of these studies, wall fouling was either minimized or avoided through prudent selection of operating conditions, such as water cut and construction material of pipe. In pilot-scale and industrial operations, attempts to operate CAF pipeline usually required expensive mitigation strategies to handle wall fouling. In most published cases, it was impossible to avoid wall fouling (see, for example [15, 33]).

The self-lubricated flow (SLF) and continuous water assisted flow (CWAF) are the commonly applied forms of LPF in the industry. As mentioned earlier, the SLF refers to the water lubricated pipeline transportation of a viscous mixture known as bitumen froth containing approximately 60% bitumen, 30% water, and 10% solids by volume [12, 19, 20]. The water fraction in the froth lubricates the flow; additional water is usually not added. In a SLF pipeline, water assist appears to be intermittent, and the oil core may touch the pipe wall at times [10, 12, 29]. Continuous water assisted flow denotes the pipeline transportation of heavy oil or bitumen when the water lubrication is more stable and the oil core touches the pipe wall infrequently [10, 13, 34]. Approximately 20–30 vol% water required to produce lubricated flow is supplier from an external source to a CWAF pipeline. Both SLF and CWAF involve wall fouling. For example, the thickness of fouling layer was measured from 5.5 to 8.5 mm in a 150-mm SLF pipeline transporting bitumen froth at 25°C [19]. Similar thicknesses in a 100-mm CWAF pipeline were found to vary from 1 to 5 mm depending on the operating temperature and mixture velocity [10, 23].

1.3 Modeling LPF pressure losses

Lubricated pipe flow has been applied in a specific industrial context for transporting viscous oils like heavy oil and bitumen with limited success in many cases [1, 3, 9, 18, 20, 21]. A challenge to the broader application of LPF technology is the lack of a reliable model to predict frictional pressure losses, even though numerous empirical (e.g., [12, 13]), semi-mechanistic or phenomenological (e.g., [10, 11, 15]) and idealized models (e.g., [14, 32, 33, 35–37]) have been proposed to date. The existing models were developed based on either single-fluid or two-fluid approach. A critical analysis of these models is important to underscore their limitations and to realize the scope of developing new approach to model LPF frictional losses.

1.3.1 Single-fluid approach

Single-fluid models are also known as equivalent fluid models. This kind of models generally takes an engineering approach to predict the pressure gradients. The flow system is modeled by considering the flow of a hypothetical fluid under comparable LPF process conditions. In some cases, this hypothetical fluid is water [10, 12, 13, 15]. In other cases, the properties of this fluid are determined using the mixture properties [11]. The flow regime in a single-fluid model is assumed to be in turbulent state, and the friction factor is recognized as inversely proportional to the nth power of a representative Reynolds number (Re), that is, $f = K/Re^n$. The constants K and n are either determined empirically or simply assigned. The Reynolds number is defined with respect to the properties of the hypothetical liquid and the pipeline conditions: an equivalent density (ρ) and viscosity (μ) of the hypothetical liquid, the pipe diameter (D), and the average mixture velocity (V). The famous Blasius formula ($f = 0.079/Re^{0.25}$) is often the basis of single-fluid modeling approach. This empirical law was originally proposed for the turbulent flow of water in a smooth pipe. The value of K in Blasius formula ($K = 0.079$) can be tweaked to take into account the equivalent hydrodynamic roughness produced by the pipe-wall and/or wall fouling layer. Thus, single-fluid models take an empirical approach to predict pressure loss for lubricated pipe flow; the actual physical mechanisms governing pressure losses in a water lubricated pipeline are mostly disregarded.

1.3.2 Two-fluid approach

There are a few two-fluid models available in the literature [23, 32, 33, 38]. However, most of these were proposed for smooth pipe CAF, that is, this kind of models does not take into account the hydrodynamic roughness. As a result, these models are not suitable for SLF or CWAF. However, these models do have an advantage over single-fluid models. The actual mechanism of frictional pressure loss is addressed to some extent while developing a two-fluid model. The modeling approach is described in details with two examples as follows.

Oliemans et al. [32] described the mechanism of frictional losses in their pioneering model developed for a CAF system. They identified the shear in the turbulent water annulus as the major contributing factor to pressure losses. However, they had to empirically address two important aspects of core annular flow: physical roughness on the oil core and water holdup. They also used a couple of idealized concepts like Reynold's lubrication theory and Prandtl's mixing length. This two-fluid model systematically underpredicted the CAF pressure losses. Also, the implementation of the model is not straightforward.

Ho and Li [31] adapted the key features of Shi et al. [33] to develop another two-fluid model. They recognized the major source of frictional pressure loss in CAF to be the shear in the turbulent water annulus and modeled the turbulence based on the concept of Prandtl's mixing length. They also considered the oil core to be a plug having a rough surface. However, instead of empirically quantifying this roughness like [33], the complexity of physical roughness was simplified in [32] based on the concept of hydrodynamic roughness. An idealized core annular flow regime was subdivided into four hypothetical zones as presented in **Figure 2**, which also depicts the dimensionless distances of these zones from the stationary pipe wall. The velocity profiles in the sublayers are usually presented using these nondimensional terms. The relationships of flow rate and pressure drops were obtained by integrating these velocity profiles with respect to the dimensionless distance. The equations are presented in **Table 1**.

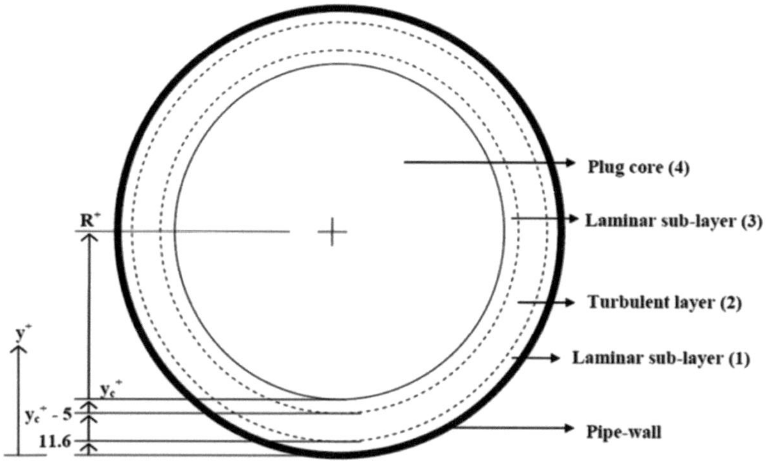

Figure 2.
Hypothetical subdivision of perfect or ideal core annular flow into four zones showing dimensionless distances from the pipe wall [31].

Zone (Figure 2)	Equations	Range
Laminar sublayer (1)	$u_1^+ = y^+$	$0 \le y^+ \le 11.6$
Turbulent layer (2)	$u_2^+ = 2.5\ln(y^+) + 5.5$	$11.6 \le y^+ \le y_c^+ - 5$
Laminar sublayer (3)	$u_3^+ = 2.5\ln(y_c^+ - 5) - y_c^+ + 10.5 + y^+$	$y_c^+ - 5 \le y^+ \le y_c^+$
Plug core (4)	$u_4^+ = 2.5\ln(y_c^+ - 5) + 10.5$	$y_c^+ \le y^+ \le R^+$
(1) + (2) + (3)	$Q_w = 2\pi(\nu_w^2/v^*)[(2.5R^+y_c^+ - 1.25y_c^{+2})\ln(y_c^+ - 5) + 3R^+y_c^+ - 2.125y_c^{+2} - 13.6R^+]$	$0 \le y^+ \le y_c^+$
(4)	$Q_o = \pi(\nu_w^2/v^*)(R^+ - y_c^+)^2[2.5(\ln y_c^+ - 5) + 10.5]$	$y_c^+ \le y^+ \le R^+$

Table 1.
Velocity profiles and equations relating flow rates and pressure losses [31].

The principal focus of Ho and Li [31] was the water annulus in a CAF pipeline. The annular thickness was the most important parameter in the two-fluid model. However, they had to determine this thickness empirically. Moreover, they used the idealized concept of perfect CAF, that is, the perfectly concentric orientation of the oil core in the pipe, even though the orientation is more likely to be eccentric [29, 33]. The eccentricity of the oil core has a consequential effect on the CAF pressure losses [39].

Even after involving a number of simplifications, the Ho and Li model very closely addresses the physical mechanism of CAF pressure losses. This model allows predicting the pressure gradients using the values of oil and water flow rates. As expected, this two-fluid model underpredicts the CWAF friction losses consistently. This is because a CWAF system involves considerable wall fouling and oil core eccentricity, while the two-fluid model was developed for the perfect CAF in a hydrodynamically smooth pipe.

Adapting the modeling methodology described in [32], a physics-based approach to model CWAF pressure losses was proposed in [38]. Please refer to Ref. [23] for the details of the development. It is a semi-mechanistic two-fluid model, which requires simulating the turbulent flow of annular water on the fouling oil layer in a lubricated pipeline. The turbulence in the water annulus is modeled with the anisotropic ω-RSM model instead of the standard isotropic models. It can capture the effects of the thickness of the wall fouling layer, the equivalent hydrodynamic roughness produced by the viscous oil layer on the pipe wall, and the water holdup. The model was validated using actual CWAF data collected by varying pipe diameter, oil viscosity, water fractions, and flow rates. Compared to existing CFD models, this model is more robust as it not only produces better predictions but also requires significantly fewer computing resources. Although a promising development, the current version of the model involves some simplifications and is difficult to implement.

2. Descriptions of selected models

2.1 CAF model

Arney et al. [11] performed a comprehensive study on the core annular flow in a horizontal pipeline involving both experiments and theoretical analysis. Their primary objective was to enrich the CAF database and introduce a simple approach to calculate the frictional pressure losses.

For the experiments, Arney et al. [11] used two oils: waxy crude oil (ρ = 985 kg/m^3 and μ = 0.6 Pa s) and No. 6 fuel oil (ρ = 989 kg/m^3 and μ = 2.7 Pa s). The experimental setup consisted of three pipeline segments made of a glass pipe having 15.9 mm inner diameter (ID). The 6.35-m long first part was used for flow visualization using a Spin-Physics SP2000 high-speed video system and a 35 mm camera. The second part of the pipeline was used to connect two pressure tap. 1.42 m apart. The last part of the pipe was 1.47 m long and utilized to measure the *in situ* volume fraction of water, that is, water holdup.

Two important parameters used for this study were the water holdup (H_w) and input water fraction of (C_w). It was observed that the H_w was consistently larger than the C_w. That is, the oil core in CAF was moving faster than the annular water phase. Similar experimental finding was also reported in [23]. Areny et al. [11] then collated all the previous CAF experimental data from the literature with their own measurements to propose the following correlation between H_w and C_w:

$$H_w = C_w[1 + 0.35(1 - C_w)] \quad (1)$$

They also measured the pressure losses for a variety of flow conditions. Based on the data, they proposed a single-fluid model. The friction factor (*f*) was correlated to a system specific Reynolds number (Re_a):

$$f = 0.079/Re_a^{0.25}, Re_a > 4000 \quad (2)$$

$$Re_a = \rho_c DV/\mu_w \quad (3)$$

$$\Delta P/L = f\rho_c V^2/2D \quad (4)$$

where $\Delta P/L$ is the pressure gradient, μ_w is the water viscosity, and ρ_c is an equivalent fluid density. The viscosity of the equivalent liquid was considered to be equal to that of water (μ_w). Empirical expression used to correlate the density of this hypothetical liquid (ρ_c) to the densities of oil (ρ_o) and water (ρ_w) is as follows:

$$\rho_c = H_w\rho_w + (1 - H_w)\rho_o \quad (5)$$

Using this model, it was possible to predict a large number of CAF pressure drop data sets with a reasonable accuracy. The model showed good conformance with friction factor values at high Reynolds number. However, there was significant under prediction when Reynolds number was low. This was due to the fact that at low Reynolds number, the core annular flow was slightly unstable.

2.2 SLF model

Joseph et al. [12] investigated the "Self-Lubrication" phenomenon of Bitumen froth (approximately 60% bitumen, 30% water, and 10% solids by volume), which was extracted using Clark's hot water extraction process from the oil sands of Athabasca. The water in the froth, while transporting through the pipelines, was released due to high shear resulting in a lubricating layer near the wall. This is just another form of CAF where the annular water comes from the mixture itself.

Two different setups were used to experimentally study the phenomenon of self-lubrication. First, a setup of 6 m long 25 mm ID pipe loop was used at the University of Minnesota. The froth was continuously recirculated. The duration of the experiments varied from 3 to 96 hours. The velocities for which pressure gradients were measured ranged from 0.25 to 2.5 m/s. Water volume fractions were kept within 20–40%, and the froth temperature ranged between 35 and 55°C. From the collected pressure gradient data at different flow rates, it was observed that there was a critical velocity range (from 0.5 to 0.7 m/s) below which the self-lubrication was being lost. The longest experimental time in this setup was 96 hours. This test was conducted in an already fouled pipeline. Despite the residual fouling, no further fouling was observed during the experiment. The authors suggested that this may be due to the clay particles (Kaolinite) from the released water protecting the oil core from accumulating to the pipe wall. The researchers also observed that heating the froth to a higher temperature would not necessarily improve the lubrication. This was due to the opposing effects of lowered bitumen viscosity at high temperature and reabsorption of the released water into the core. The other test setup (0.6 m ID and 1000 m long pipeloop) was at Syncrude (Canada), where pilot-scale tests were performed.

Based on the experimental data, a single-fluid model for the SLF of bitumen froth was proposed. In this model, a "Blasius-type" equation was used to correlate the f with a water equivalent Reynolds number (Re_w):

$$f = 0.079K_j/Re_w^{0.25} \tag{6}$$

In Eq. (6), the complex flow behavior of self-lubricated flow is addressed with an empirically determined value of K_j. It was assumed to be a function of temperature only (K_j = 23 when temperature ranges 35–47°C and K_j = 16 when temperature ranges 49–58°C). Water content was considered to have negligible effect on K_j. Frictional pressure losses are 15–40 times greater when predicted using the above model than those for water flowing alone under identical flow conditions. The application of this model for predicting LPF pressure losses is extremely limited according to previous researches [10, 13].

2.3 CWAF model 1

McKibben et al. [13] carried out the investigation to examine free water-crude oil flows and, specifically, to establish a correlation for predicting the pressure gradients in continuous water assisted flow.

The experiments were conducted using the followings:

a. A 53-mm ID pipeline consisting of approximately 60 m long horizontal insulated section;

b. The water fractions between 0.10 and 0.36;

c. The temperatures ranging from 18 to 39°C

d. The average velocities of the mixtures between 0.5 and 1.2 m/s;

e. Four different oils with the viscosities of 91.6, 24.9, 7.1, and 5.8 Pa s.

On the basis of the CWAF data sets at different water equivalent Reynolds number, the correlation of Fanning friction factor was found as:

$$f = \frac{1410}{Re_w} \tag{7}$$

$$f = \frac{\left(-\frac{dP}{dL}\right)D}{2\rho_w V^2} \tag{8}$$

$$Re_w = \frac{DV\rho_w}{\mu_w} \tag{9}$$

The inverse relationship between friction factor and water Reynolds number suggested that friction was controlled by a very thin water layer. A water layer of this type formed the lubrication region surrounding the oil core and provided the lubricating force required to overcome the effect of natural buoyancy.

2.4 CWAF model 2

Rodriguez et al. [15] mainly focused on lab- and pilot-scale experimental measurements. They also took a semi-mechanistic approach to model frictional pressure losses of horizontal core annular flows in pipes having both fouled and clean walls. That is, the model was actually developed for the CWAF systems.

Lab-scale tests were conducted with a 27-mm ID PVC pipe and a crude oil having a viscosity of 0.5 Pa s at 20°C. For the pilot-scale experiments, a steel pipeline (77 mm ID and 274 m length) was used to pump a highly viscous crude oil (36.95 Pa s and 972.1 kg/m^3 at 20°C). A freshwater network was used to control the water injection. A piston pump pumped the water, and its flow rate was adjusted via a calibrated frequency inverter. The water superficial velocity was kept constant at 0.24 m/s, and three oil superficial velocities, 0.80, 1.00, and 1.10 m/s, were tested.

In the experiments, a wavy core of viscous oil was observed, and the annular flow of water was mostly turbulent (Reynolds number for the water flow: $1000 < \text{Re}_2 < 14,500$, $\text{Re}_2 = \frac{\rho_2 V_2 D}{\mu_2}$). The proposed model first defined the irreversible hydrodynamic component of the frictional pressure gradient ($\frac{\Delta P}{L}$):

$$\frac{\Delta P}{L} = b\left(\frac{\rho_m VD}{\mu_m}\right)^{-n} \frac{\rho_m V^2}{2D} \tag{10}$$

where D is the pipe ID, V is the mixture velocity, ρ_m is the mixture density, μ_m is the mixture viscosity, and b is an empirical constant. The μ_m was obtained by evaluating the ratio between the wall shear stress in core-annular flow (τ_o) and the wall shear stress if the annular water was flowing alone in the pipe at mixture flow rate (τ_w). Assuming the phases have the same density and use the same power law to express the friction factors in both flows, the shear stress ratio (R_τ) was expressed as:

$$R_\tau = \left|\frac{\tau_o}{\tau_w}\right|_{\rho_0=\rho_2} = \frac{b\left(\frac{\rho_2 V_2 D}{\mu_2}\right)^{-n} \frac{\rho_2 V_2^2}{2}}{b\left(\frac{\rho_2 VD}{\mu_2}\right)^{-n} \frac{\rho_2 V^2}{2}} = \frac{1}{(1-\varepsilon)^n [1+(s-1)\varepsilon]^{2-n}} \tag{11}$$

$$\varepsilon = \frac{1}{1+s\frac{V_2}{V}} \tag{12}$$

where V_2 is the average in-situ water velocity. Also, R_τ is the ratio between the corresponding pressure drops, and from Eq. (11), one obtains

$$R_\tau = \left|\frac{\frac{\Delta P}{L}}{\frac{\Delta P}{L}_{2,o}}\right|_{\rho_1=\rho_2} = \left(\frac{\mu_2}{\mu_m}\right)^{-n} \tag{13}$$

where $\frac{\Delta P}{L}_{2,o}$ is the extrapolated pressure drop for the annulus fluid alone in the pipe at mixture flow rate:

$$\frac{\Delta P}{L}_{2,0} = \lim_{\varepsilon \to 0} \frac{\Delta P}{L} = b\left(\frac{\rho_2 VD}{\mu_2}\right)^{-n} \frac{\rho_2 V^2}{2D} \tag{14}$$

From Eqs. (13) and (14), the mixture viscosity can be expressed as:

$$\mu_m = \frac{\mu_2}{(1-\varepsilon)[1+(s-1)\varepsilon]^{(2-n)/n}} \tag{15}$$

Eq. (15) shows that the mixture viscosity is affected by the slip ratio: the faster the core moves relative to the annulus, the lower the mixture viscosity and pressure drop. For the simple case of perfect core annular flow (PCAF), putting n = 1 and s = 2, Eq. (15) can be transformed into:

$$\frac{1}{\mu_m} = \frac{1-\varepsilon^2}{\mu_2} \tag{16}$$

The final form of the pressure drop model was obtained by introducing a two-phase multiplier defined as

$$\varnothing_{2,o} = \frac{\frac{\Delta P}{L}}{\frac{\Delta P}{L}_{2,0}} \tag{17}$$

Using Eq. (10), (13), and (14), the above equation becomes

$$\varnothing_{2,o} = \left(\frac{\rho_m}{\rho_2}\right)^{1-n} R_\tau \tag{18}$$

which becomes using Eq. (12) and $\rho_m = \varepsilon\rho_1 + (1-\varepsilon)\rho_2$:

$$\varnothing_{2,o} = \left[1 - \left(1 - \frac{\rho_1}{\rho_2}\right)\varepsilon\right]^{1-n} (1-\varepsilon)^{-n}[1+(s-1)\varepsilon]^{n-2} \tag{19}$$

The hydrodynamic component of frictional pressure gradient can be expressed as

$$\frac{\Delta P}{L} = b\left(\frac{\rho_2 JD}{\mu_2}\right)^{-n} \frac{\rho_2 J^2}{2D}\left[1 - \left(1 - \frac{\rho_1}{\rho_2}\right)\varepsilon\right]^{1-n} (1-\varepsilon)^{-n}[1+(s-1)\varepsilon]^{n-2} \tag{20}$$

or

$$\frac{\Delta P}{L} = \varphi Q^{2-n}\left[1 - \left(1 - \frac{\rho_1}{\rho_2}\right)\varepsilon\right]^{1-n} (1-\varepsilon)[1+(s-1)\varepsilon]^{n-2} \tag{21}$$

where $\varphi = \frac{b}{2}\left(\frac{\pi}{4}\right)^{n-2} \rho_2^{1-n} \mu_2^n D^{n-5}$, and Q is the mixture flow rate.

The proposed model can be used to analyze, correlate, and generalize pressure drop data. Along with that, the model allows for the satisfactory representation of different annulus flow regimes, kinematic effects, and wall conditions, including fouling. The model accounts for effects of buoyancy on the core. However, it cannot provide reliable predictions without regressing the values of b and n on the basis of reliable data set.

2.5 CWAF model 3

In continuation of the previous research [13], McKibben et al. [10] carried out an extensive experimental investigation of CWAF. The tests were conducted at the Saskatchewan Research Council (SRC), Saskatoon, Canada using 25, 100, and 260 mm steel pipe flow loops. The average thickness of wall fouling (t_c) was estimated using a special double-pipe heat exchanger [19]. The estimations were validated by using a hot-film probe to measure the physical thickness of the fouling oil layer. Heavy oils having viscosities in the range of 0.62–91.6 Pa s were used for the experiments. The input water fraction was within the range of 30–50%. Additional details of the experimental facilities are available in [23].

Based on the experimental study, a new empirical correlation for the Fanning friction factor was proposed as follows:

$$f(CWAF) = 15\left(V/\sqrt{gD}\right)^{-0.5} f_{CF}^{1.3} f_{OIL}^{0.32} C_{CF}^{-1.2} \tag{22}$$

where V is the average mixture velocity, g is the gravitational acceleration, D is the pipe diameter, f_{CF} is the friction factor of aqueous phase, f_{OIL} is the friction factor of oil phase, and C_{CF} is the total volume fraction of water in the mixture. It is a phenomenological model, which is claimed to take into account the effects of inertia, gravity, lubricating water, wall fouling, and viscous oil in CWAF. A large data set comprising more than 300 data points were used for the empirical derivation of the model constants.

2.6 CFD models

A scientific methodology of modeling single phase turbulent flow is to use computational fluid dynamics (CFD) [40]. In general, this modeling approach decomposes the turbulent flow into two parts: (i) time-averaged mean motion; (ii) time-independent fluctuations. The product of such decomposition is the transformation of Navier-Stokes (NS) equations into Reynolds Averaged Navier Stokes (RANS) equations [41]. In course of the mathematical transformation, additional terms of turbulent stresses are produced to make the matrix of equations “unclosed”; that is, the number of unknowns is higher than the number of equations. Various turbulent stresses in RANS equations are necessarily modeled empirically for the “closure” of the matrix [42]. The continuity and RANS equations can be presented with the following simplified differential equations:

$$\frac{\partial U_i}{\partial x_i} = 0 \tag{23}$$

$$\frac{\partial U_i}{\partial t} + U_j \frac{\partial U_i}{\partial x_j} = -\frac{1}{\rho}\frac{\partial p}{\partial x_i} + \frac{\partial}{\partial x_j}\left(\frac{\mu}{\rho}\frac{\partial U_i}{\partial x_j} - \tau_{ij}\right) + S_i \tag{24}$$

where x_i represents the coordinate axes, U_i is the mean velocity, p is the pressure, ρ is the density, μ is the viscosity, S_i is the sum of body forces, and τ_{ij} represents the components of the Reynolds stress tensor. The available models for τ_{ij} can be divided in the categories of eddy-viscosity models and Reynolds stress models [43, 44].

Eddy-viscosity models were developed based on the concept of a hypothetical term known as eddy-viscosity (μ_t), which is considered to produce turbulent stresses caused by macroscopic velocity fluctuations [41]. These models can further be divided into three major groups, namely zero-equation, one-equation, and two-equation models [43, 44]. Two-equation models, instead of zero- and one-equation models, are generally used at present to solve complex engineering problems [43, 45]. The most commonly used two-equation models are the k-ε and k-ω models [44]. A significant limitation of this group of models is that they are meant to describe isotropic turbulence [46, 47]. That is, only the significant components of the Reynolds stresses can be computed with the two-equation models. As a result, the group of two-equation models is practically limited to flows where anisotropy is not important [47, 48]. It should be mentioned that the turbulent water annulus in a CWAF pipeline can experience both anisotropy and rough surfaces [10, 12, 15, 30]. These models are also not suggested for turbulent flow in narrow channels and over very rough surfaces [49, 50].

Anisotropic turbulence can be addressed using Reynolds Stress Models (RSM), in which the hypothetical concept of eddy-viscosity is discarded [46]. An example of anisotropic model is ω Reynolds Stress Model (ω-RSM) [45]. In this model, the closure for Reynolds stresses is obtained by using seven differential transport equations [42]. It is a higher level and more elaborate modeling approach compared to the isotropic two-equation models. This kind of models is more widely applicable compared to eddy-viscosity models [43–46]. However, this flexibility is gained through a high degree of complexity in the computational system. The solution of an increased number of transport equations requires significantly higher computational resources compared to the applications of different two-equation models. Even so, a Reynolds stress model was successfully applied to simulate flow conditions that involve anisotropy and rough surfaces [47–51].

To acknowledge the superiority of a Reynolds stress model, a study of the equivalent hydrodynamic roughness (k_s) produced by a wall fouling/coating layer of viscous oil ($\mu_o \sim 21$ kPa.s) was conducted using a rectangular flow cell [23]. The oil surface became rough when turbulent water ($Re_w > 10^4$) was pumped through the flow cell. The rough viscous surface produced very large values of k_s compared to the similar values produced by a clean surface. The relative performance of k-ω model and ω-RSM is presented in **Figure 3**. The ω-RSM can provide reliable predictions of the measured values of friction losses, while the k-ω model yields significant under predictions. This is because the process conditions involved turbulent flow, a hydrodynamically rough surface, and a narrow flow channel, which produced anisotropic turbulence. Comparable analysis was also conducted involving various rough surfaces like solid wall, sandpapers, wall-biofouling layers, and wall-coating layers of heavy oils in different flow cells. Invariably, the ω-RSM always allowed for reliable predictions, while the k-ω model failed to do so. This analysis along with the supporting literature evidently prove that a RSM would be a better choice than a two-equation model to simulate flow conditions, which involve anisotropy and hydrodynamically rough surface.

It should be mentioned that turbulence is a complex subject. Even though RANS methodology is feasible to computationally resolute the phenomenon of turbulence, it averages the process variables with a steady-state assumption in course of solving NS equations. The minor scale unsteady features of turbulence are usually neglected in this kind of averaging [44]. Most important of these features is the turbulent eddies. The scale of these eddies can vary over orders of magnitude [46]. The CFD solution of taking the effect of these eddies into account is computing differential NS equations without any kind of modeling. The available methods for the purpose are Large Eddy Simulation (LES) and Direct Numeric Simulation (DNS). However, these two simulation techniques demand extremely high computational resources [44]. At a computing rate of 1 gigaflop, the requirement of computational time for DNS is of the order of the Reynolds number to the third power (Re^3). Similar

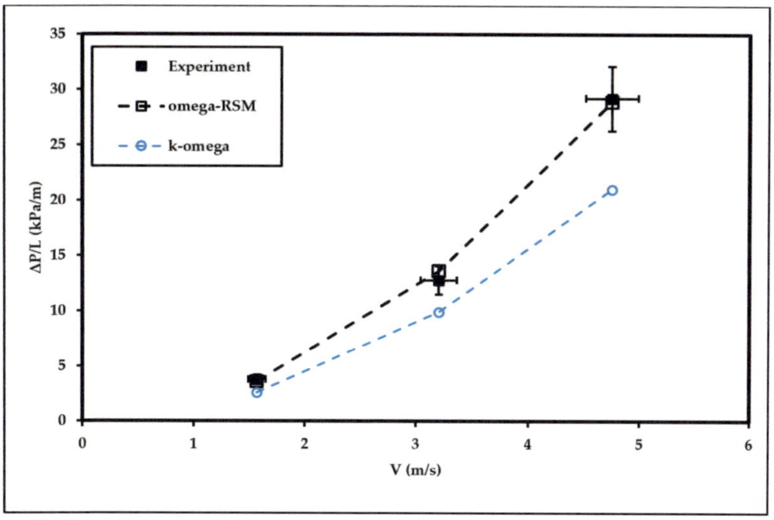

Figure 3.
Comparison of experimental pressure gradients with simulation results: 25.4 × 15.9 × 2000 mm rectangular flow cell; average coating thickness, t_c *= 1.0 mm; equivalent hydrodynamic roughness,* k_s *= 3.5 mm;* $10^4 < Re_w < 10^5$ *[23].*

requirement in LES is generally ten times less than DNS. As an industrial flow system like CWAF pipeline can involve *Re* in the order of 10^5 or higher, application of LES or DNS is not realistic for CWAF at this point.

Two-fluid CFD modeling approach to predict frictional pressure losses of core annular flow was used in different studies [14, 35, 36]. They considered the water annulus to be turbulent and the viscous core to be a laminar plug. Usually, the annular turbulence was modeled with standard k-ε and k-ω models using commercial CFD packages like ANSYS CFX. CFD simulations were also conducted using FLUENT for horizontal oil-water flow with viscosity ratio $\left({}^{\mu_o}/_{\mu_w}\right) = 18.8$ in different flow regimes, namely core annular flow, oil plugs/bubbles in water, and dispersed flow [37]. In FLUENT, the volume of fluid (VOF) model of multiphase flow and the SST k-ω scheme of turbulence closure was applied to simulate the oil-water flow. The SST k-ω turbulence scheme at the interface provided better predictions than the standard k-ε and re-normalization group (RNG) k-ε models. Although these turbulence models might show some superiority over Prandtl's mixing length model used in [32], they are meant for isotropic turbulence and are not suggested for the turbulent flow that involves anisotropy or very rough surfaces [49–51]. In addition, this modeling approach is also expensive computationally as it requires solving the governing equations for both phases of oil and water. Using an anisotropic model makes this modeling approach even more expensive from a computational perspective. Moreover, the interphase transfer of mass and momentum is modeled in this methodology by using the default mixture model ANSYS CFX or FLUENT. The correlations used for these models are not validated for the interfacial mixing of LPF systems.

3. Comparative analysis

The performance of existing models in predicting friction losses is analyzed by comparing the experimental results collected for a lab-scale LPF system with the corresponding results obtained using five different models in **Figure 4**. The water equivalent friction factor (f_w) and Reynolds number (Re_w) defined as follows are presented in this figure:

$$f_w = \frac{\Delta P}{L}\frac{D}{2\rho_w V^2} \tag{25}$$

$$Re_w = \frac{DU_w\rho_w}{\mu_w} \tag{26}$$

where $\Delta P/L$ is the pressure gradient, D is the internal pipe diameter, V is the bulk velocity, U_w is the water superficial velocity, and ρ_w and μ_w represent water density and viscosity, respectively. The experiments were conducted in a 26-mm horizontal PVC pipeline to collect data under typical CWAF operating conditions [7, 34]. The heavy oils used for the experiments had densities and viscosities in the ranges of 900–950 kg/m^3 and 3.3–16.0 Pa s, respectively. Although different flow patterns were observed in course of the experiments, the core annular flow with oil fouling on pipe wall, that is, the CWAF was the dominant regime under a wide range of flow conditions. The results shown in **Figure 4** reveal the major limitation of the existing models to be their system specificity.

The CAF model proposed by Arney et al. [11] significantly under predicts the experimental results for CWAF tests. The model was developed based on the experiments conducted in a 15.9-mm glass pipeline, which was selected for the purpose of controlling wall fouling and visualizing the flow regime. Compared to the CAF model, higher predictions of f_w-values by the SLF model and CWAF model 2 can be seen in **Figure 4**. These three models were developed for CAF systems having different degrees of wall fouling and intermittent lubrication. It is interesting to note that the trends of the results produced by all of these three models are similar to that of Blasius law predictions, which represent the frictional losses as only water flows through a pipe. This is because these models were developed based on the Blasius correlation. Clearly, the methodology of modeling CWAF friction losses by modifying Blasius law is not a successful approach. On the other hand, the CWAF model 1 also fails to provide satisfactory predictions. It was developed by modifying the standard f correlation for laminar flow regime (f = 64/Re).

Among the models presented in **Figure 4**, the CWAF model 3 demonstrates superiority in predicting both the magnitude and the trend of CWAF friction factors. The performance of the model tends to be better at higher flow rates. The reason is, most likely, the difficulty in establishing the lubricated flow regime at lower flow rates. The improved predictions of this model can be attributed to the following facts:

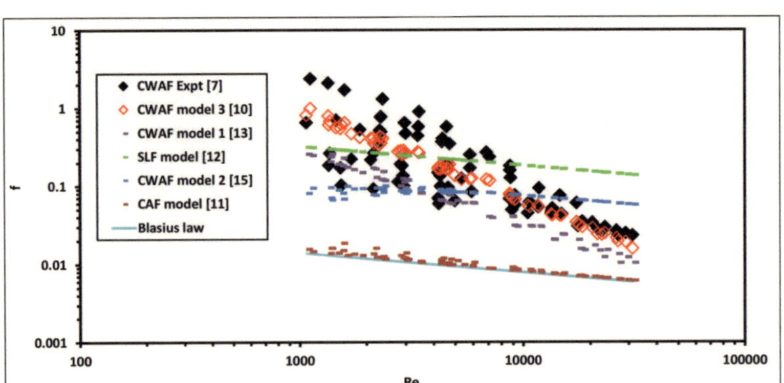

Figure 4.
Comparison of experimental and predicted values of water equivalent friction factor with respect to an equivalent Reynolds number.

i. Instead of a purely empirical reasoning, the model was developed based on phenomenological observations. The physics of frictional pressure losses in a CWAF pipeline dominated the modeling approach. It can be considered as an empirical two-fluid model.

ii. Special attention was paid to the contribution of wall fouling in increasing CWAF pressure drops.

iii. A large data set covering a wide variety of process conditions was used to regress out the model constants.

However, it should be emphasized that the CWAF model 3 needs further validation with multiple independent data sets. Producing more quality data of CWAF pressure losses would be essential for the purpose.

As demonstrated here, the existing models generally adopt an empirical approach to predict LPF friction losses. The effects of operating conditions in these models are usually accounted for with empirical constants. The actual physical mechanisms that govern the pressure losses in an LPF pipeline are almost entirely disregarded. As a result, the models developed for a CAF system cannot be applied to a CWAF system and vice versa. More rigorous studies are essential to develop a fully mechanistic approach to model the LPF friction losses. The CFD can be expected to play a significant role in the process.

4. Challenges and opportunities

The lab-scale applications reported in the literature and a few commercial successes prove that continuous water assisted flow is a reliable method for the long-distance transportation of heavy oil. One of the major barriers to spread commercial applications of this flow technology is the lack of a viable model to predict frictional losses on the basis of operating parameters, such as pipe diameter, flow rates, fluid properties, and water fraction. A new model capable of dealing with the hydrodynamic effects produced by the "wall fouling" layer in a CWAF pipeline is required to facilitate wider industrial implementation of this pipeline transportation technology. The phenomenon of wall fouling has not been thoroughly probed till now, although it is an important characteristic of CWAF technology. Therefore, the future researches should be focused on the investigation of the hydrodynamic effects produced by the wall fouling layer.

The oil core touching the pipe wall in a large water-assisted pipeline is another unaddressed phenomenon. Experiments conducted at Saskatchewan Research Council suggest that this phenomenon is significant for intermittent water assist when the bulk velocity is less than 1 m/s and the water fraction is less than 30% [10]. More devoted research works are necessary to determine the contribution of intermittent core/wall contact to the LPF pressure loss.

The presence of solids like sands or clays in a CWAF pipeline is another important issue. The solids embedded on the wall fouling layer and the oil core may increase the equivalent roughness. In some cases, the fine particles in the lubricating water can change its apparent viscosity and the nature of contact between the oil-covered wall and the oil core (see [12] for additional details). Future work in this field would help to characterize the effects of solid fraction on the pressure losses in CWAF pipelines.

5. Conclusions

The objective of the current chapter is to provide a brief introduction to the water lubricated transportation of heavy oil. The contents are summarized as follows:

i. Lubricated pipe flow is an alternative technology for long distance transportation of heavy oil. This kind of water assisted pipeline transportation is more economic and environmentally friendly than the conventional viscous oil transportation technologies. Its applications can be categorized as core annular flow, self-lubricated flow, and continuous water assisted flow. From an engineering perspective, CWAF is more significant than CAF or SLF.

ii. A technical challenge to the field-scale application of CWAF is the absence of a reliable model to predict friction losses.

iii. The models proposed to date for CWAF friction losses can be categorized as single-fluid and two-fluid models. In general, the methodology followed to develop a two-fluid model is more mechanistic, while a single-fluid model is an empirical development.

iv. Applicability of an existing model for a specific set of flow conditions cannot be justified at present without a comparative analysis based on a reliable data set.

v. The most important research opportunities to develop a more reliable model for CWAF friction losses are related to the following subjects:

  a. Wall fouling

  b. Intermittent water lubrication

  c. Solid laden CWAF

  d. CFD modeling

  e. Enrichment of the CWAF database

Acknowledgements

A part of the manuscript was the outcome of a research project that was funded by the *NSERC Industrial Research Chair in Pipeline Transport Processes* (RSS). We are grateful to Canada's Natural Sciences and Engineering Research Council (NSERC) and the Industrial Sponsors: Canadian Natural Resources Limited, CNOOC-Nexen Inc., Saskatchewan Research Council Pipe Flow Technology Centre™, Shell Canada Energy, Suncor Energy, Syncrude Canada Ltd., Total, Teck Resources Ltd. and Paterson & Cooke Consulting Engineers Ltd. We are especially thankful to Dr. Sanders (Professor, Department of Chemical and Materials Engineering, University of Alberta) for his kind advisory role in the research project.

Conflict of interest

The authors certify that they have no affiliations with or involvement in any organization or entity with any financial interest (such as honoraria, educational grants, participation in speakers' bureaus, membership, employment, consultancies, stock ownership, or other equity interest, and expert testimony or patent-licensing arrangements), or nonfinancial interest (such as personal or professional relationships, affiliations, knowledge or beliefs) in the subject matter or materials discussed in this manuscript.

Other declarations

A part of the manuscript is adapted from the corresponding author's PhD dissertation.

Author details

Sayeed Rushd[1]*, Rasel A. Sultan[2] and Shahriar Mahmud[3]

1 King Faisal University, Al Ahsa, Saudi Arabia

2 Del Tech Manufacturing, Prince George, BC, Canada

3 Bangladesh University of Engineering and Technology, Dhaka, Bangladesh

*Address all correspondence to: mrushd@ualberta.ca

References

[1] Nunez GA, Rivas HJ, Joseph DD. Drive to produce heavy crude prompts variety of transportation methods. Oil and Gas Journal. 1998;**96**(43):59-63

[2] Canadian Association of Petroleum production (CAPP). Crude Oil: Forecast, Market and Transportation [Internet]. 2018. Available from: https://www.capp.ca/publications-and-statistics/publications/320294 [Accessed: September 02, 2018]

[3] Saniere A, Henaut I, Argillier JF. Pipeline transportation of heavy oils, a strategic, economic and technological challenge. Oil & Gas Science and Technology–Revue d'IFP Energies nouvelles. 2004;**59**:455-466

[4] Martínez-Palou R, de Lourdes MM, Zapata-Rendón B, Mar-Juárez E, BernalHuicochea C, de la C, et al. Transportation of heavy and extra-heavy crude oil by pipeline: A review. Journal of Petroleum Science and Engineering. 2011;**75**:274-282

[5] Ashrafi M, Souraki Y, Karimaie H, Torsaeter O, Bjorkvik BJ. Experimental PVT property analyses for Athabasca bitumen. In: Proceedings of the Canadian Unconventional Resources Conference; 15–17 November 2011; Calgary. Richardson, Texas: SPE; 2011. SPE-147064-MS. DOI: 10.2118/147064-MS

[6] Bjørnseth F. Heavy oil production technology challenges and the effect of nanosized metals on the viscosity of heavy oil: A literature review and an experimental study [Thesis]. Trondheim: Norwegian University of Science and Technology; 2013

[7] Shi J. A study on high-viscosity oil-water two-phase flow in horizontal pipes [Thesis]. Cranfield: Cranfield University; 2015

[8] Hart A. A review of technologies for transporting heavy crude oil and bitumen via pipelines. Journal of Petroleum Exploration and Production Technologies. 2014;**4**:327-336

[9] Institute for Prospective Technological Studies, European Commission. Prospective Analysis of the Potential Non-conventional World Oil Supply: Tar Sands, Oil Shales and Non-conventional Liquid Fuels from Coal and Gas [Internet]. 2005. Available from: http://ipts.jrc.ec.europa.eu/publications/pub.cfm?id=1365 [Accessed: September 02, 2018]

[10] McKibben M, Sanders S, Gillies R. A new method for predicting friction losses and solids deposition during the water-assisted pipeline transport of heavy oils and co-produced sand. In: Proceedings of the SPE Heavy Oil Conference-Canada; 11–13 June 2013; Calgary. Richardson, Texas: SPE; 2013. SPE-165480-MS. DOI: 10.2118/165480-MS

[11] Arney MS, Bai R, Guevara E, Joseph DD, Liu K. Friction Factor and Holdup Studies for Lubricated Pipelining – I, Experiments and Correlations. International Journal of Multiphase Flow. 1993;**19**(6):1061-1076

[12] Joseph DD, Bai R, Mata C, Sury K, Grant C. Self-lubricated transport of bitumen froth. Journal of Fluid Mechanics. 1999;**386**:127-148

[13] McKibben MJ, Gillies RG, Shook CA. Predicting pressure gradients in heavy oil-water pipelines. The Canadian Journal of Chemical Engineering. 2000;**78**:752-756

[14] Crivelaro KCO, Damacena YT, Andrade THF, Lima AGB, Farias Neto SR. Numerical simulation of heavy oil flows in pipes using the core-annular

flow technique. WIT Transactions on Engineering Sciences. 2009;**63**:193-203

[15] Rodriguez OMH, Bannwart AC, de Carvalho CHM. Pressure loss in coreannular flow: Modeling, experimental investigation and full-scale experiments. Journal of Petroleum Science and Engineering. 2009;**65**:67-75

[16] Vuong DH, Zhang HQ, Sarcia C, Li M. Experimental study on high viscosity oil/water flow in horizontal and vertical pipes. In: Proceedings of SPE Annual Technical Conference and Exhibition; 4–7 October 2009; New Orleans. Richardson, Texas: SPE; 2009. SPE-124542-MS. DOI: 10.2118/124542-MS

[17] Strazza D, Grassi B, Demori M, Ferrari V, Poesio P. Core-annular flow in horizontal and slightly inclined pipes: Existence, pressure drops, and hold-up. Chemical Engineering Science. 2011;**66**: 2853-2863

[18] Joseph DD, Bai R, Chen KP, Renardy YY. Core-Annular Flows. Annual Review of Fluid Mechanics. 1997;**29**:65-90

[19] Schaan J, Sanders RS, Litzenberger C, Gillies RG, Shool CA. Measurement of heat transfer coefficients in pipeline flow of Athabasca Bitumen froth. In: Proceedings of 3rd North American Conference of Multiphase Technology; 6–7 June 2002; Banff. Cranfield, UK: BHR Group. 2002. pp. 25-38

[20] Sanders RS, Ko T, Bai R, Joseph DD. Factors governing friction losses in self lubricated transport of bitumen froth: 1. Water release. The Canadian Journal of Chemical Engineering. 2004;**82**:735-742

[21] Bannwart AC, Rodriguez OMH, Biazussi JL, Martins FN, Selli MF, de Carvalho CHM. Water-assisted flow of heavy oil in a vertical pipe: Pilotscale experiments. International Journal of Chemical Reactor Engineering. 2012;**10**: A32. DOI: 10.1515/1542-6580.2965

[22] Gadelha AJF, de Farias Neto SR, Swarnakar R, de Lima AGB. Thermohydrodynamics of core-annular flow of water, heavy oil and air using CFX. Advances in Chemical Engineering and Science. 2013;**3**:37-45

[23] Rushd S. A new approach to model friction losses in the water-assisted pipeline transportation of heavy oil and bitumen [Thesis]. Edmonton: University of Alberta; 2016

[24] Charles ME, Govier GW, Hodgson GW. The pipeline flow of equal density oil water mistures. Canadian Journal of Chemical Engineering. 1961;**39**:27-36

[25] Bannwart AC, Rodriguez OMH, de Carvalho CHM, Wang IS, Vara RMO. Flow patterns in heavy crude oil–water flow. Journal of Energy Resources Technology. 2004;**126**:184-189. DOI: 10.1115/1.1789520

[26] Taitel Y, Dukler AE. A model for predicting flow regime transitions in horizontal and near horizontal gas-liquid flow. AIChE Journal. 1979;**22**(1): 47-55

[27] Holland FA, Bragg R. Fluid Flow for Chemical Engineers. 2nd ed. New York: Elsevier; 1995

[28] Ooms G, Segal A, Van der Wees AJ, Meerhoff R, Oliemans RVA. A theoretical model for core-annular flow of a very viscous oil core and a water annulus through a horizontal pipe. International Journal of Multiphase Flow. 1984;**10**(1):41-60

[29] Shook CA, Gillies RG, Sanders RS. Pipeline Hydrotransport with Applications in the Oil Sand Industry. SRC Publication No. 11508-1E02. Saskatoon: Saskatchewan Research Council; 2002

[30] Arney MS, Ribeiro GS, Guevara E, Bai R, Joseph DD. Cement-lined pipes for water lubricated transport of heavy oil. International Journal of Multiphase Flow. 1996;**22**(2):207-221

[31] Ho WS, Li NN. Core-annular flow of liquid membrane emulsion. AIChE Journal. 1994;**40**(12):1961-1968

[32] Oliemans RVA, Ooms G, Wu HL, Duijvestijn A. Core-annular oil/water flow: The turbulent-lubricating-film model and measurements in a 5 cm pipe loop. International Journal of Multiphase Flow. 1987;**13**(1):23-31

[33] Shi J, Al-Awadi H, Yeung H. An experimental investigation of high-viscosity oil-water flow in a horizontal pipe. The Canadian Journal of Chemical Engineering. 2017;**95**(12):2423-2434

[34] McKibben MJ, Gillies RG, Shook CA. A laboratory investigation of horizontal well heavy oil–water flows. The Canadian Journal of Chemical Engineering. 2000;**78**:743-751

[35] de Andrade THF, Crivelaro KCO, de F Neto SR, de Lima AGB. Numerical study of heavy oil flow on horizontal pipe lubricated by water. In: Öchsner A, da Silva LFM, Altenbach H, editors. Materials with Complex Behaviour II, Advanced Structured Materials, Vol. 16. Berlin Heidelberg: Springer-Verlag; 2012. pp. 99-118. DOI: 10.1007/978-3-642-22700-4_6

[36] Sakr AL, Sultan GI, Safan MG, Tolba MM. Flow structure development due to injection water into the annulus of heavy oil pipe flow. In: Proceedings of the North Africa Technical Conference and Exhibition; 20–22 February 2012; Cairo. Richardson, Texas: SPE; 2009. SPE-151402-MS. DOI: 10.2118/151402-MS

[37] Shi J, Gourma M, Yeung H. CFD simulation of horizontal oil-water flow with matched density and medium viscosity ratio in different flow regimes. Journal of Petroleum Science and Engineering. 2017;**151**:373-383

[38] Rushd S, Sanders RS. A new approach to model friction losses in the water-assisted pipeline transportation of heavy oil and bitumen. Canadian Journal of Chemical Engineering. 2019 (Accepted on 5th November 2018)

[39] Huang A, Christodoulou C, Joseph DD. Friction factor and holdup studies for lubricated pipelining – II: Laminar and k-ε models for eccentric core flow. International Journal of Multiphase Flow. 1994;**20**(3):481-491

[40] White FM. Fluid Mechanics. 4th ed. Boston: McGraw-Hill; 1999. p. 349

[41] Bird RB, Stewart WE, Lightfoot EN. Transport Phenomena. 2nd ed. New York: John Wiley & Sons; 2001

[42] Pope SB. Turbulent Flows. Cambridge: Cambridge University Press; 2000

[43] Wallin S. Engineering turbulence modelling for CFD with a focus on explicit algebraic Reynolds stress models [Thesis]. Stockholm: Royal Institute of Technology; 2000

[44] University of Ljubljana (Department of Physics). Turbulence Models in CFD [Internet]. 1999. Available from: http://mafija.fmf.uni-lj.si/seminar/files/2006_2007/Turbulence_models_in_CFD.pdf [Accessed: September 02, 2018]

[45] Chalmers University of Technology (Department of Thermo and Fluid Dynamics). An Introduction to Turbulence Models, Publication 97/2 [Internet]. 2011. Available from: http://www.tfd.chalmers.se/~lada/postscript_files/kompendium_turb.pdf [Accessed: September 02, 2018]

[46] Aupoix B, Arnal D, Bezard H, Chaouat B, Chedenergne F, Deck S, et al. Transition and turbulence modeling. Aerospacelab. 2011;**2** (AL02-01):1-13. Available from: https://hal.archives-ouvertes.fr/hal-01181225

[47] Fletcher DF, Geyer PE, Haynes BS. Assessment of the SST and omega based Reynolds stress models for the prediction of flow and heat transfer in a square section u-bend. Computational Thermal Sciences. 2009;**1**:385-403

[48] Amano RS, Guntur K, Lucci JM. Computational study of gas turbine blade cooling channel. In: Proceedings of the 14th International Heat Transfer Conference (Volume 5); 8–13 August 2010; Washington. New York: ASME; 2010. IHTC14-22920: pp. 239-247. DOI: 10.1115/IHTC14-22920

[49] Mothe KK, Sharif MAR. Assessment of the omega based models in the prediction of incompressible turbulent flows over flat rough surfaces. In: Proceedings of the 44th AIAA Aerospace Sciences Meeting and Exhibit; 9–12 January 2006; Reno. Reston: AIAA; 2010. AIAA 2006–0718: pp. 8707-8715. DOI: 10.2514/6.2006-718

[50] Bonkdari H, Lipeme-kouyi G, Asawa GL. Developing turbulent flows in rectangular channels: A parametric study. Journal of Applied Research in Water and Wastewater. 2014;**1**(2):53-58

[51] Zhang XW, Jin XL, Wang Y, Deng K, Ren ZM. Comparison of standard k-ε model and RSM on three dimensional turbulent flow in the SEN of slab continuous caster controlled by slide gate. ISIJ International. 2011;**51**(4): 581-587. DOI: 10.2355/isijinternational.51.581

Chapter 6

Ablation of Oil-Sand Lumps in Hydrotransport Pipelines

Mahshad Pazouki and Sayeed Rushd

Abstract

Oil-sand ore is a kind of heavy crude oil found primarily in Canada. The surface mining of this petroleum resource requires expensive 400-ton capacity trucks to transport the ore to the slurry plant. The slurry prepared with the crushed ore is usually conditioned in a hydrotransport pipeline prior to extracting bitumen. As the elimination of the mammoth trucks has a tremendous economic and environmental incentive, it is of industrial interest to employ new processes capable of conditioning oil-sand right at the mine face. This would demand an accelerated rate of conditioning compared to what is achieved at present in the industry. One of the significant steps of the conditioning process is oil-sand lump ablation (OSLA). An understanding of the fundamental concepts associated with OSLA is essential to achieve any industrial-scale change in the current conditioning method. A number of parameters such as temperature, lump size, pipe diameter, pipe length, flow rate, and shear influence the ablation process. The current chapter introduces the concept of OSLA. It also includes a comprehensive review of the most important models available to predict the ablation rate and the scope of future works.

Keywords: heavy oil, bitumen, transportation, conditioning, ablation rate, modeling

1. Introduction

About 15–20% of the Canadian oil-sand reserve containing 140 billion barrels of bitumen can be found at a depth less than 75 m, where surface mining is economically feasible [1]. The remaining 80–85% buried at higher depths can possibly be recovered with underground-type mining employing in situ techniques. The Great Canadian Oil Sands, now known as Suncor Energy Inc., developed an open-pit mine, a hot water extraction plant and an upgrading complex in 1967. Their operation was followed in 1979 by Syncrude Canada Ltd.'s open-pit mine at Mildred Lake. Currently, Suncor Energy Inc., Syncrude Canada Ltd., Albian Sands Energy Inc., and Canadian Natural Resources Ltd. employ surface mining [1–3].

A typical surface mining operation, as shown in **Figure 1**, includes the following steps [1, 2]:

1. Removal of overburden using shovels and trucks

2. Mining oil-sand with hydraulic or electric shovels

3. Transport of oil-sand ore from the mine face to the crushers with trucks

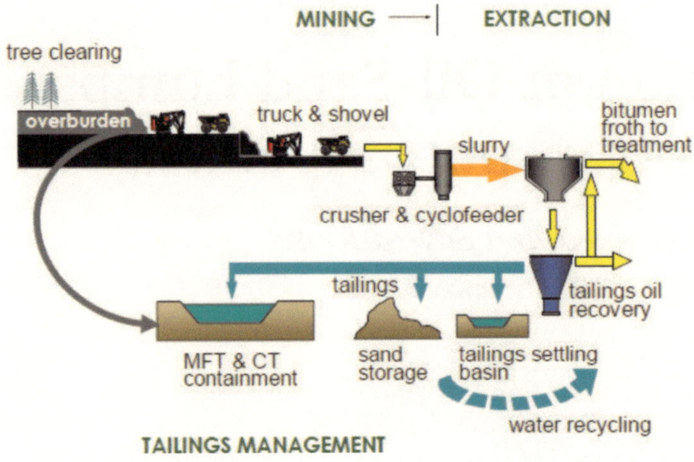

Figure 1.
Major steps for bitumen recovery in surface mining operation [1].

4. Crushing of large oil-sand lumps into smaller parts

5. *Oil-sand conditioning*

6. Gravity separation of bitumen froth

7. Diluted froth treatment to separate water and solids

8. Supplemental solvent recovery from tailings

9. Dewatering and concentrating the tailings

Among the above mentioned steps, *oil-sand conditioning* is probably the most important phase that can further be divided into three stages [1, 3]:

i. Oil-sand lump ablation or size reduction

ii. Liberation of bitumen from sand grains

iii. Aeration of bitumen droplets

Oil-sand ore carried from the mine with conveyors was originally conditioned with rotating drums or tumblers [1]. Investigations into the possibility of replacing conveyors and tumblers with pipelines began at the Syncrude Research Center in the early 1980s [3, 4]. The idea was based on the examination of the pipelines transporting tailings from the separation vessels to the tailing ponds. Subsequently, a large-scale prototype of an oil-sand hydrotransport system known as the extraction auxiliary production system was commissioned in 1993. It became a successful commercial unit that could digest up to 5000 ton of oil-sand per hour. Since then, hydrotransport pipelines are being used in other commercial extraction plants to simultaneously transport and condition oil-sand ore. Roughly 60, 000 ton of oil-sand flowing as slurry of about 60 wt% solids is digested per hour to produce ½ million barrels of bitumen per day at present [3, 5]. Commercial applications of hydrotransport pipelines enabled the conditioning process to be carried out at considerably lower temperatures [6]. Syncrude Aurora now operates their 5 km hydrotransport line at 35–40°C [7].

The use of hydrotransport pipelines improved the oil-sand conditioning at lower temperatures [1]. As the efficiency of the conditioning process significantly influences the final recovery of the bitumen, achieving an efficient method for the conditioning has always been the target among oil-sand producers [1, 3]. In addition, reducing the production cost and greenhouse gas emissions has also been ongoing targets in the industry [1]. Continued process improvements have led to more economic bitumen production process and reduced environmental footprint [8]. Presently, oil-sand companies aim to eliminate the use of expensive trucks to transport the ore to the slurry plant by conditioning the oil-sand slurry at the mine face [2]. This kind of onsite conditioning would demand a consequent reduction of pipeline length, which would potentially cause incomplete conditioning and could especially affect lump ablation [2]. In order to resolve the problem, an accelerated conditioning process would be required. Before implementing any significant changes in the oil-sand conditioning, a better understanding on the fundamentals of OSLA must be developed.

As the first step of conditioning, the crushed and screened oil-sand lumps (size range, 50–150 mm) are ablated or digested inside the hydrotransport pipeline [9]. The ablation occurs due to the dual effects of heat transfer and mechanical energy [6]. Generally, bitumen acts as a glue to hold the matrix of sand grains together. It should be mentioned that the viscosity of bitumen is a strong function of temperature. As a result, the viscosity of the bitumen on the surface layer of the lump reduces considerably as an oil-sand lump is exposed to the hot slurry medium. The softened surface layer is then sheared away due to the shearing inside the hydrotransport pipeline, and, thereby, a new lump surface is exposed to the hot medium. The new surface undergoes the same mechanism. This process repeats itself to the point that the entire lump is ablated. The heat transfer from the slurry and the contacts of a lump to neighboring lumps as well as the pipe wall control the rate of ablation [10]. The parameters that mostly affect OSLA can be identified as temperature and composition of the slurry, size and temperature of the initial lump, and mechanical shear imparted to the lump [1, 11].

The current chapter aims to contribute to the better understanding of the lump ablation process. It would assist to identify the important parameters that affect the OSLA and to recognize the way in which each one of these parameters influences the ablation process. The oil-sand conditioning process can be improved by changing one or more of these factors to achieve accelerated ablation and conditioning as required. Previous studies of oil-sand lump ablation are also described, and important areas not covered by those investigations are identified. In addition, some background theory needed in the development of improved ablation models is introduced.

2. Effects of operating parameters on ablation

As mentioned previously, the ablation of oil-sand lumps is the result of two important phenomena: heat transfer and shear forces [1]. Any factor that affects these phenomena would have an effect on the ablation process. The most important of this kind of parameters, viz., slurry temperature, initial lump size, pipeline diameter, and pipeline velocity or shear stress, are discussed here.

2.1 Slurry temperature

Since the commissioning of the first commercial oil-sand mining and bitumen extraction operation, attempts to reduce bitumen production costs and environmental impacts have driven process improvements and flowsheet changes [6]. One way to reduce both simultaneously is to operate the process at a lower slurry temperature. Slurry temperature is effectively the most important parameter in the extraction of bitumen from Athabasca oil-sand, as it affects all three steps of oil-sand conditioning, i.e., oil-sand lump ablation, bitumen liberation, and air attachment.

Bitumen, because of its high viscosity, holds the mixture of sand grains and fine minerals together within an oil-sand lump [6]. The viscosity of bitumen decreases sharply with increasing temperature as shown in **Figure 2**. Since the viscosity is above 10^5 mPa.s at room temperature (T = 20°C), bitumen looks like a solid, and it is essentially impossible to separate bitumen from the sand grains at this temperature. When the temperature increases to 50°C, bitumen viscosity reduces by more than one order of magnitude, i.e., to the order of 10^3 mPa.s. The separation of bitumen from sand grains occurs relatively quickly at such temperature. Lower viscosity of bitumen must be attained in order to reduce the lump size and liberate bitumen from sand grains efficiently [6, 12]. It should be mentioned that froth quality and bitumen recovery are also dependent on slurry temperature as the temperature influences the air bubble-bitumen attachment. Although oil-sand producers ultimately wish to operate hydrotransport pipelines at low temperatures due to the reduced operating costs and environmental impacts, they are well aware that a balance between bitumen recovery and temperature reduction must be attained. At this point, operating temperatures between 40 and 55°C are common in the industry [6].

2.2 Initial lump size

The heat transfer to an oil-sand lump is an important factor in ablation process [13]. The thickness of the layer softened and ablated away is equal in essence for all lump sizes under comparable thermal conditions. Therefore, the fraction of lump mass that is removed decreases as the lump size increases. In other words, the time necessary for complete digestion of a lump increases with the initial lump size [1].

2.3 Pipeline diameter

Pipeline diameter is a key parameter in the ablation of an oil-sand lump. For a specific slurry flow rate, more energy is dissipated in a pipe having smaller

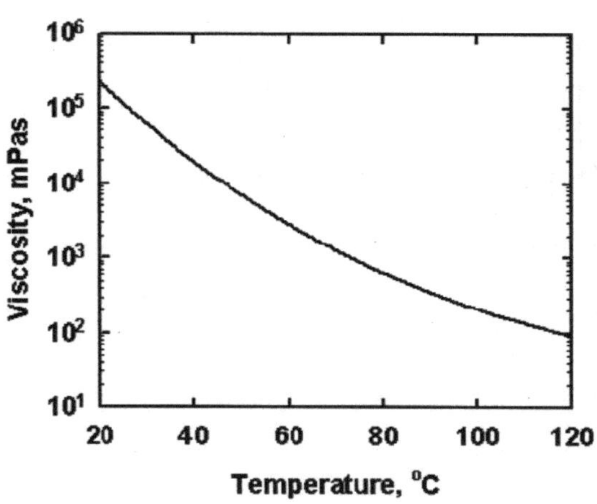

Figure 2.
Viscosity of Athabasca bitumen as a function of temperature [1].

diameter, and this leads to the higher interparticle shear stresses [1]. The relationship between energy dissipation and pipe diameter can be expressed as follows [14]:

$$\bar{\varepsilon} = \left(2f_f V^3\right)/D \tag{1}$$

where,

$\bar{\varepsilon}$ is the volumetric average of energy dissipation in the pipeline (m^2/s^3),V is the bulk velocity (m/s), D is the pipe diameter (m), and f_f is the Fanning friction factor.

According to Eq. (1), energy dissipated in a pipe inversely varies with the diameter of the pipe. Consequently, higher energy dissipation, i.e., higher ablation rate, is expected in smaller pipelines. On the other hand, the residence time of an oil-sand lump inside a smaller pipe will be shorter as the bulk velocity is higher in the pipe for a constant slurry flow rate. A shorter residence time is likely to result in a lower ablation rate [1]. A balance must be sought between the residence time and the energy dissipation in order to achieve an acceptable ablation rate. That is, the pipeline diameter should be optimized for a specific flow condition.

2.4 Shear stress

An oil-sand lump moves at a different velocity than the surrounding slurry and contacts with the pipe wall as well as other lumps present in the slurry. The difference in the velocities results in a shear stress on the lump surface [1]. In a pipeline, the shear force acting on the lump surface is expected to increase with increasing mixture velocity and concentration. Higher slurry concentrations promote particle-particle interactions and particle-wall interactions [1, 3]. Thus, all steps of the oil-sand slurry conditioning process depend on the slurry velocity in a hydrotransport pipeline [5]. However, more research has been conducted to study the effect of shear exposure on bitumen liberation and bitumen aeration than lump ablation. Clearly, this area demands further investigations.

3. Examples of previous studies

Important examples of previous experimental and modeling studies on lump ablation are briefly discussed in this section. The parameters considered in each study are summarized, and the parameters that need further investigations are identified.

3.1 Traynis wheel test stand

Traynis [15] studied the grinding or crushing of coal particles using a wheel test stand (WTS). It was experimentally demonstrated that the pressure losses for slurry in a WTS agree quite well with that in a horizontal pipeline. This agreement was a result of the similarities in energy dissipation mechanisms of moving solid particles in both systems. The mechanism of energy dissipation was found to determine the process of the particle crushing.

Most of the tests reported in [15] were completed using a WTS made of a 200-mm diameter pipe. Three runs were repeated with another WTS made of 300-mm diameter pipe to investigate the effect of pipe size on particle crushing or grinding. Pipes were polished using abrasive materials like quartz to ensure that the pipe wall was smooth when the experiments were started. The apparatus was filled up to 1/3 of the total volume with a mixture of water and coal. For each run using the smaller WTS, 10–40 kg of coal was loaded. Coal particles from two different hydraulic mines were used for the experiments. At certain time intervals, the degree of size reduction of the coal particles was determined by measuring the particle size distribution of the remainder of the particles. To reconfirm that the WTS system was representative of the horizontal pipeline, a number of experiments were replicated by circulating slurry in 4 and 1.3-km long pipelines. The effects of slurry velocity and concentration, pipe diameter, coal particle size, existence of abrasive rocks, pipe length, and mechanical properties (strength and hardness) of the coal particles on the size reduction of the coal particles were investigated. The outcomes of the experimental investigation can be summarized as follows:

1. The results obtained using short pipelines (~10 km) agreed with those of the wheel test stand experiments. However, transporting the coal particles for a long distance (>20 km) resulted in faster crushing. This is probably because the mixture passed through the feed pump many times. It caused more rapid size reduction of the coal particles. The effect was more evident when the initial coal particle size was large.

2. The slurry velocity was changed from 1.8 to 6 m/s for experiments with different coal types. In all cases, velocity had an insignificant effect on the crushing of coal particles.

3. Slurry concentrations were varied from 1:16 to 1:2 (mass of solid, mass of liquid). These experimental runs showed that size reduction of the coal particles was independent of slurry concentration.

4. The pipe size of the wheel test stand did not affect the extent of size reduction of the coal particles.

5. For coal particles having initial diameters in the ranges of 3–6 or 50–100 mm, the degree of size reduction was only affected by the initial particle size. For particles smaller than 3 mm, increasing the initial particle size resulted in more visible increase in the intensity of crushing.

6. The existence of abrasive rocks, which were 50–100 mm in diameter initially but were crushed to the 3–6 mm size range in the slurry, caused more rapid size reduction of particles. Considerable crushing was observed within the first 15 km.

7. The crushing rate was higher in the first few kilometers of the pipe, and it decreased as coal particles moved along the pipeline. This must be because of the fact that shear stress decreases as particle size decreases, which would be expected when the particle slip velocity decreases. Rounding of the edges of the particles within the first kilometers of the pipe might be another reason for reduction of the crushing rate with pipeline length.

8. Experiments using coal particles with initial size in the range of 6–13 mm showed that as the strength factor of the coal particles increased, the crushing rate decreased. Strength factor is an indicator of the grindability of the coal particles.

One of the strengths of this study is that it introduces a new experimental method for studying the mass loss of solid particles. In addition, this is the only study done on the effect of slurry velocity and concentration on the particle mass loss in slurry pipelines. However, one cannot directly apply the results of this research to oil-sand hydrotransport pipeline. This is because the nature of coal particles is very different from oil-sand lumps. Coal is a brittle organic sedimentary rock that contains varying amounts of carbon, hydrogen, nitrogen, oxygen, and sulfur [16]. On the other hand, oil-sand contains bitumen, sand grains, clays, and small amount of water, and the viscosity of bitumen highly varies with temperature [1]. As coal is brittle, coal particles tend to break down into smaller particles when they are exposed to the shear forces. However, mass loss from oil-sand lumps occurs by gradual mass removal from the surface of the lump. The amount of mass loss from an oil-sand lump depends on many parameters, although temperature seems to be the most important factor. Thus, one cannot study oil-sand lump ablation without considering the slurry temperature. Additionally, because the nature of the two materials is different, the effect of slurry concentration and velocity on their mass loss can be expected to be different.

3.2 Law et al. experimental study

Law et al. [17] examined the ablation of frozen mixtures of water and paraffin wax (octadecane) with solid particles such as kaolinite clay, titanium oxide, aluminum powder, and sand. Because the ablation of oil-sand samples was complex, they chose to study the ablation of less complex materials. This investigation was conducted with the purpose of obtaining information from a well-controlled system and applying it for designing the rotating drums, which were used for oil-sand lump ablation at that time. A turbulent axisymmetric water jet, whose velocity varied between 1.7 and 2.8 m/s, was used. The temperature of the jet was changed from 32 to 60°C depending on the material being tested. The water temperature was chosen based on the sample's melting point, which was 26–29°C for octadecane. Cylindrical samples (L = 150 mm, d = 11 mm) were manufactured and immediately frozen in liquid nitrogen. The samples were then placed in front of the jet using a sliding platform. The sliding platform was moved up and down by using a stepping motor. Before the start of the experimental run, the front of the sample was placed in-line with a certain point called the melt front pointer. During an experimental run, the sample was never moved from this point. Instead, the sliding platform was lowered at a speed equal to the ablation rate of the frozen sample. The downward movement of the sliding platform against time was recorded and plotted. The slope of this line, for each set of the experiments, was considered to be the ablation rate. It was observed that for each operating condition, the slope of the plotted line remained constant with time. The results of this study can be summarized as follows:

1. For all samples, an increase in the jet temperature increased the ablation rate. For instance, ablation rate of lumps made from octadecane and 60 (vol%) sand at V = 2.8 m/s and T = 60°C was equal to 4.4×10^{-3} m/s, whereas it was equal to 3×10^{-3} m/s at T = 50°C.

2. Increasing the jet velocity from 1.7 to 2.8 m/s increased the ablation rate equivalent to that of raising the jet temperature by approximately 10°C (from 50 to 60°C). This is because surface shear stress on the sample is proportional to V^2.

3. Addition of solid particles to the samples affected the ablation rate in a complex way. The effect depended on the type of the solid particles and the lump material because thermal conductivity of the solid particles was different, so the heat transfer coefficient within the sample differed depending

on the solid particles. In the case of octadecane-kaolinite samples, at a certain jet velocity and temperature (V = 2.06 m/s, T = 50 and 60°C), ablation rate gradually increased for solid content up to 17 vol%, and, for solids content beyond 17%, the ablation rate increased significantly. The reasons for this observation were mentioned to be the increase of the heat transfer area due to the roughening of the melting surface and the ablation of the sample as clusters instead of layers at high solid contents. That is, ablation can also depend on the type and components of the oil-sand ore.

Although this experimental study could provide a good indication of the way different parameters influence ablation, the study had some limitations such as:

i. Viscosities of the samples were significantly different from that of bitumen.

ii. Sample was stationary and was exposed to water only from one face.

iii. Ablation with only water was investigated.

iv. The number of experimental runs using samples manufactured with sand particles was very limited.

More investigations need to be conducted on actual OSLA using a system that better represents the actual conditioning medium. These experiments are necessary to develop a predictive model applicable for all operating conditions.

3.3 SRC experimental study

In 1996, an experimental study on the ablation of actual oil-sand lumps was conducted at the Saskatchewan Research Council (SRC) Pipe Flow Technology Centre™ [10]. Lump ablation at various operating conditions were investigated by loading a certain amount of lumps to a 264 mm pipe loop using a feeder close to the pump discharge. At certain time intervals, lumps were trapped with a basket before discharging into the storage tank. The trapped lumps were weighed and put back into the loop. The SRC experiments showed the following results:

1. For all types of the oil-sand lumps, the time or pipeline length required to reach a certain ablation rate was strongly dependent on the slurry temperature. As shown in **Figure 3**, the time required for the complete ablation at T = 50°C was one third of that at T = 30 and 18°C for soft lumps. It is worth noting that the effect of temperature on the ablation rate was found to be qualitatively similar for various oil-sand ores [1, 10].

2. The ablation rate was faster for lumps with lower initial temperature at higher slurry temperatures (T = 30 and 50°C)

This experimental study is very valuable, as it is the first available research on the ablation of actual oil-sand lumps. However, the research did not include the effects of important parameters such as slurry velocity and concentration on ablation.

3.4 Masliyah ablation model

Masliyah ablation model (model 1) was developed for the ablation of oil-sand lumps based on the fact that crushed oil-sand lumps and sand particles form a moving layer at the bottom of the pipe, while the fine solids—water blend known as the carrier fluid, exist within the whole pipe cross section [1, 6, 10]. The height of the bottom layer reduces with axial position along the pipeline when lump size reduces. It was assumed that the heated surface layer of the lump peeled off because of the shear stress inside the pipe. This process was presumed to repeat up to the point when 95% of the lump mass would be removed. The SRC two-layer model was used to estimate the velocities of top and bottom layers of the flow. An average value of the shear stress on the lumps was calculated using the slurry viscosity and the axial flow velocity [1, 6]. The model was validated with the data available in [10]. The predicted effects of mixture velocity, pipeline diameter, slurry temperature, initial lump size, slurry density, and pipeline length are discussed as follows [1]:

1. Raising the slurry velocity does not cause any significant enhancement in the oil-sand lump ablation rate. As shown in **Figure 4**, the beneficial effects of increasing velocity are offset by the reduced residence time of the lump.

2. The rate of lump digestion reduces with the increasing pipe diameter at T = 25°C (**Figure 5A**). As a result, longer pipelines are needed for complete ablation. At T = 50°C, however, the effect of pipe diameter is not very

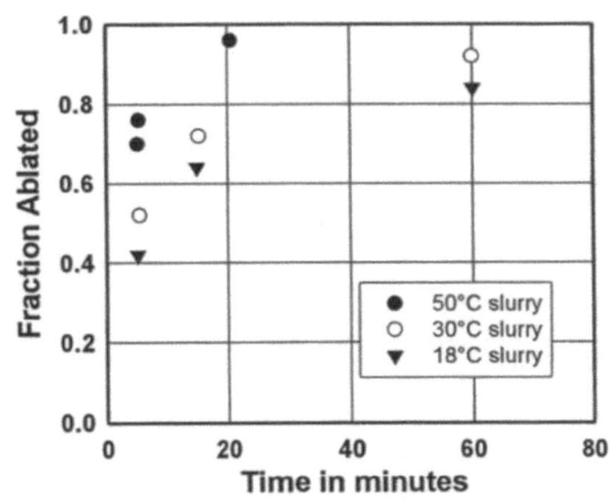

Figure 3.
SRC Experimental results showing the effect of slurry temperature on the ablation time of soft lumps [10].

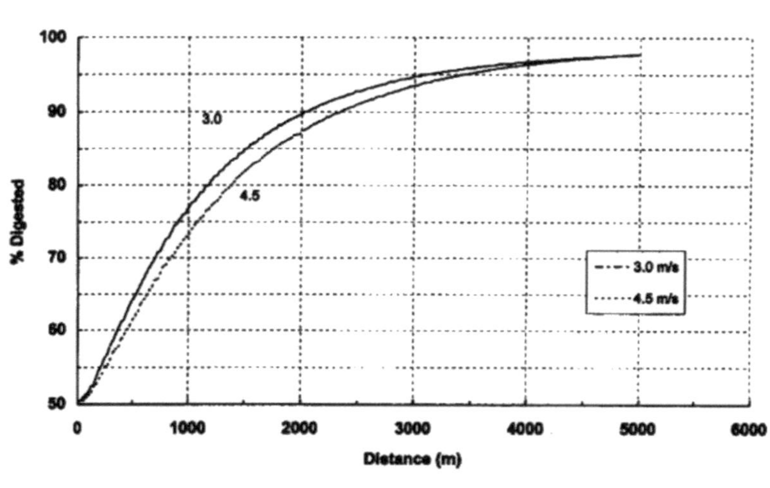

Figure 4.
Effect of mixture bulk velocity on the oil-sand lump ablation (T = 50°C; T_{oL} = 5°C; d = 20 mm) [1].

significant (**Figure 5B**). This is probably because heat transfer is the dominant factor in the ablation of an oil-sand lump. At a higher temperature, the viscosity of the bitumen and the surface shear stress required for consequent removal of the soft surface layer reduce significantly.

3. Increasing the slurry density increases the ablation rate at a fixed temperature (**Figure 6**). This is because an increment in slurry density is caused by a reduced water flow rate, i.e., an increased residence time. For the same reason, ablation rate reduces with decreasing slurry density.

4. Smaller oil-sand lumps ablate more rapidly (**Figure 7**). A given mass of oil-sand ore in the form of smaller lumps is digested much faster than the same mass of ore in the form of larger lumps due to the larger surface area of the former.

Although model 1 is one of the pioneer models that can be used for industrial-scale ablation of oil-sand lumps, it has a number of limitations. Some of the limitations are identified as follows:

i. Only one set of experimental data was available for validating this model. The data was also limited to a certain pipeline size and a set of velocities.

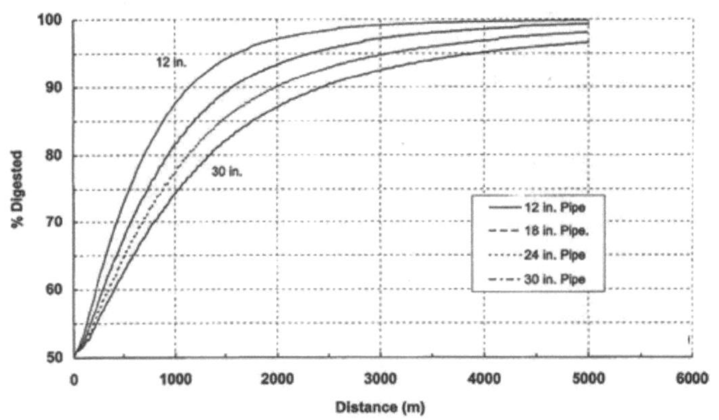

(A)

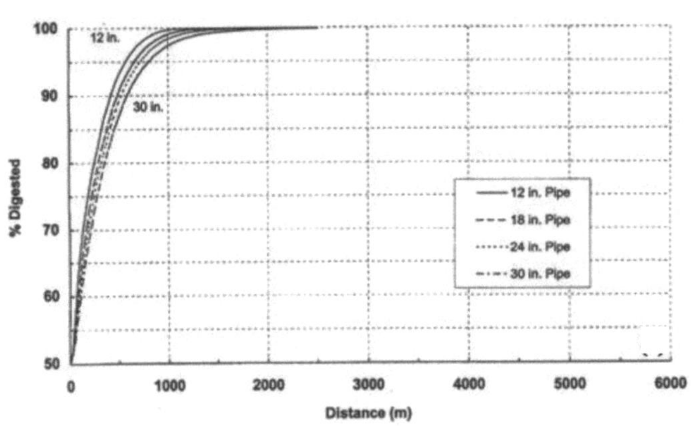

(B)

Figure 5.
Effect of pipe diameter on the ablation of oil sand lumps (T_{oL} = 5°C; V = 3.8 m/s; ρ_m = 1550 kg/m^3; d = 20 mm): (A) T = 25°C; (B) T = 50°C [1].

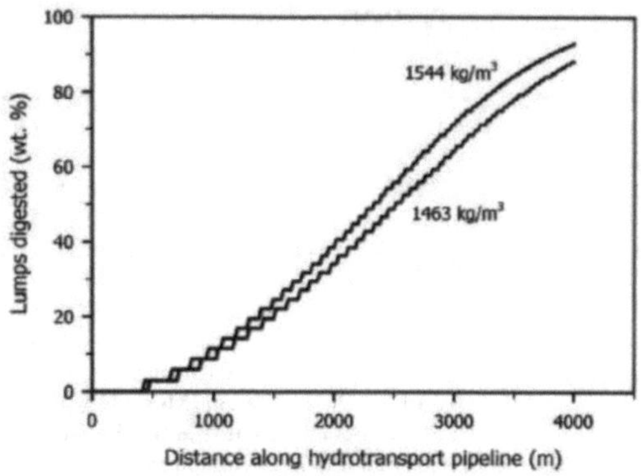

Figure 6.
Effect of slurry density on the ablation of oil sand lumps (T_{oL} = 0°C; D = 710 mm; d = 100 mm and T = 40°C) [1].

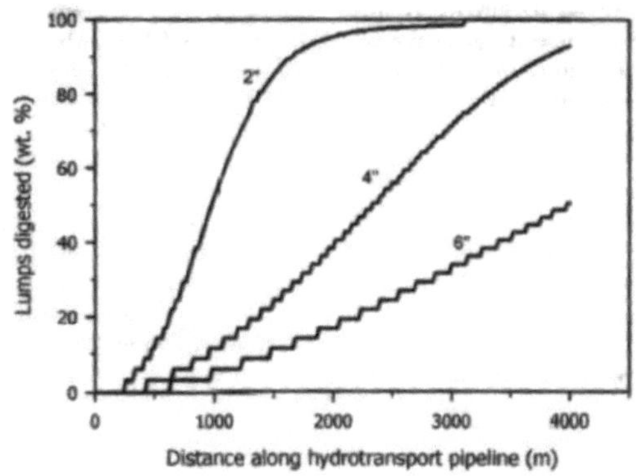

Figure 7.
Effect of initial lump size on the ablation of oil-sand lumps (T_{oL} = 0°C; D = 710 mm; ρ_m = 1548 kg/m³ and T = 40°C) [1].

ii. A part of the model, namely, the method of estimating surface shear force on the lumps is yet to be verified.

iii. The version of the SRC two-layer model used for the modeling is out-of-date. The SRC model has been improved lately. However, the improvements were never integrated into model 1.

iv. According to model 1, the bulk velocity was used to calculate the shear stress acting on the oil-sand lumps. However, the velocity of the lumps in slurry is about 90% of the bulk velocity. This means, the relative velocity causing the shear stress on the lump is likely to be only 10% of the velocity. Therefore, the shear stress used in Model 1 might be overestimated.

3.5 Eskin et al. ablation model

Eskin et al. model (Model 2) was developed for the ablation of a spherical oil-sand lump using a hypothesis similar to that of Masliyah model [13]. The following assumptions were used for the purpose:

1. The effects of shear stresses and heating were integrated with respect to a surface critical temperature. The critical temperature remained constant

during the ablation and was related to the minimum adhesive strength of the bitumen. This model-specific temperature was determined using the experimental results available in [17].

2. The lump was considered to retain its spherical shape during ablation.

Based on the assumptions, the complex phenomena were simplified as a one-dimensional heat conduction problem for a shrinking sphere. The critical temperature was used as an input to the applied convection boundary condition. A correlation between the reduction of sphere size and the critical temperature was found by simplifying the boundary condition-based equations. The heat conduction equations were also included in the model. The model was applied to study the effect of lump size and initial temperature on OSLA.

Although the fundamental concepts used for model 2 agree with general hypothesis of how individual oil-sand lumps ablate, the way of integrating shear stress to a critical temperature is debatable. Moreover, the effects of slurry velocity and slurry concentration on ablation cannot be investigated with the model. Most importantly, the model was never validated against any experimental data.

3.6 Pazouki study

Pazouki [18] applied an innovative method by using strain gauge technology for online tracking of the mass loss of the anchored oil-sand lumps. Four small strain gauges looped in a full-bridge circuit were used for the measurements. The new method also allowed measuring the drag force on ablating and non-ablating objects. The accuracy of the drag force measuring technique was evaluated by comparing the measured drag force with the calculated drag force on the number of smooth spheres in water.

As part of the experiment, an idealized oil-sand lump was anchored in a basket at the height of $40D_L$, where D_L is the diameter of the cylindrical lump. Artificially manufactured oil-sand lumps were used for the experiments. Slurries ($C = 0.15$ and 0.30) were prepared by mixing pre-weighed industrial quartz ($d_{50} = 0.190$ mm) with water. The flow temperature was adjusted using a double-pipe heat exchanger located in the vertical section of the pipeline loop. DASYlab 10.0 software was used to the readings of strain gauges, temperature, and flow meter.

In addition to experimentation, Pazouki [18] also developed a new modeling approach. The overall outcomes of this study can be summarized as follows:

1. The online measurement method and the experimental apparatus built at SRC provide the opportunity to test OSLA at many different operating conditions. The strain gauge measurement method also enables measuring the drag force on the ablating oil-sand lump.

2. The data obtained using the artificially produced oil-sand lumps were repeatable.

3. The OSLA was enhanced significantly with the increasing flow temperature. Heat transfer played a more important role for ablation in the slurry where surface shear stresses can be expected to be lower.

4. The slurry velocity was found to affect OSLA appreciably. The ablation rate increased with V^n, where $n = 2–4.7$. The value of n was a function of temperature.

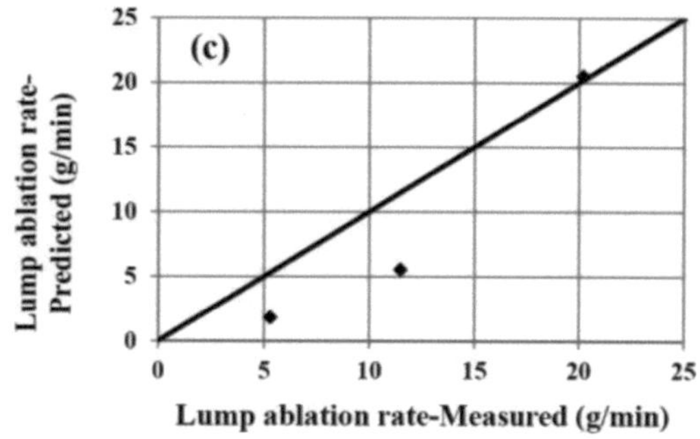

Figure 8.
Comparison of measured and predicted ablation rate for slurry (C = 0.30) at T = 45°C [18].

5. Ablation in water occurred at a slower rate than that in the slurry. However, an increase of slurry concentration from 15 to 30% at T = 30°C reduced the ablation rate. This reduction was most likely related to the turbulence modulation in slurries that could affect both shear force and heat transfer.

6. For the range of concentrations used for the study, the slurry concentration did not have a significant influence on the ablation of oil-sand lumps.

7. The drag force acting on a lump depended on the slurry concentration.

8. The equivalent fluid model was found to be most appropriate for simulating the drag force. In this model, the slurry is treated like a single-phase fluid with density and viscosity related to the solid concentration.

9. The proposed model demonstrated the ablation rate to be a primary function of surface shear force and temperature. In the model, the effect of temperature was implemented as the change in the bitumen viscosity.

10. The model was validated with respect to the experimental measurements. An example of the validation results is presented in **Figure 8**. In course of the validation, it was found to better predict the ablation rate than other existing models.

Even though the study advanced both experimentation and modeling of OSLA, it suffers from the similar limitations of previous studies. It neither clarifies the procedure to apply the experimental findings to the actual hydrotransport pipelines nor verifies the model with industrial-scale data.

4. Theory and modeling

In order to estimate the ablation rate of the oil-sand lumps inside the hydrotransport pipeline, the shear stress acting on the lumps and the temperature profile of the lump at different times must be determined. The most relevant models for the estimation are follows:

1. SRC two-layer model

2. Shear stress decay law

The benefits and drawbacks of each model are discussed in the subsequent sections.

4.1 SRC two-layer model

Recall that an early version of SRC model was used to estimate the shear stress acting on the oil-sand lump(s) in model 1. If one intends to use a similar approach, it would be advisable to use a more recent version of the model. The theory behind the development of the SRC two-layer model is explained in this section. Also, the validity of this approach for calculating the shear stress acting on a lump is discussed.

In a slurry pipeline similar to the hydrotransport pipeline, fine particles (particles <0.074 mm) are considered to augment the viscosity and density of the suspending liquid, i.e., the carrier fluid. The coarse particles that are suspended by fluid turbulence are assumed to be at a constant volume fraction throughout the flow domain. The other part of the coarse particles, that is, the fraction not effectively suspended by fluid turbulence, is supposed to transmit the immersed weight to the pipe wall. These particles are found in the lower layer and contribute Coulombic or sliding bed friction [11]. The continuous concentration profile of the coarse particles was simplified to a step-change, i.e. two layers for the purpose of writing force balance equations. The velocity within each layer was assumed to be constant. **Figure 9** shows the idealized concentration and velocity distributions.

The SRC two-layer model was developed using mass and force balances for the two layers of the slurry. The force balance produces an equation for the axial pressure gradient in horizontal slurry as a function of friction losses in top layer, bottom layer, and the interface between the layers [19].

To estimate the ablation rate of an oil-sand lump, the actual shear stress on the lump needs to be estimated. If one assumes an oil-sand lump is located at the interface between the two layers, the surface shear stress acting on this lump can be assumed to be equal to the shear stress at the interface. The shear stress at the interface is calculated using the following correlation [11]:

$$\tau_{12} = \frac{1}{2} f_{12}(V_1 - V_2)|V_1 - V_2|\rho_1 \tag{2}$$

where ρ_1 denotes the slurry density in the upper layer and f_{12} is the interfacial friction factor that can be estimated from a modified Colebrook friction factor equation:

$$f_{12} = \frac{2(1+Y)}{\left[4\ \log_{10}(D/d) + 3.36\right]^2} \tag{3}$$

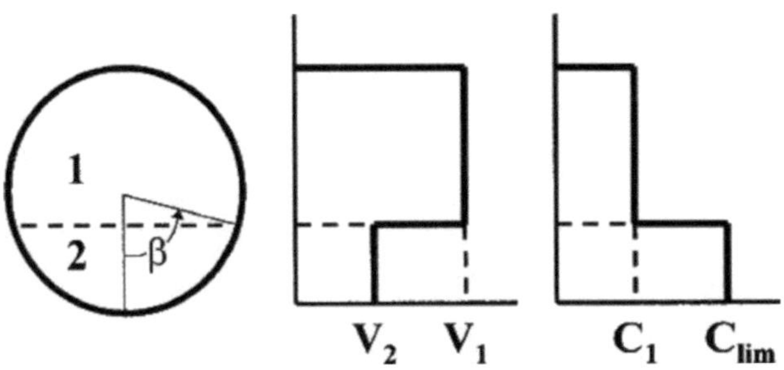

Figure 9.
Idealized concentration and velocity distributions used in the SRC two-layer model [19].

where Y is the 0 for d/D < 0.0015 and Y is calculated using the following equation when 0.0015 < d/D < 0.15:

$$Y = 4 + 1.42 \log_{10}(d/D) \quad (4)$$

Equation (4) has been formulated based on the data taken at Archimedes number, Ar $< 3 \times 10^5$.

It is inferred from Eq. (2) that the shear stress acting on a lump is proportional to $(V_1 - V_2)^2$. Calculating the velocity of the layers using the SRC two-layer model shows that V_2 is substantially small compared to V_1 and $(V_1 - V_2)$ is approximately equal to the slurry bulk velocity (V). On the other hand, other research conducted at the Saskatchewan Research Council showed that the velocity of a large particle in horizontal slurry flow was about 0.9 V. That is, the shear stress on the oil-sand lump would be proportional to $(0.1\ V)^2$, which is considerably smaller than $(V_1 - V_2)^2$. In other words, the shear stress calculated using Eq. (2) might overestimate the shear stress acting on an individual lump.

4.2 Shear stress decay law

In order to estimate the effects of slurry velocity and concentration on OSLA, a simplified approach can be taken instead of using complex two-layer model. The following assumptions are necessary to apply the method:

1. The oil-sand lump is stationary compared to the slurry flow in the hydrotransport pipe.

2. Slurry flow has reached steady state.

3. Slurry density is constant throughout the pipe.

Based on the assumptions, the local shear stress where the solid particle is located in the hydrotransport pipeline can be estimated using the shear stress decay law [11]:

$$\tau_{rz} = \frac{2s\ \tau_w}{D} \quad (5)$$

where s is the distance from pipe axis (m), τ_{rz} is the shear stress at y (Pa), τ_w is the pipe wall shear stress (Pa), and D is the pipe diameter (m).

For a known value of the wall shear stress, the shear stress at any radial position of the pipe can be calculated using Eq. (5). However, calculating the wall shear stress for a hydrotransport pipeline is complex as wall shear stress and flow density are not constant around the pipe.

Flow is not uniform throughout the cross section in a hydrotransport pipeline at all. Significant concentration and velocity gradients can exist, particularly, if operating velocity is just higher than the deposition velocity (V_c) [10]. However, the concentration profile was found to be nearly uniform for highly concentrated settling slurries at velocities significantly higher than V_c, and, for these slurries, Coulombic friction was also found to be negligible relative to kinematic friction [19]. Moreover, Coulombic friction is typically negligible as long as the particle diameter is not too large ($d_{50} \leq 0.3$ mm), and the mixture velocity is high (say, $V > 2V_c$) [11]. It is therefore worthwhile to consider the so-called kinematic friction loss component of the SRC pipe flow model [Eqs. (6) and (7)] to calculate the wall shear stress for such systems [10, 11]. Using this model, the kinematic friction loss component is determined so that it accounts for the friction associated with the flow of the carrier fluid, the friction related to particle collisions and the tempering effect of near-wall lift. If the Coulombic friction can be assumed to be negligible, then only the kinematic friction is important, and the τ_w can be calculated using the following equations [10, 11, 20]:

$$\tau_w = 0.5\, V^2 \left(f_f \rho_f + f_s \rho_s\right) \tag{6}$$

$$f_s = \lambda^{1.25} \left[A\, Ln\left(d^+\right) + B\right] \tag{7}$$

for d + ≤ 21: $A = -1.1 \times 10^{-4}$ *and* $B = 4.2 \times 10^{-4}$
for d + ≥ 21:$A = -5.6 \times 10^{-5}$ *and* $B = 2.6 \times 10^{-4}$

$$d^+ = \frac{d\, v^* \rho_f}{\mu_f} = \frac{d\left(f_f/2\right)^{0.5} V\, \rho_f}{\mu_f} \tag{8}$$

$$\lambda = \left[\left(\frac{C_{max}}{C}\right)^{1/3} - 1\right]^{-1} \tag{9}$$

The estimation of kinematic friction provides a tool for realistic assessment of the effects of slurry velocity and concentration on the local shear stress. To

Term	Value
D (mm)	103
s (mm)	20
ρ_f (kg/m^3)	1000
μ_f (Pa.s)	0.001
d_{50} (mm)	0.120
ρ_s (kg/m^3)	2650

Table 1.
Input parameters for estimating shear stress inside a pipe.

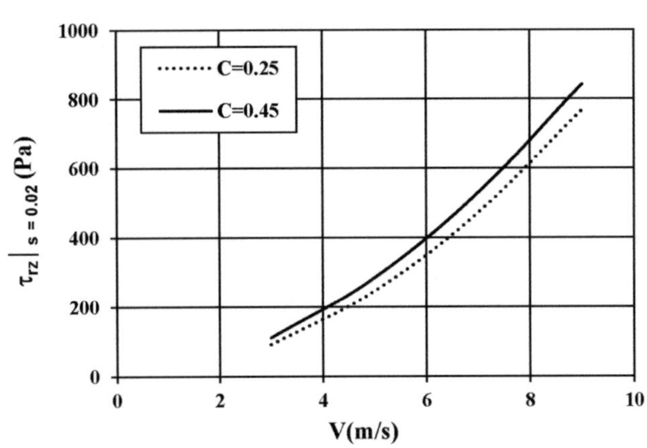

Figure 10.
Pipeline local shear stress at different slurry velocities and concentrations.

demonstrate the application, a model system is considered here. The system properties are chosen on an arbitrary basis and outlined in **Table 1**.

The local shear stress is shown in **Figure 10** as a function of mixture velocity for two different solid concentrations. The graph indicates that, for a solid particle placed 20 mm from the center of a 103 mm pipe, the surface shear stress on the particle increases substantially at higher mixture velocities. It should be noted that the graph indicates the effect of velocity and concentration on shear stress qualitatively. The results do not represent the actual shear stress on the oil-sand lumps. This example shows the application of shear stress decay law for a simplified case, where the solid particle is stationary. However, in the actual hydrotransport pipelines, oil-sand lumps move along the pipe axis. In order to estimate the shear stress acting on the lump, the slip velocity of the lump must be considered.

5. Conclusions

A limited number of studies on the ablation of large particles exist in open literature. Among these studies, only a few looked at the effect of velocity and concentration on the ablation. Slurry concentration and velocity were experimentally demonstrated to have minimum effect on the grinding or crushing of the coal particles [15]. The effect of slurry velocity on OSLA was also shown to be insignificant with the application of model 1 [1]. Interestingly, increasing velocity was found to cause a considerable increase in the ablation rate for stationary samples [17, 18]. An estimate of pipe local shear stress based on a simplified application of shear stress decay law likewise shows that increasing slurry velocity enhances the shear stress inside the pipeline. That is, the velocity is likely to have a positive impact on OSLA. However, an increase in slurry concentration appears to have a more substantial effect on the increase of local shear stress at low slurry velocities. Rigorous investigations are required to clarify the impacts of slurry velocity and concentration on OSLA. To the best of authors' knowledge, no significant research has been done until now to study the effect of slurry velocity and concentration on the ablation of actual oil-sand lumps. Also, a publicly available model that can estimate the ablation rate of the oil-sand lump as a function of shear stress forces is not available to date, although such a model is highly required for engineering usage in the industry.

Acknowledgements

A part of the manuscript was the outcome of a research project that was funded by the NSERC Industrial Research Chair in Pipeline Transport Processes (RSS). We are grateful to Canada's Natural Sciences and Engineering Research Council (NSERC) and the Industrial Sponsors: Canadian Natural Resources Limited, CNOOC-Nexen Inc., Saskatchewan Research Council Pipe Flow Technology Centre™, Shell Canada Energy, Suncor Energy, Syncrude Canada Ltd., Total, Teck Resources Ltd., and Paterson & Cooke Consulting Engineers Ltd. We are especially thankful to Dr. Sanders (Professor, Department of Chemical and Materials Engineering, University of Alberta) for his advisory role in the research project.

Other declarations

A part of the manuscript is adapted from the first author's PhD dissertation.

Nomenclature

A	area (m^2)
Ar	Archimedes number (−)
C	solid volume concentration (−)
D	pipe diameter (m)
D_L	lump diameter (m)
d_p	particle diameter (m)
d_{50}	mean particle diameter (m)
d^+	dimensionless particle diameter (−)
f_f	Fanning friction factor (−)
f_s	solid friction factor (−)
L	length (m)
R	radius (m)
s	distance from pipe center (m)
s^+	dimensionless distance from pipe center (−)
T	temperature (°C)
T_{0L}	lump initial temperature °C
t	time (s)
V	bulk velocity (ms^{-1})
V_s	solid object velocity (ms^{-1})
V_c	deposition velocity (ms^{-1})
v	local velocity (ms^{-1})
μ	viscosity (Pa.s)
ρ	density (kgm^{-3})
τ	shear stress (Pa)
τ_w	wall shear stress (Pa)
τ_{wp}	wall shear stress on solid object (Pa)

Author details

Mahshad Pazouki[1] and Sayeed Rushd[2]*

1 TransCanada, Calgary, Alberta, Canada

2 King Faisal University, Al Ahsa, Saudi Arabia

*Address all correspondence to: mrushd@ualberta.ca

References

[1] Masliyah JH, Xu Z, Czarnecki JA. Handbook on Theory and Practice of Bitumen Recovery from Athabasca Oil Sands. Alberta, Canada: Kingsley Knowledge Pub.; 2011

[2] Flint L. Bitumen Recovery Technology: A Review of Long-Term R&D Opportunities. Calgary, AB, Canada: LENEF Consulting Ltd; 2005

[3] Sanders RS, Schaan J, McKibben MM. Oil sand slurry conditioning tests in a 100 mm pipeline loop. The Canadian Journal of Chemical Engineering. 2007; **85**(5):756-764. DOI: 10.1002/cjce.5450850521

[4] Sanders RS, Ferre AL, Maciejewski WB, Gillies RG, Shook CA. Bitumen effects on pipeline hydraulics during oil sand hydrotransport. The Canadian Journal of Chemical Engineering. 2000;**78**(4):731-742. DOI: 10.1002/cjce.5450780416

[5] Qiu L. Effect of oil sands slurry conditioning on bitumen recovery from oil sands ores [thesis]. University of Alberta; 2010. DOI: 10.7939/R3BS65

[6] Masliyah J, Zhou ZJ, Xu Z, Czarnecki J, Hamza H. Understanding water-based bitumen extraction from Athabasca oil sands. The Canadian Journal of Chemical Engineering. 2004; **82**(4):628-654. DOI: 10.1002/cjce.5450820403

[7] Schaan J, Cook N, Sanders RS. On-line wear measurements for commercial-scale, coarse-particle slurry pipelines. In: 17th International Conference on the Hydraulic Transport of Solids. Cape Town, South Africa; 7-11 May 2007. pp. 291-300

[8] Friesen WI, Dabros T, Kwon T. A bench-scale study of conditioning behaviour in oil sands slurries. The Canadian Journal of Chemical Engineering. 2004;**82**(4):743-751. DOI: 10.1002/cjce.5450820413

[9] Sanders RS, Schaan J, Hughes R, Shook C. Performance of sand slurry pipelines in the oil sands industry. The Canadian Journal of Chemical Engineering. 2004;**82**(4):850-857. DOI: 10.1002/cjce.5450820427

[10] Shook CA, Gillies RG, Sanders RS. Pipeline hydrotransport: With applications in the oil sand industry. SRC Publication No. 11508-1E02. SRC Pipe Flow Technology Centre; 2002

[11] Shook CA, Roco MC. Slurry Flow: Principles and Practice. Amsterdam, Netherlands: Elsevier; 2015

[12] Wallwork V, Xu Z, Masliyah J. Processibility of Athabasca oil sand using a laboratory hydrotransport extraction system (LHES). The Canadian Journal of Chemical Engineering. 2004;**82**(4):687-695. DOI: 10.1002/cjce.5450820407

[13] Eskin D, Leonenko Y, Lezhnin S, Vinogradov O. A model of oil sand lump digestion. Minerals engineering. 2002; **15**(6):469-472. DOI: 10.1016/S0892-6875(02)00065-1

[14] Hesketh RP, Fraser Russell TW, Etchells AW. Bubble size in horizontal pipelines. AIChE Journal. 1987;**33**(4): 663-667. DOI: 10.1002/aic.690330414

[15] Traĭnis VV. Parameters and Flow Regimes for Hydraulic Transport of Coal by Pipelines. Rockville, MD, USA: Terraspace, Incorporated; 1977

[16] Speight JG. Handbook of coal analysis. New Jersey, USA: John Wiley & Sons; 2015

[17] Law HS, Masliyah JH, Nandakumar K. Ablation of ice-solids and wax-solids

mixtures in turbulent axisymmetric water jets. The Canadian Journal of Chemical Engineering. 1987;**65**(3):420-429. DOI: 10.1002/cjce.5450650309

[18] Pazouki M. An experimental study on oil sand lump ablation [thesis]. Edmonton, AB, Canada: University of Alberta; 2013. DOI: 10.7939/R3D378

[19] Gillies RG, Shook CA, Xu J. Modeling heterogeneous slurry flows at high velocities. The Canadian Journal of Chemical Engineering. 2004;**82**(5): 1060-1065. DOI: 10.1002/cjce.5450820523

[20] Gillies DP. Particle contributions to kinematic friction in slurry pipeline flow [thesis]. Edmonton, AB, Canada: University of Alberta; 2012. DOI: 10.7939/R3W31B

Chapter 7

Evaluation of Asphaltenes Deposition Inhibition Factors in Heavy Crude Oil Pipelines

Hazlina Husin, Mysara Eissa Mohyaldinn Elhaj and Colin D. Wood

Abstract

Asphaltenes deposition is considered as Achilles's heel in the oil industry. The nucleation, precipitation and deposition of asphaltenes reduce the production rate significantly in affected wells and sometimes it can completely block the flow by plugging the flowlines, tubing and process facilities, in severe cases. This chapter evaluates the extrinsic and the intrinsic (thermodynamic) factors within the heavy crude oil production system. The main consequences of asphaltenes deposition are discussed such as the solvent-to-crude oil dilution ratio, crude oil physical properties (cloud point, pour point and API gravity), chemical solvent type (carbon number, functional group), agitation time and temperature changes. This chapter is expected to become the means for understanding the factors affecting the asphaltenes nucleation, precipitation and deposition.

Keywords: heavy crude oil, flow assurance, asphaltenes, deposition, inhibition

1. Introduction

Oil industry has been increasing its activities of crude oil production in subsea environments, deep waters and cold environments [1]. However, the physical and chemical conditions of these unconventional reservoirs have been challenging in which oil industry needs to overcome flow assurance issue such as asphaltenes deposition [2].

Asphaltenes deposition will affect the oil production rate at all oil production stages [3]. For example, when asphaltenes deposit in reservoir rock, they change the rock wettability hence decrease the chances of having an effective oil recovery [4, 5]. In some enhanced oil recovery (EOR) methods such as carbon dioxide injection [6] and chemical (cEOR) [7], the guest chemical would interact with asphaltenes in crude oil and subsequently result in asphaltenes deposition. Crude oil which flows in flowlines, tubing or process facilities will create a temperature gradient within a specified boundary [8, 9]. This temperature change will weaken the molecular interactions of resins which hold the asphaltenes molecule in crude oil. Due to its polarity, asphaltenes molecule will accumulate and subsequently start to nucleate. In general, asphaltenes deposition is considered as the major problem for oil industries as it occurs in upstream, midstream and downstream.

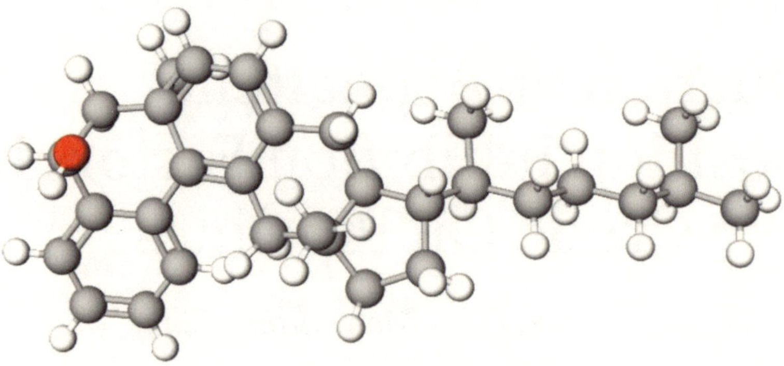

Figure 1.
Asphaltenes molecular structure.

Asphaltenes are the heaviest, polar hydrocarbon substances containing heteroatoms such as nitrogen, sulfur and oxygen. Asphaltenes have an average molecular weight, MW_{ave} of 750 g/mol. **Figure 1** shows a schematic structure of the continental-model asphaltenes which obtained using MolView software v2.4.

There are several approaches for inhibiting deposition of asphaltenes such as via physical removal, use of dispersant and use of chemical solvent. Physical removal is done by using a pigging tool. It is however uneconomical because the collected solid asphaltenes must be properly disposed and the production must halt for the workers to conduct pigging. The use of dispersant is also uneconomical as the oil industry will need to use a large amount of dispersant for a long period of time. Chemical solvent such as benzene, toluene, chlorate or xylene as asphaltenes inhibitor is widely used as the base component of commercial chemical inhibitor. But, due to their low flash point and high toxicity level, these chemical solvents require extra safety precaution during handling and storage.

Since asphaltenes have a complex structure, differ depending on oil reservoirs and are not in any well-defined chemical group, it is difficult to predict the reason why asphaltenes became unstable and deposited out from crude oil [10]. Previous studies have revealed that asphaltenes deposition is a function of various parameters [11, 12]. Hence, this chapter evaluates the effect of asphaltenes deposition inhibition factors such as solvent-to-crude oil dilution ratio, crude oil physical properties (cloud point, pour point and API gravity), the chemical solvent type (carbon number, functional group), agitation time and temperature change on two types of Malaysia crude oils. The findings from this study are desirable to develop prevention and remediation methods for asphaltenes deposition problem.

2. Materials and methods

Pre-treated crude oil samples were collected from Malaysia oilfield with properties listed in **Table 1**. The crude oil samples contain negligible water and solids (such as sand, clay and corrosion debris).

Different types of liquid solvent were bought from Fisher Chemicals with a reported purity of 98–99%. The molecular structures of the solvents are shown in **Table 2**, used to control asphaltenes precipitation. Mixtures of solvent-to-crude oil dilution ratio of 1:1, 5:1, 10:1 and 15:1 were prepared for each solvent.

The method used to deposit the asphaltenes particles from crude oil is called "standard direct experimental" method. This method follows the modified ASTM 2007/ASTM D6560 standard procedure [13, 14]. Once the solvent and the crude

Properties	Sample 1	Sample 2
Cloud point (°C)	33.8	15.0
Pour point (°C)	−2.0	−18.0
API gravity (°API)	37.4	12.5
Wax content (%)	4.0	10.0
Phase	Liquid	Liquid
Color	Dark brown	Dark brown

Table 1.
Crude oil properties for the study.

Solvent	Molecular formula (MW)	Molecular structure
n-Pentane	C_5H_{12} (72.15 g/mol)	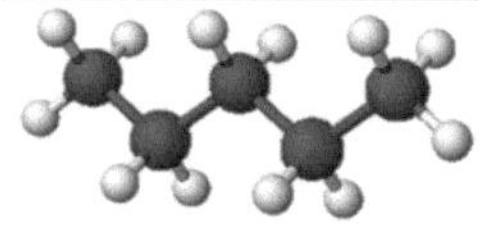
Benzene	C_6H_6 (78.11 g/mol)	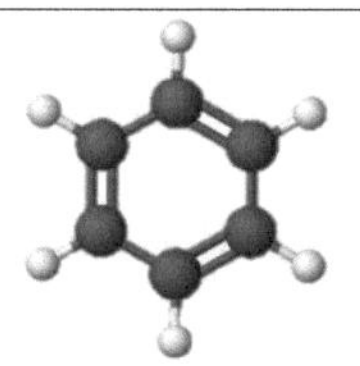
n-Hexane	C_6H_{14} (86.18 g/mol)	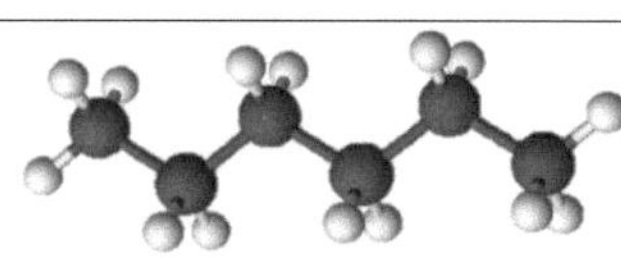
Toluene	C_7H_8 (92.14 g/mol)	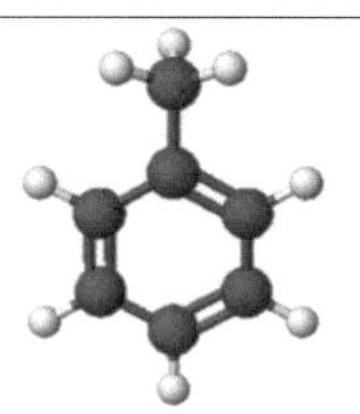
Heptane	C_7H_{16} (100.21 g/mol)	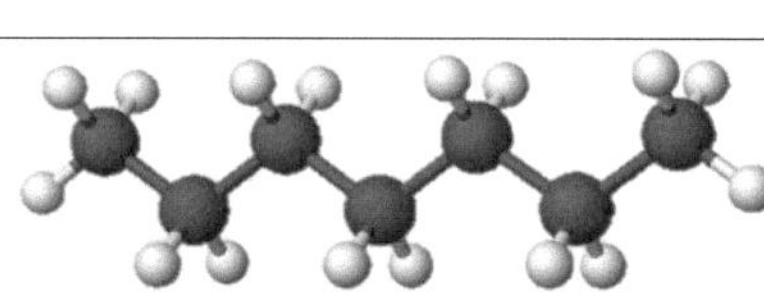

Table 2.
Molecular structure and molecular weight (MW) of solvents used for the study, in increasing carbon number.

oil were mixed together according to the desired dilution ratio, the resulted mixture was then placed in a water bath at room temperature and left for 3 h under agitation condition. Then, the aliquot was placed in a container and centrifuged at 10,000 rpm (SIGMA 3-18K, United Kingdom) with relative centrifugal force (RCF) of 16,000 times gravity. The deposited asphaltenes were filtered by using a Whatman® filter paper, grade number 42. Finally, the asphaltenes particles were dried in a vacuum oven at 100°C for 30 min and then weighed (with digital instrument precision: ±0.001 g) to determine the amount of asphaltenes precipitated. The entire procedure was repeated for different agitation time (6 and 9 h) and temperature (50 and 75°C).

3. Results and discussion

The observable deposited asphaltenes particle is shown in **Figure 2**. There have been a wide range of asphaltenes particle size as reported in literatures; 2–5 nano-micron [11] and 34–113 micron [13]. It is proposed that asphaltenes tend to self-aggregate by π-π interactions and originate the asphaltenes "nucleus" which consist of stacked nanoaggregates.

Figure 3a and **b** shows the plots for the deposited asphaltenes particles when n-pentane was used as the solvent type. Systematically, n-pentane is selected as the first stage of this study as recommended by the modified ASTM 2007 procedure.

At 1:1 dilution ratio, the amount of asphaltenes deposited was 0.860 g when the agitation time was 3 h (as shown in **Figure 3a**). At 15:1 dilution ratio, the amount of asphaltenes deposited was 3.100 g, at the same agitation time. Result showed that the amount of deposited asphaltenes particle is increased with solvent-to-crude oil dilution ratio. This is due to the increasing formation of unstable asphaltenes aggregates. It is expected that a maximum asphaltenes yield would be produced at a dilution ratio of 40:1 [15].

The same trend was found at longer agitation time, i.e., 6 and 9 h. The increasing amount of deposited asphaltenes particles at higher dilution ratio was verified with published results [16, 17]. In the presence of n-pentane solvent, the asphaltenes particles became unstable and precipitated as primary particles. At higher volume of the solvent and longer agitation time, more resins which responsible for asphaltenes solubility in crude oil are "washed off" that lead to rapid aggregation of asphaltenes molecules. Consequently, macroparticles (i.e., heavier particles) asphaltenes are formed.

Interestingly, at lower region of dilution ratio (1:1 and 5:1), the amount of asphaltenes deposited from sample A crude oil having agitation time of 3 h is the same as that of 6 h (as in **Figure 3a**). As one may see, **Figure 3b** shows the same behavior despite having different physical properties. This indicates that the

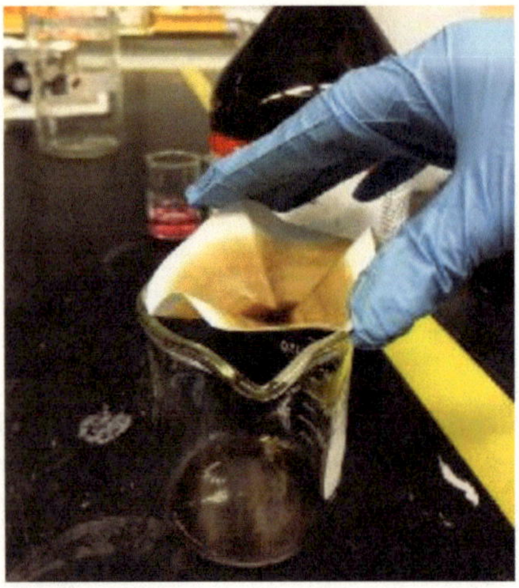

Figure 2.
Filtered asphaltenes particles.

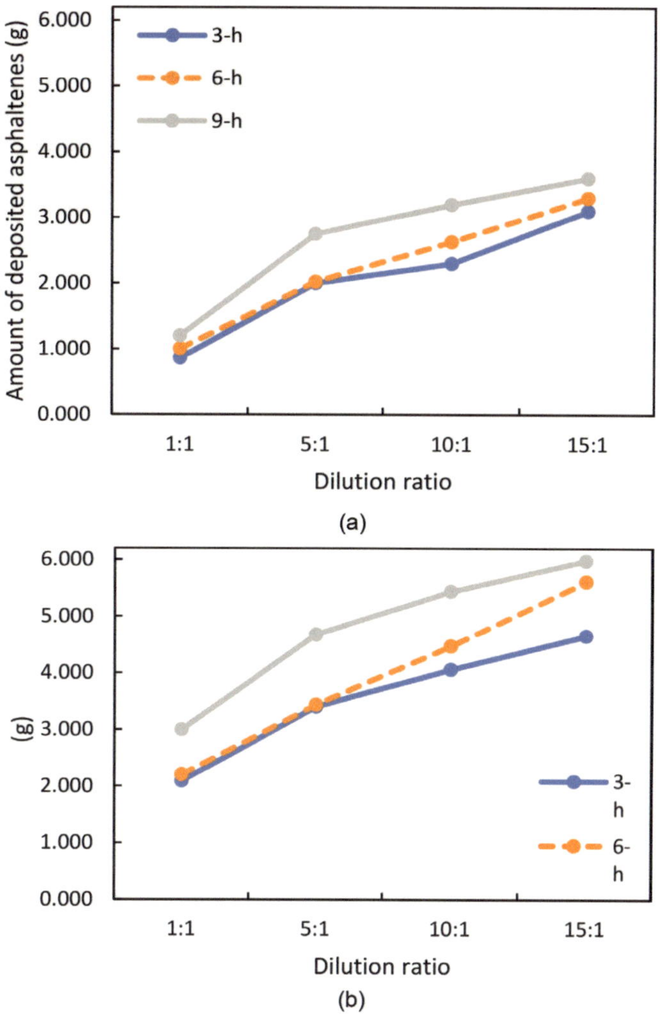

Figure 3.
Amount of deposited asphaltenes at selected solvent-to-crude oil dilution ratio and different agitation time. The conditions of experiment were: chemical solvent type = n-pentane and T = 30°C. (a) Sample A and (b) sample B.

solubility of asphaltenes in n-pentane solvent is approximately constant at low solvent concentration or in this case lower region of solvent dilution ratio. We proposed that this observation is due to the rapid diffusion and aggregation of asphaltenes molecules hence resulting in high precipitation rate at short mixing times. It is worth highlighting that future asphaltenes deposition studies should distinct between the lower and higher regions of solvent dilution ratio. Note that the kinetic effect is ignored here since the agitation time is constant (either at 3 or 6 h).

To evaluate the effect of the properties of the crude oil such as the cloud point, pour point and density (reflected by the API gravity of crude oil), plots in **Figure 3a** and **b** were compared. Although sample B having higher wax content and heavier, i.e., more dense than sample A, both plots showed similar trend of increment of deposited asphaltenes particles at higher volume of solvent (i.e., higher region of solvent dilution ratio). At 1:1 dilution ratio, we found that the magnitude of increment for both sample A and sample B is 2.2–2.5, when the agitation time is 3–9 h. Surprisingly, at 5:1, 10:1 and 15:1 dilution ratio, the magnitude of increment is constant, i.e., ~1.7 when the agitation time is 3–9 h. Previous study [13] showed that the amount of deposited asphaltenes was either reached a plateau value or reduced, when the dilution volume is higher than 15:1.

Figure 4a–d shows the plots for the deposited asphaltenes particles when different types of liquid solvent were added to sample A. For the molecular structure of these solvents, see **Table 2**. This study shows the effect of increment carbon number (n-pentane, C5, n-hexane, C6 and n-heptane, C7) and the effect of functional group (i.e., alkane group and benzene-ring group). Firstly, we observed the influence of n-pentane, n-hexane and n-heptane at 30°C (as in **Figure 4a**). At dilution ratio 1:1, 0.860 g of asphaltenes were deposited when n-pentane and n-heptane were added. A slight reduced (0.710 g) amount was observed when n-hexane was added. This result suggests that the solubility of asphaltenes decreases when higher solvent carbon number was used. As the viscosity of the solvent increases with carbon number, asphaltenes diffusion is slower. Subsequently, fewer asphaltenes-asphaltenes collision which then leads to lesser amount of deposited asphaltenes.

Meanwhile, at dilution ratio from 5:1 to 15:1 (as in **Figure 4b–d**), we observed that the highest amount of asphaltenes was obtained when n-pentane was added, followed by n-hexane and lastly n-heptane. Interaction between the long, planar molecules of n-heptane and the planar, supramolecular asphaltenes compound would have resulted in lower steric effect, and thus leading to a decrease in aggregation size [18].

Secondly, we found that both benzene and toluene do not significantly influence the deposition of asphaltenes, having measurements of less than 0.005 g. Despite typically reported by authors worldwide, the asphaltenes deposition was not affected by thermodynamic temperature variation. The reason anchoring this observation is due to the solvent chemical properties. Asphaltenes are known to be miscible in benzene (1 benzene ring) and toluene (1 benzene ring and 1 methyl branch), or other similar structure compounds such as ethylbenzene (1 benzene

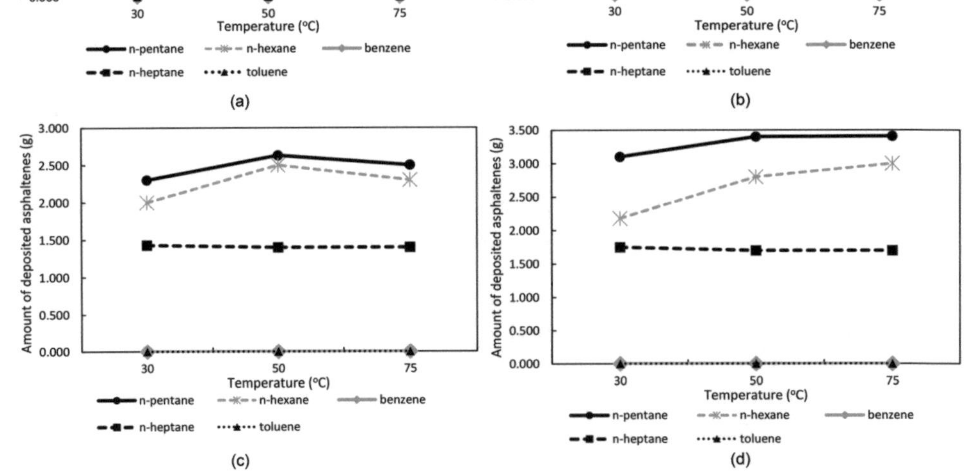

Figure 4.
Amount of deposited asphaltenes as a function of solvent type and temperature. The conditions of experiment were: crude oil type = sample A, agitation time = 3 h. (a) 1:1, (b) 5:1, (c) 10:1, and (d) 15:1 dilution ratio.

ring and 1 ethyl branch) and xylene (1 benzene ring, 2 methyl branches). These solvents would interact with asphaltenes and give similar value of enthalpy of mixing ($+\Delta H_{mix} = 0.5 \pm 0.2$ kJ) [19]. In asphaltenes mitigation strategy application, these benzene-based chemicals are applied as dispersants.

Temperature factor plays an important role in preventing or reducing asphaltenes deposition [20]. **Figure 4a** illustrates that 0.860, 0.980 and 0.960 g of asphaltenes particles were deposited at 30, 50 and 75°C, respectively when n-pentane was added. Consecutively, n-hexane follows the same trend, while n-heptane had resulted in 0.710, 0.690 and 0.550 g of asphaltenes deposits at 30, 50 and 75°C, respectively. In general, it was noted that a discontinuity increment in the amount of deposited asphaltenes occurs at increasing temperature. Heating process or existence of sufficient thermal energy hinders the mechanism of asphaltenes aggregation. This is due to the weak force of interparticle attraction between asphaltenes molecules and between asphaltenes and constituents (such as saturates and resins) [21] hence lowers the ability for the asphaltenes particles to coalesce. The influence of temperature effect on asphaltenes deposition was similar with earlier study [13].

4. Conclusions

For the two samples of Malaysia crude oil, the amount of deposited asphaltenes increased as the solvent-to-crude oil dilution ratio and the agitation time increased. Similar trend was observed for both samples despite having distinctive physical properties (cloud point, pour point and API gravity). Results also showed that the amount of deposited asphaltenes is solvent-dependent (carbon number and functional group) and temperature-dependent. Also, we suggest that future studies on asphaltenes deposition should be linked to low/high region of solvent dilution ratio.

Acknowledgements

The authors would like to thank Universiti Teknologi PETRONAS (UTP) for the financial support given through the YUTP 015-LC0-064 grant.

Nomenclature

MW	molecular weight, g/mol
CP	cloud point, °C
PP	pour point, °C
API	The American Petroleum Institute gravity
RCF	relative centrifugal force, g
ΔH_{mix}	enthalpy of mixing, J/mol

Abbreviations

cEOR	chemical enhanced oil recovery
API	The American Petroleum Institute gravity
ASTM	American Society for Testing and Materials

Author details

Hazlina Husin[1*], Mysara Eissa Mohyaldinn Elhaj[1] and Colin D. Wood[2]

1 Universiti Teknologi PETRONAS, Ipoh, Malaysia

2 Commonwealth Science and Industrial Research Organization, Kensington, Australia

*Address all correspondence to: hazlina.husin@utp.edu.my

References

[1] Hammami A, Ratulowski J. Precipitation and deposition of asphaltenes in production systems: A flow assurance overview. In: Mullins OC, Sheu EY, Hammami A, Marshall AG, editors. Asphaltenes, Heavy Oils, and Petroleomics. New York: Springer; 2007. pp. 617-660

[2] Theyab MA. Fluid flow assurance issues: Literature review. Scientific Federation Journal of Petroleum. 2018;**2**:1-11

[3] Soleymanzadeh A, Yousefi M, Kord S, Mohammadzadeh O. A review on methods of determining onset of asphaltenes precipitation. Journal of Petroleum Exploration and Production Technology. 2019;**9**:1375-1396

[4] Zanganeha P, Dashtib H, Ayatollahic S. Comparing the effects of CH_4, CO_2, and N_2 injection on asphaltenes precipitation and deposition at reservoir condition: A visual and modeling study. Fuel. 2018;**217**:633-641

[5] Li X, Guo Y, Sun Q, Lan W, Liu A, Guo X. Experimental study for the impacts of flow rate and concentration of asphaltenes precipitant on dynamic asphaltenes deposition in microcapillary medium. Journal of Petroleum Science and Engineering. 2018;**162**:333-340

[6] Alrashidi H, Afra S, Nasr-El-Din HA. Application of natural fatty acids as asphaltenes solvents with inhibition and dispersion effects: A mechanistic study. Journal of Petroleum Science and Engineering. 2019;**172**:724-730

[7] Curren M, Kaiser A, Adkins S, Qubian A, Al-Enezi H, Sana H, et al. Effects of cEOR chemicals on asphaltenes precipitation. In: Proceedings of the SPE Annual Technical Conference and Exhibition; Texas, USA: SPE; 2018

[8] Hoepfner MP, Limsakoune V, Chuenmeechao V, Maqbool T, Fogler HS. A fundamental study of asphaltenes deposition. Energy and Fuels. 2013;**27**:725-735

[9] Sedghi M, Goual L, Welch W, Kubelka J. Effect of asphaltenes structure on association and aggregation using molecular dynamics. The Journal of Physical Chemistry B. 2013;**117**:5765-5776

[10] Pan H, Firoozabadi A. Thermodynamic micellization model for asphaltenes precipitation from reservoir crudes at high pressures and temperatures. SPE OnePetro. 2000;**15**:58-65

[11] Mullins OC, Sabbah H, Eyssautier J, Pomerantz AE, Barré L, Andrews AB, et al. Advances in asphaltenes science and the Yen-Mullins model. Energy and Fuels. 2012;**26**:3986-4003

[12] Ilyin S, Arinina M, Polyakova M, Bondarenko G, Konstantinov I, Kulichikhin V, et al. Asphaltenes in heavy crude oil: Designation, precipitation, solutions, and effects on viscosity. Journal of Petroleum Science and Engineering. 2016;**147**:211-217

[13] Husin H, Aman Z, Chyuan OH. Correlation between rate of deposition and temperature of asphaltenes particles. Materials Today: Proceedings. 2018;**5**:22128-22136

[14] Santosa D, Amaral M, Filho EBM, Dourado RS, Coutinho JAP, Borges GR, et al. Revisiting the methodology for asphaltenes precipitation. Journal of Petroleum Science and Engineering. 2019;**178**:778-786

[15] Wang S, Liu J, Zhang L, Masliyah J, Xu Z. Interaction forces between asphaltenes surfaces in organic solvents. Langmuir. 2010;**26**:183-190

[16] Groenzin H, Mullins OC. Asphaltenes molecular size and structure. The Journal of Physical Chemistry A. 1999;**103**:11237-11245

[17] Akbarzadeh K, Dhillon A, Svrcek WY, Yarranton HW. Methodology for the characterization and modeling of asphaltenes precipitation from heavy oils diluted with n-alkanes. Energy and Fuels. 2004;**18**:1434-1441

[18] Chaisoontornyotin W, Haji-Akbari N, Fogler HS, Hoepfner MP. Combined asphaltenes aggregation and deposition investigation. Energy and Fuels. 2016;**30**:1979-1986

[19] Miadonye A, Evans L. The solubility of asphaltenes in different hydrocarbon liquids. Petroleum Science and Technology. 2010;**28**:1407-1414

[20] Hong E, Watkinson P. A study of asphaltenes solubility and precipitation. Fuel. 2004;**83**:1881-1887

[21] Ferworn KA, Svrcek WY, Mehrotra AK. Measurement of asphaltenes particle size distributions in crude oils diluted with n-heptane. Industrial and Engineering Chemistry Research. 1993;**32**:955-959

Chapter 8

Flow of Heavy Oils at Low Temperatures: Potential Challenges and Solutions

Emad Soliman

Abstract

Global demand for non-conventional heavy and extra-heavy oil has been marginal until the end of twentieth century because of their composition complexity and high viscosity that cause many operational difficulties in the production with decline of their economic viability. However, growing energy demands in the beginning of the twenty-first century motivate many countries to handle such non-conventional resources. Heavy extra-heavy crudes usually have higher pour points due to high content of high molecular weight components, such as waxes, asphaltenes, and resins. The structural changes for these components cause abrupt rise in oil viscosity and simultaneous deposition of wax and asphaltene on the inner walls of pipelines. This can cause clogging of pipes accompanying oil flowability reduction with extra burden on the pumping system and consequently increases its power requirement and cost. This chapter presents technological challenges in transportation, describing the different mitigation strategies that have been developed to improve the low-temperature flow properties of heavy crude oils (heating, dilution, oil-in-water emulsion, and upgrading and core annular flow).

Keywords: heavy crude oils, cold flow, heating, dilution, oil-in-water emulsion, upgrading, core annular flow

1. Introduction

The continual global demand for petroleum fuels led to the decrease in supply from conventional reservoirs. Where, the conventional light oils had typically been produced at a high rate and a low cost. Therefore, the plateau in conventional oil production and the corresponding increase in the demand for liquid fuels have motivated markets to respond with higher oil prices and subsequently have stimulated in the global oil industry the advancement in technologies for the exploitation of reservoirs of transitional and unconventional oils [1, 2]. An array of these new oils including oil sands, tight oil, heavy oils (including extra-heavy oils and bitumen), deep water oil, and eventually oil shale are projected to fill the gap through the next few decades and that could continue to be revised upward as new technologies are developed so that they could dominate the liquid-fuel supplies through the endings of the twenty-first century especially regarding the global abundance of these new oil supplies.

 IntechOpen

The oil industry is expected to invest huge sums in petroleum production and oil infrastructure in the years ahead, up to an estimated $1 trillion over the next decade alone. Without a concerted policymaking effort, these investments will likely continue to flow disproportionately toward unconventional oils. The involvement of unconventional crude oils in the international energy markets faces serious difficulties that need certain technological developments in the production, refining and transportation [3].

As conventional crude oil supplies have peaked and leveled off globally in recent years, oil has begun to transition, the makeup and geography of the new or tomorrow's oil, however, will be dramatically different from the current ones. Generally, the International Energy Agency defines conventional oil as "a mixture of hydrocarbons that exist in liquid phase under normal surface conditions". Meanwhile, unconventional oils are defined as those oils obtained by unconventional production techniques because they cannot be recovered through pumping in their natural state from an ordinary production well without being heated or diluted. By other words, unconventional oils require new, highly energy intensive production techniques and new processes to deal with their inaccessible placements or unusual compositions. In this context, the U.S. Department of Energy divides unconventional oil into four types: heavy oil (HO), extra-heavy oils (EHO), bitumen, and oil shale. Moreover, some analysts include oils produced from natural gas or coal using gas-to-liquids (GTL) processes and coal-to-liquids (CTL) processes in the unconventional oil category. These unconventional oil-processing techniques broaden the feedstock of unconventional oils to include unconventional natural gas, such as tight gas, shale gas, coal-bed methane, and methane hydrates [4] (**Figure 1**).

These new oils are an abundant untapped potential energy source, which is expected to be a large contributor to the world's energy needs in the future.

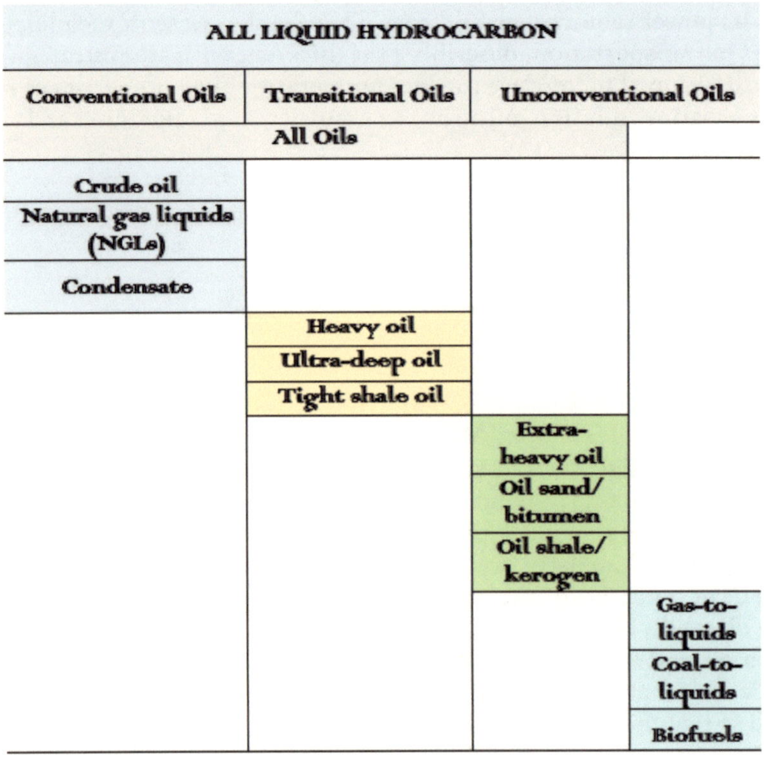

Figure 1.
Transformation of liquid fuels.

However, these oils tend to be less valuable than conventional crude in addition to the technological costs per barrel that are currently much higher than for conventional resources, which is readily transformed into the most marketable petroleum products by today's standards.

2. Physico-chemical properties of new crude oils

Crude oil is a mixture of different compounds consisting of hydrocarbons, heteroatoms, and metals. The various compounds can differ widely in molecular structure, volatility, density and viscosity. These compounds are usually divided into fractions: saturates, aromatics, resins and asphaltenes (SARA). The saturate fraction is non-polar and includes linear, branched and cyclic alkanes. Long-chain linear alkanes ($>C_{20}$) are known as waxes, or paraffinic waxes, and can make oil recovery challenging as the wax becomes solid at low temperatures. Aromatics are slightly more polar and consist of one or more aromatic rings connected by double bonds. Saturates and aromatics constitute the majority of the crude oil. Resins and asphaltenes are the most polar heavy molecules and complex components in crude oil. The majority of heteroatoms and metals are found in these fractions. Conventional oils are hydrogen-rich compounds with relatively short hydrocarbon chains, fewer carbon atoms (C_1–C_{60}), and lower molecular weights than most unconventional oils (around 200) where hydrogen packs all of the energy while carbon goes along for the ride [5]. The physical properties and grade of oil depends on the dominance of one of the hydrocarbons or its fractions in its composition. The light conventional crude oil has low density where several of the molecules are volatile, while the unconventional heavy oils have high density and high viscosity where resins, paraffins and asphaltenes are found at a higher amount than those in light crude oil [6, 7]. These oils require special treatments for their extraction and transportation according to their content of these compounds. Regarding the close relation between the physical properties of crude oil with these content, it is worthy to mention that the crude oils can be categorized based on the paraffin content into; less paraffin-oils (paraffins < 1.5%), paraffinic or waxy oils (paraffins = 1.5–6%) and high paraffin- oils (paraffins > 6%). Also, the crude oils can be categorized based on their content of resins into; less resins-oils (resins < 8%), resinous oils (resins = 8–25%) and high resins-oils (resins > 25%). Additionally, sulfur and vanadium are present in high concentrations in heavy oils and lead to reduce its quality and cause many problems through extraction, transportation, storage, and refining of these oils as a result of the side reactions of these compounds causing the corrosion effects on pipelines, boilers, and storage tanks. In this context, the crude oils can be also categorized based on their content of the sulfur compounds into; sweet oils (total sulfur level < 0.5%), sour oils (total sulfur level > 0.5–1.9%) and highly sour oils (total sulfur level > 1.9%). These sour crude oils containing larger amounts of the impurity sulfur, an extremely corrosive element that is difficult to process, and deadly when released hydrogen sulfide gas.

Generally, transitional oils have compositions comparable for that of conventional oils; however, they are extracted by unconventional means. As conventional oils become less accessible, new, more technical, energy-intensive methods are being developed for their recovery, from ultra-deep wells drilled miles below the sea to fracturing shale rock in order to tap oil trapped in low-permeability siltstones, sandstones, and carbonates deep in the earth. Meanwhile, heavy oils are tricky to extract, requiring gas injection and other invasive techniques due to their high, molasses-like viscosities that approach those of unconventional oils. On other hand, the unconventional oils are typically much heavier and sourer than even the

Location	High Quality Range	Low Quality Range
Africa	Nigeria (Agbami Light) 47°, 0.04%	Angola (Kuito) 19°, 0.68%
Asia	Indonesia (Senipah Condensate) 54°, 0.03%	China (Peng Lai) 22°, 0.29%
Australia	Bayu Undan 56°, 0.07%	Enfield 22°, 0.13%
Europe	Norway (Snohvit Condensate) 61°, 0.02%	UK (Alba) 19°, 1.24%
Middle East	Abu Dhabi (Murban) 39°, 0.8%	Saudi Arabia (SA heavy) 27°, 2.87%
North America	US (Williams Sugarland Blend) 41°, 0.20%	Canada (Albian) 19°, 2.1%
Latin America	Columbia (Cupiaga) 43°, 0.14%	Venezuela (Bascan) 10°, 5.7%
Central Asia	Kumkol (Kazakhstan) 45°, 0.81%	Russia (Espo) 35°, 0.62%

Benchmark Crudes: Brent 38°, 0.37%; WTI (West Texas Intermediate) 40°, 0.24%; Dubai 31°, 2.0%.

Table 1.
The composition of crude barrel originating from various regions.

lowest-quality conventional oil, where the heavier oils such as, oil sand (bitumen) and oil shale (kerogen), are the more carbon laden, higher in sulfur, and filled with toxic impurities. And therefore, conventional light oils tend to deliver more productivity with less waste than unconventional oils.

Crude oil make-up is highly dependent on the origin and depth of reservoir. Whereat no two sources are alike, no two crudes are exactly alike or have the same make-up. However, there can be a great deal of variation within a range. The natural resources range from high-quality "light, sweet" crudes to lower-quality "heavy, sour" crudes based on their region (**Table 1**).

The physical properties of crude oils are highly related to their make-up. Furthermore, the technological conditions applied during exploration, drilling, transportation and storage of these oils can affect strongly their physical properties. On the other side, the physical parameters have to be in the allowable ranges to achieve the highest potential yield technologically and economically. Density and viscosity are the most important physical properties affecting the flowability of the crudes. Such flowability has a great and clear impact through the transportation and storage of these liquid hydrocarbons, especially for new unconventional ones. Where, some of these new hydrocarbon resources are effectively solid and must be removed through mining or heated in place (in situ) until they flow. Extra-heavy, impure oils also require very large energy inputs to upgrade and preprocess into synthetic crude oil that is then processed by a refinery (known as feedstock). Therefore, such key properties determine the economics of a heavy oil field development. Generally, heavy oil sells at a lesser price than lighter hydrocarbons, as it will have to go through an energy intensive upgrading process before use. On the other hand, high viscosity values lead to lower production and more expensive enhanced oil recovery (EOR) investments.

The density of the oils is measured on a scale known as API gravity developed by the American Petroleum Institute and the National Bureau of Standards developed a scale of the density of liquid petroleum products. The gravity scale is calibrated in terms of API degrees, which equals:

$$\left(141.5/\text{Specific gravity (SG) at } 60^{\circ}\text{F}\right) - 131.5 \quad (1)$$

The lighter crudes have the higher API gravity. If the API gravity is greater than 10, the oil is lighter and floats on water; if less than 10, it is heavier and sinks. Generally, API gravity of light crudes exceeds 38°. However, it is commonly below 22° for heavy crudes. Intermediate crudes fall between 22 and 38°. Extra-heavy oils are below 10; the API gravity of bitumen approaches zero.

Viscosity is measured in centipoises (cP) that represents the oil's resistance to flow; the higher the value, the higher the viscosity. Viscosity is the property that most affects production and recovery operations where it complicates its production and pipeline transportation, due to its poor fluidity and high pressure differentials generated with values between 20 cP and more than 1×10^6 cP [8, 9]. The heavy oil in terms of viscosity was defined as the class of oils ranging from 50 to 5000 cP. Among the different compounds of crude oil, asphaltenes are usually most responsible for the high viscosities in HO and EHO.

The high viscosity restricts the easy flow of oil at the reservoir temperature and pressure [10]. The viscosity of heavy crudes is strongly affected by temperature variations. For this reason, thermal recovery methods are commonly used in heavy oil production. **Figure 2** shows the relationship between viscosity and temperature for two Athabasca bitumen samples. There is no universal relationship between oil density and viscosity. However, oils are generally found to be more viscous when density increases. This relation can be greatly attributed to the presence of asphaltenes, which are high molecular weight polycyclic hydrocarbons that tend to aggregate. Indeed, it has been shown that the viscosity of oils increases exponentially with asphaltene content [11].

Figure 3 is a graph relating viscosity and API ratings and it can be observed that the heavy oil region lies in the high viscosity range.

The high density and viscosity of unconventional oils at atmospheric conditions has traditionally made their recovery very energy demanding compared to lighter crudes and has resulted in very low recovery factors, which means that these oils tend to result in higher carbon emissions and other societal impacts. Mnemonically, when other factors being equal, the lower the API gravity, the oil will be more expensive to extract and process, and simultaneously the obtained oil will have lower price.

The rheological properties are highly controlling all processes in which fluids are transferred from one location to another such as in, the migration of crude oil within the oil reservoir and transportation of the crude oil from oil wells and the refining units. Therefore, the unconventional oils are required to have the fluid viscosity of the migrating conventional oils [13]. The viscosities of the unconventional crude oils (i.e. heavy oil and bitumen/tar sands) at 25°C might reach more than 10^5 cP. The extremely low flowability due to high viscosity at reservoir

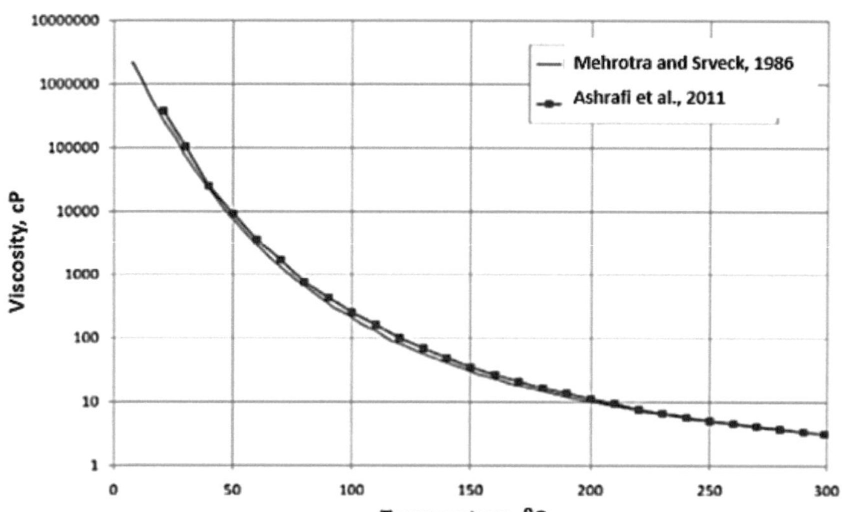

Figure 2.
Viscosity—temperature relationship of an Athabasca bitumen [12].

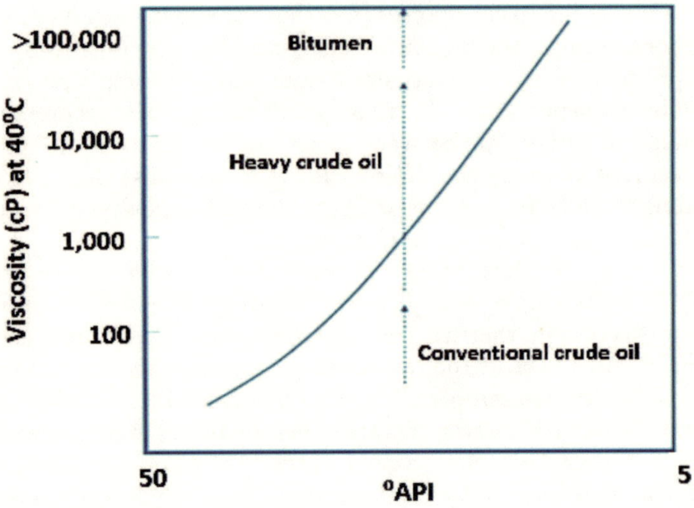

Figure 3.
General relationship of viscosity to API gravity [2].

conditions, especially in offshore conditions alongside asphaltene deposition, heavy metals, sulfur and brine or salt content making these kinds of oils more challenging to produce, transport as well as refine by conventional means without firstly upgrading them to meet conventional light crude oil properties [14, 15]. It is well known that pipelines are the most effective means of transporting crude oil from the producing field to the refinery because of their low expense and environmentally convenient. As such, transportation of the heavy crude oil via pipeline is a major challenge for petroleum industries, where conventional pipelining is not adequate because of the huge energy (i.e. high pumping power) is required to overcome the high pressure drop in the pipeline, owing to their high viscosity. Therefore, to transport heavy oils economically, the pressure drop in the pipeline must be lowered to minimize the pumping power via reducing the viscosity [16]. Furthermore, the pumping temperature is an important affecting factor of the flowability and consequently the pipeline transportation of crude oil. Heavy crudes usually have higher pour points due to high content of high molecular weight components, such as waxes, asphaltenes and resins. Under conditions in which the atmospheric temperature is below the pour point, gelation of the crude oil occurs lowering its flow completely and causing severe transportation problems. Especially in the cold offshore environment, waxes and asphaltenes deposit over inner surfaces of pipelines and eventually clog the pipelines, which decreases the accessible cross-sectional area for oil flow that causes reduction in flow rate and rise in pressure drop and multiphase flow, may occur resulting to further increases in the pumping cost [17, 18]. Here, it is worthy to mention that the desired pipeline viscosity of crude oils might not exceed 400 cP at 25°C [19–21]. However, it should be less than 200 cP at 15°C [22]. On the other side, the presence of brine or salt in the heavy crude stimulates corrosion problems in the pipeline [17]. In some cases, the formation of emulsion such as the oil-water mixture produced from the reservoir poses transportation difficulty.

3. Mitigation technologies for low flowability

Sync of the ever-growing world energy demand with the decline of conventional middle and light crude oil reserves and the limited supply and rising price

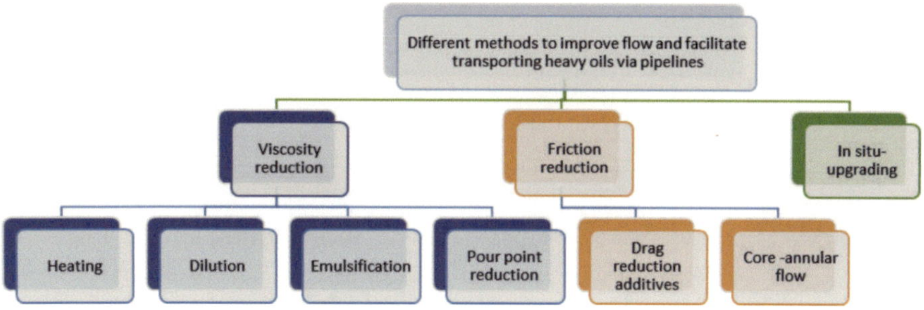

Figure 4.
Schematic diagram presenting different methods for improving flow of heavy crude oil and bitumen and facilitating their transporting via pipelines.

of crude oil led to attract the interest the petroleum industry growingly in heavy crude oil and bitumen/tar sands exploitation. Nevertheless, the exploitation of these crudes is still low because of the technical challenges that face it at all stages from recovery/production from the reservoir to transportation and refining [23, 24]. To reduce this high-pressure drop and cost of transportation, several technologies have been proposed to improve the flow properties of the heavy crude oil and bitumen through pipelines and thus to meet the production volumes projected by the market. Generally, these different technologies can be divided into three categories (**Figure 4**) including (1) viscosity reduction [e.g. preheating of the heavy crude oil and bitumen and subsequent heating of the pipeline, blending and dilution with light oils or organic solvent, emulsification through the formation of an oil-in-water emulsion and lowering the oil's pour point by using pour point depressants (PPDs)]; (2) drag/friction reduction [e.g. pipeline lubrication through the use of core-annular flow, drag reducing additive]; and (3) in situ partial upgrading of the heavy crudes to produce synthetic crudes with lower viscosity [23, 25].

In this chapter, various technologies available for transporting heavy crude oil and bitumen from the production site to the processing facilities including viscosity and friction reduction and in-situ upgrading are reviewed with extensive discussion. The author provide a review of typical methods such as heating and dilution, and also point out lubrication solutions to move heavy and extra-heavy oils, such as core-annular flow and o/w emulsions. The advantages and disadvantages of each technology are highlighted with the view that the chapter will provide direction for improvement and development of novel economically viable technologies to improve the transportation of heavy oils via pipelines. Other approaches as drag reduction additives and pour point depressants (PPDs) may be complementarily discussed in a later section in this work.

4. Heating

Heating technique is commonly utilized to overcome the difficulties related to the transportation of the heavy crude oil. This technique involves preheating the heavy crude oil followed by subsequent heating of the pipeline. Such thermal treatments are based on the strong viscosity-temperature relationship, since the viscosity of the heavy oils and bitumen is reduced by several orders of magnitude with increasing temperature and subsequently the flowability of such oil is improved and it will be easier to pump. The response of viscosity to changes in temperature for some heavy oil and bitumen is illustrated in **Figure 5** [26]. Heat should be applied

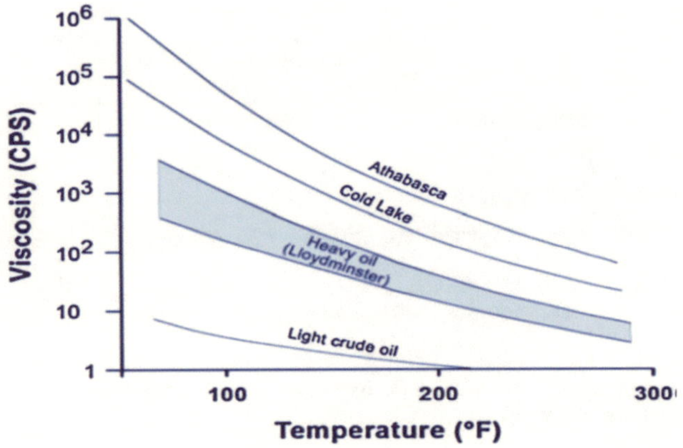

Figure 5.
Response of viscosity to increase in temperature [26].

to the oil to guarantee that its viscosity reaches acceptable values for transport in pipelines. These values typically refer to a maximum viscosity of 500 cP, below which many crude oils can be economically pumped [27, 28].

Equations proposed to represent the viscosity temperature relationship are commonly of logarithmical or double logarithmical forms. Many of these equations are based on the Eyring relationship that has been proposed in 1936:

$$\eta = \frac{N.h}{V}.\exp\left(\frac{\Delta G'}{RT}\right) \quad (2)$$

where, η, $\Delta G'$ and V represent the absolute viscosity, the Gibbs's activation energy and the molar volume, respectively. The parameters h, N and R are the Planck, Avogadro and ideal gas constant parameters, respectively. Eyring's equation was the base for proposing many others exponential types of viscosity-temperature relations. Among the large number of equations proposed, Walther's equation has been widely applied to represent the viscosity of the oil and its fractions.

$$\text{Log}\,(\log \eta + C) = A + B.\log\,(T) \quad (3)$$

where, η is the dynamic viscosity, A and B are constants that depend on the nature of the liquid, C is a fixed constant for most oils (C = 0.6 for viscosities above 1.5 cSt (cP/SG) and varies slightly with smaller viscosities) and T is the absolute temperature.

The effect of temperature upon viscosity relies greatly on the composition or volatility of the oil [29]. For pure compounds and single systems, the temperature effect on the global system is dictating by the solvent properties. However, for complex systems, such as crude oil, a raise in temperature affects mainly the petroleum macromolecular structures, promoting disruption of their aggregate and maintaining monomer units scattered. Therefore, the dispersed system should be more favorable than those of the organized macrostructures because these organized structures of the latter system enhancing its flow resistance [30]. A well-documented example is the pipeline Alyeska in Alaska, which transports the crude oil at approximately 50°C. A project involving heated pipelines is not an easy task. However, there are many considerations have to be taken into account on the design of a heated pipeline including; (a) instability in the flow of oils as a result to change of their rheological properties that may be possibly induced by heating the pipeline, (b) many number of heating and pumping stations are required over long distances of pipelines posing an additional cost, and (c) the heat losses along the pipeline because of the low flow of the crude oil i.e., heat loss is present during oil flow (the cooling effect). Therefore, heating stations should be planned anticipating gradual cooling in the line where, the pipeline is often insulated to maintain an elevated temperature and reduce the heat losses to the surrounding water as well as the earth lowers the efficiency of the technique. In such case when the pipeline is vulnerable to shut-in, a heated diluent has to be injected to restart flow in the pipeline [27]. In warmer climates where the ambient temperature does not bring the heavy oil below the pour point, electrical heating may be used to boost production without the need for special restart procedures. This has been demonstrated on shorter production lines in Colombia [31]. Although widely diffused as a method for viscosity reduction for transportation, heating is expensive due to the high cost of heat generation, especially when applying it in cold regions. Therefore, this technique might not be viable for transporting crude oil when it comes to subsea pipelines, sudden expansion and contraction along the pipeline may induce challenging problems, as well as the high rate of corrosion inside the pipeline due to the high temperature. Consequently, the capital and operating costs will be significantly high especially over long distances of pipelines from the oil field to the final storage or refinery on the high side [32]. The costs of insulated pipelines or installing heaters are less than that of dilution or upgrading.

5. Dilution

Dilution/blending of heavy crude oil and bitumen with light hydrocarbons or organic solvents is the most commonly used oldest and preferred technique to reduce viscosity and to improve their transportation via pipelines almost five decades ago. The cost-effective diluents used to enhance the transportability of heavy crude oils in pipelines are relatively cheap and readily available. The widely used diluents include light crude oils and light to medium products from the upgrader or refinery processes such as naphtha, kerosene, etc. as well as light natural gas condensate, by-product of natural gas processing. However, the use of organic solvents such as alcohol, methyl tert-butyl ether (MTBE), tert-amyl methyl ether (TAME), and dimethyl ether (DME) has been investigated [33]. Generally, it is well known that blending the fluids or diluents with the lower viscosity produce the blended mixture of heavy crude and bitumen with the lower viscosity which is easier to pump allowing transportation of a large quantity or volume of these oils at reduced cost [34]. Furthermore, dilution helps the desalting and dehydration operations downstream. The viscosity of the resulting blends depends on the viscosities and densities of the heavy crude oil and bitumen and the used diluent, the dilution rate, heavy oil/diluent ratio and also the operating temperature. It is worthy to mention here that heavy crude oil and bitumen can have a viscosity of more than 10^5 mPa s. However, viscosity of the diluted or blended oil must be less than 200 mPa s to transport heavy crude oils conveniently via the classical pipelines [22]. In this sense, large volume of diluents is required to achieve this pipeline viscosity specification where, the amount of diluents required for bitumen is higher than that for heavy crude oil where, the ratio of diluents/heavy oils in their exported blends ranges from 0 to 20%, while it is in the range of 25–50% for diluents/bitumen blends. The method also suffers certain disadvantages. The dilution of heavy crude oils is employed to enhance pipelining using two pipelines, one for the oil and another for the diluents via two main strategies depending on whether, the diluent is recycled or not. In both cases, a larger pipeline diameter is needed. This demands considerable capital and operational investment in pumping and pipeline maintenance. Moreover, the difficulty in prediction of solvent/oil ratio required for achieving a reduction in viscosity owing to the large and inconsistent number of governing parameters leads to ineffectiveness of simple mixing rules and therefore, the diluent will command a significant hold-up. The non-recycling strategy is based on the availability that secure the steady supply for the diluent. As the light hydrocarbons (i.e. diluent) may be acquired from neighboring conventional oil fields that the company owns, or from a competitor which could prove costly, something that may not be the case at peripheral heavy oil fields. In case of recycling strategy, the project economics are less of a subject to the price of diluent. However, this strategy is required, as recycling facilities and constructions required to return of diluent to the production site pose an additional capital expenditure.

Also, the oil composition has to be considered on selection of the solvent. This consideration is due to the compatibility issues between the asphaltenes and paraffins present in the oil with the solvent. If due care is not taken, deposition of asphaltenes and paraffins can cause further problems as, the light condensate recovered from natural gas (C_5+ or "Pentane Plus"), a low-density and less viscous mixture of hydrocarbon liquids, has been used to dilute the heavy crude oil and bitumen in order to enhance their transportation using pipeline in Canadian and Venezuelan oil fields. Though the efficiency of this condensate in reducing the viscosity of the heavy crude oil and bitumen significantly, instability during transportation and storage is observed as a result to precipitation, segregation and aggregation of asphaltenes [35]. These findings are because of insolubility of asphaltenes in most of condensate components involving alkanes such as n-pentane and heptanes. Besides, asphaltenes have the tendency to interact and aggregate in the oil-condensate blended mixture, as the condensates are known to be paraffin rich light oil. This may result in flocculation which leads to partial plugging of pipelines. Furthermore, the availability of condensates based on natural gas demand is one of the important limitations to its use as a heavy oil thinner where, the production of condensate is not sufficient to sustain the demand due to the growing production of heavy crude oil and bitumen [27]. Thus, light crude oils with API gravity between 35 and 42 have also been employed for dilution of the heavy ones, but it is less efficient than the condensate in lowering their viscosity [36]. Because of decline of the reserves of conventional light crudes, these oils suffer similar disadvantages as condensates like availability and compatibility with asphaltenes. Also, the light hydrocarbons such as gasoline and kerosene as distillates have been found to be effective owing to their good solvent properties. Gasoline also helps to improve the octane number in downstream processing. Thus, it has been shown that 15% kerosene mixed with heavy oil at 50°C achieves the same viscosity reduction achieved by 20% kerosene at room temperature [37]. Another common diluent used is naphtha, hydrocarbons ranging from C_6 to C_{12} from naphtha fraction of crude oil distillates. Naphtha has high API gravity (low density) which leads to efficient dilution of crude oil and shows good compatibility with asphaltenes owing to the presence of aromatic content in it. It is easily recyclable and reusable. However, the mostly used light hydrocarbons for dilution of heavy crude oils are expensive and are not readily available in large quantities. Therefore, the recycling is essential in despite of the required large investments and subsequently additional operating cost.

As previously mentioned, employing of organic solvents including MTBE, TAME and DME in thinning the heavy crude oils have been considered [33]. Recovery of DME is easier than the other solvents. Indeed, alcohols have been found to be more effective in reducing the viscosity where, the addition of ethyl alcohol at 10% led to reducing viscosity of the crude oil by almost 80% at 25°C [38]. This can be due to interaction between the hydroxyl groups and

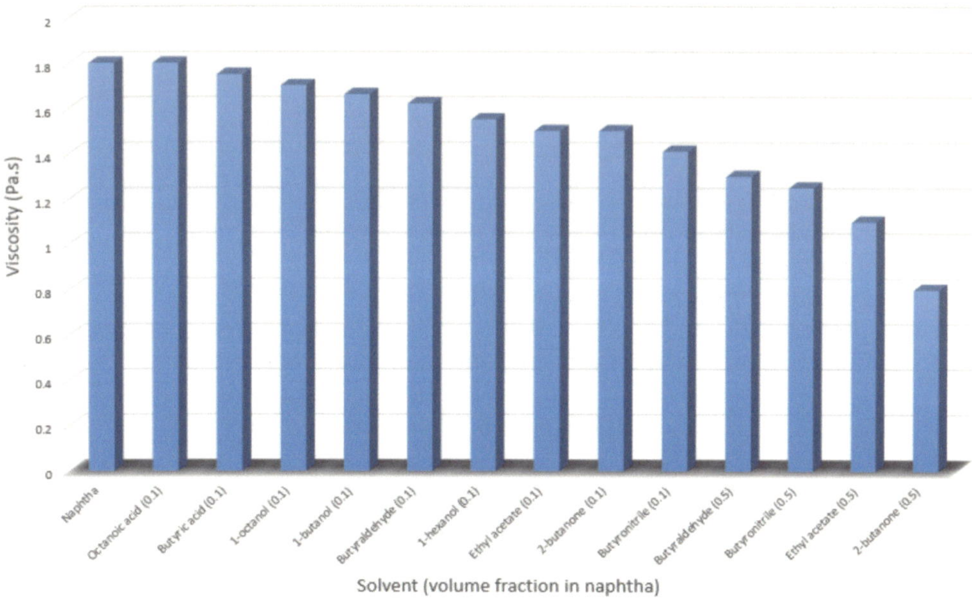

Figure 6.
Effect of naphtha/organic solvent blend at different volume fraction on the viscosity of the diluted heavy crude oil.

asphaltenes [39]. The use of these solvents is prompted based on their use in improving the octane number of gasoline. Subsequently, a mixture of hydrocarbons and organic solvents bearing polar groups onto their molecular structure has slightly exhibited efficacy in viscosity reduction of heavy crude oil at constant dilution rate [34]. Therefore, it has been proposed that a blend of naphtha and organic solvent would reduce the amount of diluents needed to lower the viscosity of heavy oil-to-pipeline transportation specifications [34]. The relative viscosities of the blend of heavy oil diluted with mixtures of naphtha and organic solvents are shown in **Figure 6**. The reduction of viscosity for resulting diluted heavy crude oil is attributed to the increasing polarity or hydrogen bonding of the solvents and the ability of the polar solvent to solubilize the asphaltenes components present in the heavy crude oil [34]. In that case, high polarity of solvents enhances their dilution efficiency causing greater reduction in viscosity of the heavy crude oil. However, the solvent can be easily recycled when its viscosity is comparable for that of the hydrocarbon as well as their boiling point.

Recently, the effect of carbon dioxide on the heavy crude oil has been studied. The findings of this studies have showed that crude oil saturated with carbon dioxide undergoes significant reduction in viscosity at a given temperature and pressure [40].

5.1 Prediction of resultant viscosity of the crude oil-diluents mixture

Generally, there is an exponential relationship between the viscosity of the resulting mixture and the volume fraction of diluent, so small fractions of diluents can cause a noticeable decline in oil viscosity. The addition of light oils or solvents resulted in lowering the frictional pressures. The rate of this reduction is greater in lower temperatures [30]. Up to date, a number of correlations have been developed for prediction of resultant viscosity of the blended mixture of heavy crude oil and diluents. But the accuracy of these relations is limited owing to the number of parameters involved in them. A few relationships have been discussed by Gateau and others [34]. In essence, viscosity of the resulting diluted heavy crude oil can be calculated from a modified correlation developed by Lederer [41] that is similar to the classic Arrhenius expression as follows:

$$\log \mu = \left(\frac{\alpha V_o}{\alpha V_o + V_d}\right) \log \mu_o + \left(1 - \frac{\alpha V_o}{\alpha V_o + V_d}\right) \log \mu_d \quad (4)$$

Where, V_o and V_d are the volume fraction of the heavy crude oil and diluents, μ_o and μ_d are the viscosity of the heavy crude oil and the diluents, respectively, and α is an empirical constant ranging from 0 to 1. Thus, an empirical formula for determining the constant α for the blend of heavy crude oil diluted with light hydrocarbons has been proposed by Shu [42]. This relation depends on the differences in densities of oil to diluents (i.e. light hydrocarbons) and their viscosity ratio.

$$\alpha = \frac{17.04\,(\rho_o - \rho_d)^{0.5237}\,\rho_o^{3.2745}\,\rho_d^{1.6316}}{\ln\left(\frac{\mu_o}{\mu_d}\right)} \quad (5)$$

where, ρ_o and ρ_d are densities of oil and solvent, respectively.

Currently, partial upgrading and dilution may be used in different stages of heavy crudes production. In the Orinoco field developments in Venezuela, the heavy oil is diluted at the production site for pipeline transport to a centralized refinery. At the refinery, the heavy oil goes through and upgrading process for further transport, while the diluent is separated from the blend and returned to the production site for recycling.

6. In-situ upgrading

In situ upgrading has been employed at surface conditions to locate the heavy crude oils and bitumen in viscosity conditions required for its production and transportation. In general, partial upgrading is often preferred with regard to entire upgrading because the cost of the process and the extension of the upgrading depend on the cost-benefit relation. This partial upgrading uses hydroprocessing to modify the relative proportion of the oil hydrocarbons (the composition of heavy oils) into a field refinery to make them less viscous and subsequently more suitable for pipeline transport without altering its refining characteristics. Hydroprocessing is a broad term that includes hydrocracking, hydrotreating, and hydrorefining. Where, the partial upgrading process is usually performed in two stages. In the first stage, called hydrocracking, the heavy oils are heated along with hydrogen under high pressure to promote the scission of macromolecular structures in the oil molecule rupture, forming smaller and simpler chemical structures as smaller paraffin and olefin molecules that change the properties of oil and the quality of its products as the viscosity of oil that reduce and it become lighter. In the second stage, called hydrotreating, hydrogen is added to promote hydrogenation without breaking structures and to remove impurities [29]. The hydrogenation process is catalyzed by metals such as nickel, palladium and platinum. Since metal catalysts are easily poisoned by sulfur containing compounds, the operation of the process requires a refined control technique. In this stage, the saturation of olefins and the conversion of aromatic compounds into naphthenics occur wherefore, the proportion of saturated carbons and aromatics increases with reducing the amount of asphaltenes and oil resins [43, 44]. Hereby, the upgrading process produces a large unwanted byproduct of coke that has to be handled and deposited. The potential advantage of upgrading over other techniques is to improve the quality for both of the oil and the residue, simultaneously with increasing its market value. Residue can represent significant portion of a crude oil barrel and its disposal treatment is not yet up to the mark [45, 46]. In this view, the conversion of residue into more consumable and valuable products is also an environmental issue. On the other hand, the increasing fuel oil demand makes the processing and utilization of bottom residue from atmospheric distillation and vacuum distillation columns unavoidable. Moreover, the partial upgrading has the advantage of lower pipeline investments, limited restart issues and no particular corrosions. Nevertheless, the investments associated with a field refinery demands a large production.

Summarily, upgrading can be considered as a process of carbon rejection and hydrogen addition [47]. Carbon rejection processes include visbreaking, thermal cracking, coking, deasphalting, and catalytic cracking, while hydrogen addition processes include catalytic hydrodemetallization, hydrodesulphurization, hydrodenitrogenation, hydrogenation, and hydrocracking in fixed bed, moving bed, ebullating bed, or slurry phase reactors [45–48]. Hereof, it was indicated that the partial upgrading is a hybrid approach involving a simultaneous use of several of these technologies. The assembly used as the strategy for upgrading depends on the product value and the SARA fraction distribution in the oil [45, 49]. Since asphaltenes and resins are the major constituents of heavy oils, these components present the highest impact on the method selection. So the heavy oils containing whether high resin and low asphaltene or low resin and high asphaltene even when both have the same API gravity, different upgrading processes must be employed [45, 50]. Farther, other considerations beside the properties of the oil have to be taken into account through choice of the upgrading treatments assembly such as, regional logistics between the well-head and the refining site, operational concern, transport distance, cost, environmental concerns and the legislation.

Example of the proper upgrading treatments developed by Association for the Valorization of Heavy Oils (ASVAHL) are deasphalting process Solvahl, thermal treatment Tervahl process and catalytic hydrotreatment Hyvahl processes [51]. Recently, many studies at Institut français du pétrole (IFP) aim to associate these different processes to optimize the heavy crude conversion. The combination of thermal cracking, solvent deasphalting and hydrocracking processes are commonly used for the processing of heavy oils. Furthermore, the present strategy in the petroleum industry is to integrate in situ upgrading to thermal enhanced oil recovery (EOR) methods for achieving the cost, environment and energy effectiveness. The Syncrude operations in Canada are an example of this, where surface mined 8° API bitumen is upgraded to a 30–32° API synthetic blend. It has also been practiced in Venezuela to export extra-heavy oil from the Orinoco belt. In situ upgrading technologies which can be achieved during thermal recovery methods include ISC, SAGD, CSS and subsequently the novel THAI and its add-on catalytic upgrading process in situ (CAPRI), collectively called THAI-CAPRITM [15, 24, 52]. These processes rely on the reduction of heavy crude oil viscosity by heat to improve its flow from the oil reservoir to the production well. This in situ thermal cracking process reduces the viscosity of the heavy oil and bitumen to a high order of magnitude, thereby improving flow and production. However, of all these processes, the THAI-CAPRI process integrates a catalytic upgrading process into the recovery. Details of the above-mentioned in situ upgrading technologies for during heavy oil recovery have been presented in a review article involving the novel extraction and upgrading technologies for heavy oil and bitumen by Shah and others [46].

7. Coverage of heavy oils by water

Coverage of heavy oils by water to facilitate their transportation via pipelines is a technology based on a physical phenomenon in which a less viscous phase migrates to the high shear region near the pipe wall, where it lubricates the flow. Once the pumping pressures are balanced by the wall shear stress, covered transport requires pressures that are comparable to pumping water alone, regardless the viscosity of oil [53, 54]. The oil and water phases can be configured in various ways during pumping. In horizontal pipes, the most common configurations are stratified flow based on the two phase's density, oil-in-water emulsions (stabilized by surfactants), and core annular flow, which are the most approaches employed in transport of heavy oils via pipelines including elevation ducts. These configurations depend strongly on the flow rate of oil and water [55].

7.1 Core-annular flow

Core-annular flow (CAF) is a technology to facilitate the flow of heavy crude oil and bitumen through pipeline by reducing the pressure drop in the pipelines owing to the friction. Where, this friction caused by the high viscosity of these fluids can make their flow in a single-phase is so difficult or undoable. This technique was first reported by Isaacs and Speeds in 1904 for the possibility of pipelining viscous fluids through the lubrication of pipe walls with water [56]. However, a commercial pipeline dedicated to transportation of heavy oil through annular flow was not in operation until the 1970s [57, 58]. Commercial establishing for core-annular flow systems involves not only technical questions but also operational methodologies to increase their feasibility and flexibility. The effectiveness of the commercial implementation of core annular flow is related to its adaptability to existing pipeline systems, in particular its capacity to share with other types of fluids that are not incorporated in the core flow regime. Core flow has attracted much industrial interest, being the subject of many patents to facilitate the pipeline transportation of heavy oils [59]. Almost, the most important industrial application among them up to date is employed in the Shell project in California, in which a 38.6 km pipeline from North Midway Sunset reservoir to the central facilities at Ten Section (California) operates for 12 years with 30% volumetric water with a flow rate of 24,000 barrels per day. Other examples include the 55 km lubricated pipelines from San Diego to Budare (Venezuela) used for transporting Zuata heavy crude oil (9.6 API°) and the self-lubricated pipelines of Syncrude's Canada Ltd. [60].

The core-annular flow technique is based on exploiting the low viscosity of water and high density of heavy oil to form a concentric flow pattern, where thin film of water or aqueous solution formed onto the inner wall of the pipe, "lubricating" the flow of core fluid consisting of heavy oil. Lining the pipe with an aqueous film is done by injecting water into the pipeline at its head (**Figure 7**). Accordingly, the core-annular flow is considered one of the regimes based on flow of two-phase via pipelines. Somewhere further down the pipeline, formation of an aqueous coating around the core of heavy oil will stabilize the flow with reducing the friction pressure loss. Where, the reduced longitudinal pressure gradient and a total pressure drop are similar to that of volumetrically equal pure water flow [57, 58, 61]. The typical fraction of water required in this approach (10–30%) implies that the pressure drop along the pipeline depends very closely on the viscosity of water, but weakly on that of the heavy oils [23, 62]. Furthermore, the pressure drop reduction exceeds 90% when the flow of water as an annulus near the pipe wall surface is done around the core of heavy crude oil comparing with that without water lubrication [57].

If the core-annular flow is assumed to be perfect and well centered, then the pressure drop can be calculated from the following equation:

$$\frac{\Delta P}{L} = \frac{Q}{\frac{\pi}{8}\left[\frac{R^4}{\mu_w} + R_s^4\left(\frac{1}{\mu_0} - \frac{1}{\mu_w}\right)\right]} \tag{6}$$

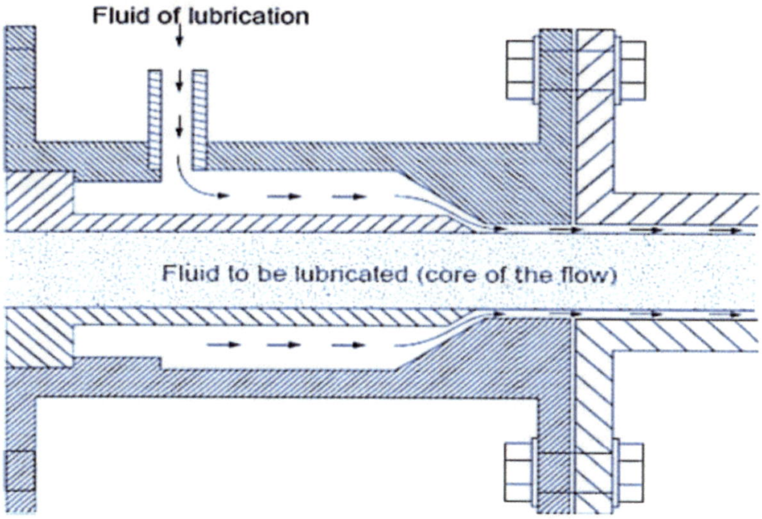

Figure 7.
Illustration of the core annular flow injector configuration [57].

where $\Delta P/L$ is the pressure drop of the centered core-annular flow (Pa/m), Q is the total flow rate (m^3/s), R is the radius of the pipe (m), R_s is the core radius (m), and μ_w and μ_o are the dynamic viscosities of water and oil, respectively (Pa s).

Despite of core annular flow capability for reducing the pressure drop to that of moving water, achieving this perfect flow having high stability is very rare because such perfection and great stability may only accomplish with density-matched and immiscible fluids. i.e., have similar densities and do not form emulsion [57, 63]. Whereas, the mechanisms of hydrodynamic destabilization of the annular flow originate from capillary forces and inertia (the difference between the interfacial velocity of the fluids), are evidenced by the deformation of the liquid-liquid interface. These flow velocity and capillary instability arising from surface tension and the density difference between the liquids break the inner core into slugs at low velocity and stratification occurs in the system. Hereby, several flow regimes may occur depending on the properties of the oil such as density, surface tension, and shear rate of the flow and fluid injection flow rate. Where, wavy core-annular flow, in which waves are created at the water and oil interface can occur (**Figure 8**). This flow regime is more likely to be present in the core fluid through the core annular flow process [64]. Furthermore, a radial movement of the oil core to the upper wall of the pipe by act of a buoyancy force can occur when the density difference between the oil and water is large (**Figure 9**). On the other side, increasing the velocity enhances the core flow stability.

Still, there are significant problems encountered in the commercial application of annular flow for heavy oil transportation as instability of the flow regime, fouling and corrosion of the pipe walls. These problems emerge the moment flow rate drops or the pipeline is shut-in (**Figure 10**) where, the liquids will segregate into two horizontal layers (stratification of the phases). If the pipeline has an elevated section; this could lead to a permanent heavy oil plug blocking the flow. Knowingly, many potential interruptions may occur in any normal pumping operation of crude oil due to mechanical failure, power interruptions, and ruptures in the pipeline or climate concerns. Ditto, interruptions occurring in the core annular flow-based pipelining even for relatively short periods of time can lead to the stratification of the two phases. This stratification of the two phases can be aggravated by the difficulties of restarting the flow in case of unscheduled downtime, where high

Figure 8.
Schematic diagram of wavy Core annular flow [54].

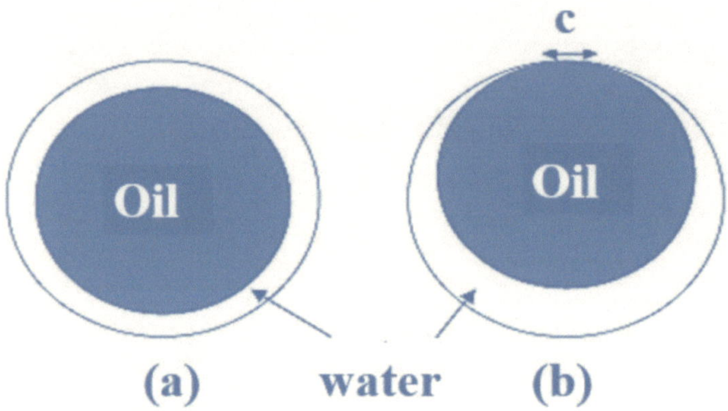

Figure 9.
Radial position of the oil core: A perfect core-annular flow and b with density difference. C is the contact perimeter between the oil phase (core) and the pipe wall.

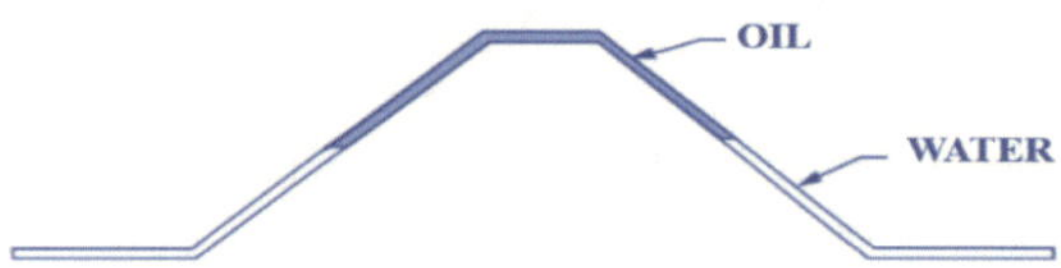

Figure 10.
Shut-in issues occurred in core-annular flow.

pressures are required to restore the system [23]. This high pressure can cause major failures in the pipeline as it may exceed the maximum allowable pressure. A basic process has been proposed for restarting core flow with heavy oils after a long standstill period in which a flow of a low viscosity fluid, water, is pumped first into an inlet portion of the pipeline with increasing the flow gradually until reaching the critical velocity required to develop annular flow in a steady state condition. Then, the heavy oil is injected into the inlet portion of the pipeline, and similarly its flow is gradually increased either by adjusting a variable speed motor to the pump or by adjusting a control valve in a viscous oil bypass line. An increase of pressure due to the pumping of heavy oil is much smaller than the pressure peaks observed during the low viscosity fluid build-up stage.

Fouling as one of the main problems in implementing core annular flow caused by the gradual adherence and accumulation of oil at the pipe walls that can cause a blockage in the pipe section, preventing flow. The tendency of the oil to adhere to the pipeline walls exhibited during annular flow of 9°API oil [65]. Even in the hydro-dynamically stable annular flow which is able to maintain its structure through various line accidents, the oil tends to embed itself in the pipe walls by means of thermodynamic effects. Although the hydrodynamic stability of the system can be achieved by adjusting the process parameters, the stability of the system is still dependent on thermodynamic aspects [54, 55, 66]. In addition, formation of the incrustations in pipes can occur due to the reversal of wettability of the system caused by asphaltenes and naphthenic acids present in the oil [81]. The effects of wettability reversal over the load loss of the system can be avoided or reduced by the addition of sodium meta-silicate to the aqueous phase [67–69].

7.2 Emulsification

Several methods have traditionally been proposed to enhance the mobility of heavy crudes for pipeline transportation; these include heating crude oils or diluting them with lighter fractions of hydrocarbons. However, each of these methods has economic, technical and logistical drawbacks especially when it comes to transportation of heavy crude oil through offshore pipelines.

Emulsions naturally occur in petroleum production and pipelining, mainly those of water-in-oil (W/O) and more complex (double) emulsion like oil-in-water-in-oil (O/W/O) emulsions (**Figure 11**). Such emulsions are detrimental for oil production since oil's viscosity raises, increment corrosion issues and are difficult to break in desalting and dehydrating units before refining. Nevertheless, O/W emulsion reduces the viscosity of heavy crude oils and bitumen and may provide an alternative to the use of diluents or heat to reduce viscosity in pipelines [70]. Thereby, emulsification of heavy oils with water with aid of active surface additives is considered to be one of the newest and most economical alternative techniques to overcome flow assurance problems associated with transportation of heavy crude oil through pipelines under the cold offshore environments. This technology is hydraulically transporting heavy crude oil via pipeline in form of oil-in-water (O/W) with the drop sizes in micron range can reduce the viscosity to values of 50–200 cP at 15°C [23, 25, 28], in which it can be easily pumped [37, 70–72]. The methods used to generate the oil droplets to create the different possible emulsions includes use of devices such as dispersing machines, mixing with rotor-stator, colloid mills, high-pressure homogenizers applying high shearing stresses, emulsification by membrane and ultrasonic waves [39, 73, 74].

These oil-in-water emulsions are thermodynamically unstable where they can be subjected to several breakdown processes like Ostwald ripening, sedimentation and creaming due to density difference and coalescence of the drops [70]. Therefore, a surfactant (or mixture of surfactants) suitable for varying conditions have to be

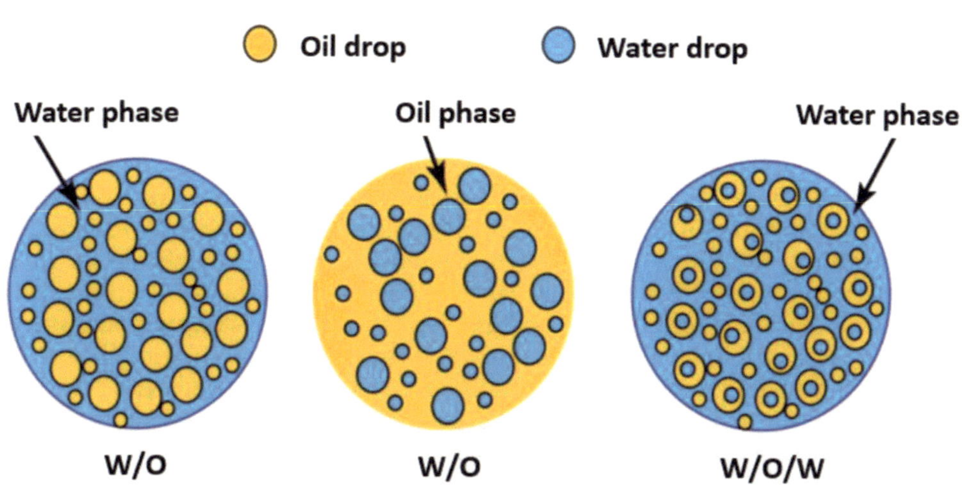

Figure 11.
The several emulsion forms used to transport heavy crude oils.

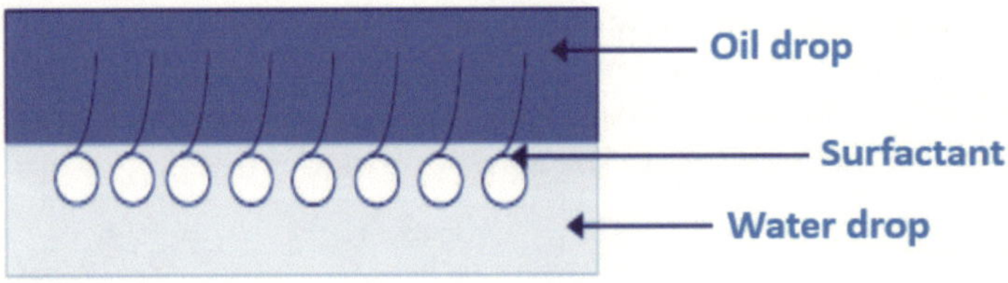

Figure 12.
Surfactant-stabilized emulsion.

added forming a strong interfacial film to enhance the kinetic stability of these emulsions and consequently improve their transportability by pipelines [75]. Besides, in some cases additional substances as stabilizing agents (high molecular weight) to avoid phase separation. Accordingly, cost of the surfactant, its ability to maintain the emulsion stability during pipeline transportation and the ease of its separating from the crude oil at the final destination (since the density of heavy oil is close to that of water) are major challenges associated with the emulsification technology for transporting heavy crude oils. Monolayer of the surfactant molecules settles onto the oil-water interface to prevent drop growth and phase separation during transportation. As shown in **Figure 12**, such interfacial monolayer is formed by directing the polar region (i.e. hydrophilic head) of the surfactant toward the water phase and non-polar tail (i.e. hydrophobic region) toward the oil phase. Hereof, the efficiency of the surfactant based on the properties of this adsorbed layer (the polar hydrophilic head and the non-polar hydrophobic tail) of surfactants that stabilize the oil-water surface against shear and decrease the interfacial tensions, and subsequently control the flow behavior of the emulsion [70]. However, heavy crude oil emulsion exhibits either Newtonian behavior at high shear rate or a shear thinning rheological behavior at low shear rate [25, 76].

Pointedly, properties of the emulsion including the rheological characteristics and stability depend mainly on many parameters such as, drop size and their polydispersity, temperature, salinity and the pH of the water, the components of the heavy crude oil, mixing energy and oil/water volume ratio [39, 77]. Drop size and their polydispersity depends on surfactant type, energy of mixing and pressure. The use of a dynamic mixer as a rotor-stator mixer, may cause the formation of small droplets with a diameter of less than 10 μm which can cause a significant increase in the viscosity of O/W emulsion and emulsion inversion to an oil continuous emulsion that are detrimental to pipelining.

The use of surfactants and water to create a stable oil-in-water emulsion with heavy crude oil has been a topic of several investigations with a series of patents. Stabilizing the O/W emulsion from Egyptian Geisum crude oil using an anionic surfactant for pipeline transportation has been studied. The findings exhibited that stability and viscosity of surfactant-stabilized O/W emulsion increased with increasing concentration of the anionic surfactant that reduces the O/W interfacial tension and size of dispersed droplets [78]. Similar results were found in a non-ionic surfactant stabilized O/W emulsion [37]. Triton X-114 is one of the commonly used non-ionic surfactants based on their ability to withstand the salinity of the produced water, they are also cheap, their emulsion is easy to separate, and they do not form undesirable organic residues that affect the oil properties [79]. However, the use of anionic and non-ionic surfactants produced a synergistic effect that allows a lower viscosity and more stable O/W emulsion. The preferred water soluble nonionic chemical surfactants for viscous crude oils are the commercially available ethoxylated alkyl phenols and ethoxylated alcohols; while the preferred water-soluble anionic chemical surfactants are ethoxylated alcohol sulfates. On the other side, as a water-soluble chemical surfactant, or, together with a bio-emulsifier can absorb onto the hydrocarbon/water interfaces. Hence, "surfactant packages" composed of water-soluble chemical and/or biological co-surfactants have been proposed to transport viscous hydrocarbons by pipeline through the formation of low-viscosity bio-surfactant-stabilized oil-in-water emulsions, or the so-called hydrocarbosols. Here, hydrocarbon droplets dispersed in the continuous aqueous phase are substantially stabilized from coalescence by the presence of bio-surfactants (Bio-emulsifiers), in particular, microbial surfactants [80]. Generally, bio-emulsifiers act by orienting their molecules at the oil/water interface, avoiding the coalescence of the oil droplets and stabilizing the resulting emulsion with maintaining their reduced viscosity over time. The hydrocarbosols viscosities were reduced by at least a factor of 10. Bio-emulsifiers, specifically extracellular microbial polysaccharides ("emulsans") produced by different strains of the *Acinetobacter* bacteria have been extensively researched [81, 82]. Among the preferred bio-surfactants are heteropolysaccharides produced by bacteria of the genus *Acinetobacter* and the genus *Arthrobacter*, and in particular, those produced by strains of *Acinetobacter calcoaceticus*. Still some heavy oils were not successfully emulsified with the surfactant packages studied. Here, it is worthy to mention that fresh water, sea water or even formation water may be available for emulsification and thus, these are very efficient oil-in-water emulsifiers possessing a high degree of specificity in both fresh water and sea water for emulsifying hydrocarbon substrates which contain both aliphatic and aromatic or cyclic components. Here, the use of formation water instead of fresh water resulted in a lower interfacial tension between crude oil and formation water and a more viscous O/W emulsion because of the formation of smaller crude oil droplets [71].

In this context, the heavy crude oil is a complex mixture of hundreds of thousands of compounds that include the asphaltenes which act as natural emulsifiers as well as other active surface components such as naphthenic acids, resins, porphyrins, etc. [70]. The presence of these component increases the complexity of crude oil emulsion, as they can interact and reorganize at the oil-water interface. On the other hand, the presence of natural hydrophilic particles such as clay and silica in the crude oil may cause instability in the emulsion [70]. Additionally, the O/W emulsion system may sometimes contain solids and gas, which increases the complexity of the process. In general, it is worthy to mention that the behavior of heavy crude oil-in-water emulsion is complex due to the interaction of several components within the system and many other factors mentioned hereinabove. By the fact that the use of surfactants can significantly increase the cost of emulsification, the activation of surfactants naturally occurring in heavy and extra heavy crude oils is a reliable option. The ionization of acidic groups present in fatty and naphthenic acids as well as asphaltenes with a strong alkali can make these surfactants more hydrophilic causing a reduction of the interfacial tension [70]. Thus, the activation of the natural molecules present in bitumen with amines serves as natural surfactants to form a bitumen-in-water emulsion or Orimulsion® (Orimulsion® is a bitumen-in-water emulsion and simultaneously the technology developed to facilitate the transportation of Cerro Negro bitumen) [83]. Moreover, some works refer to the use of the natural surfactants present in crude oil with particles such as silica, clay, iron oxides, etc., more stable emulsions can be obtained by saturation of surface of these particles by asphaltenes [70]. There are still many unresolved questions related to the peculiar behavior of these emulsions as result to the complexity of the molecular composition of oil having a wide range of chemical structures and molecular weights, the hydrophilic-lipophilic balance (HLB) values of the surfactants, the multiple interactions oil-water-surfactant and the possible molecular rearrangements at the oil/water interface.

Economically, pipelining of oil in form of O/W emulsion must transport as much oil as possible and as little water as possible (high O/W volume ratio). Notwithstanding, when the oil content in O/W exceeds 70%, the viscosity may become too high or inverse to W/O emulsion. Meta-stable and easy-to-break O/W emulsions should be produced with the minimum quantity of cost-effective surfactant and other additives where, strategy of implementing the technology of oil-in-water emulsions to transport heavy crude involves three stages such as producing the O/W emulsion, transporting the formed emulsion and separating the oil phase from the water phase. However, recovering the crude oil entails breaking the oil in-water emulsion. Hereby, different techniques including thermal demulsification, electro-demulsification, chemical demulsification, freeze-thaw method, pH modification, addition of solvent and demulsification by membranes have been developed to achieve the separation stage [73, 84].

In this technology, the surfactants should allow at the same time a simple but efficient rupture of the O/W emulsion before crude oil refining and the separated water should be treated in order to comply with environmental and industrial regulations for water discharge or recycling. The demulsification produced 0.5% BS&W for the heavy oil and less than 100 ppm oily contaminants in the separated water. Nonetheless, O/W emulsions can be considered for applications in improved recovery processes of heavy oils and for the increase in the recovery factor of mature fields [85], including the amounts not retrieved from the exhausted fields of light oils [86]. Furthermore, such emulsification technology can improve residual oil removal from mature fields that are not as efficiently recovered by traditional methods that apply heat or diluents [87]. Also, restarting a pipeline after an emergency shutdown and re-emulsification of oil may not pose major problems [88].

The potential of this technology to enhance pipeline transportation of heavy crude oil has been demonstrated in Indonesia in 1963, as well as in a 13-mile distance using 8 in. diameter pipeline in California [71]. Field-tested pipeline transportation of heavy crude oil as oil-in-water emulsions containing high fractions of oil has been proposed by Hardy and others [89]. A large number of studies, mostly experimental in nature, have been carried out on oil-water emulsions [78, 90]. However, the results of these studies are not uniform and are sometimes contradictory and thus, the results obtained from a study on a certain crude are difficult to apply to another. This can be attributed to the complex viscosity behavior of emulsions which depends on several factors such as base sediment and water (BS&W), temperature, shear rate, type and concentration of surfactant, and the components of the crude itself. Consequently, oil companies investigate on their own the specific crude that they produce the type of emulsion that is most appropriate allowing simple and economical crude oil recovery at the end of the pumping process.

Transportation of emulsified heavy oil had its technological viability clearly demonstrated by many field studies and the development of the pioneer process of large-scale fuel emulsions, a joint project between BP Canada and the Alberta Energy Company has developed TRANSOIL®, a technology that is comprised of the steps of oil emulsification, characterization of transport properties, and evaluation of storage and recovery conditions by de-emulsification [91]. The emulsions obtained were pumped continuously for 6 days at 80 m^3/day and stored for 6 days without any sign of degradation. Besides, ORIMULSIONS® process has been developed for the generation of energy in thermal plants by the state owned company PDVSA (Petr oleos de Venezuela) in Venezuela in the 1980s [17] and commercialized by its filial Bitumen's Orinoco S.A. (BITOR) [60, 70]. Ditto, emulsion flow rate tests performed with 13° API oil from a field in Shanjiasi (China) showed a reduction in pressure loss by as much as 80%, which was achieved by emulsions formulated with water fractions around 0.6–0.8. This reduction in pressure occurred especially when the tests were conducted in more drastic conditions, such as with extra-viscous oils and at low temperatures [14]. In another work, emulsification of heavy oil in Sicily and the Adriatic Sea (Italy) has been tested for application in transport and production. The trial was based on the injection of an emulsifying aqueous phase into the well without any modification to the existing system. The dispersed oil content in the well-produced oil-in-water emulsions was 70%. The emulsions showed a viscosity 30–50 times lower than that of the diluted oil and oil productivity was greater four times than obtained with the conventional diesel dilution production [92].

7.3 Heavy oil emulsions for transport in cold environments

Considering, handling difficulty for O/W emulsions in cold environments because of phase destabilization, freezing or an increase in viscosity to a level too high for pipeline transport, brine with a high salt content and freezing point depressants like ethylene glycol in sufficient concentration are suggested to be used to maintain the oil-in-brine emulsion in pipeline condition at 253.15 K or less, but insufficient to break down the emulsion. Here, it is worthy to consider the costs of emulsion rupture and further processing of produced water to discharge or reuse.

The emulsion is suggested to contain a small amount of 0.1–5% by weight, of a conventional surfactant to facilitate the formation and preservation of the emulsion. The use of common surfactants mixes has been proposed to form the emulsion and supplemented by use of the xanthan biopolymer to enhance the stability of the emulsion [93]. In another research, a novel surfactant, tri-triethanolamine monosunflower ester, was synthesized from fatty acids produced from hydrolysis of sunflower (*Helianthus annuus*) oil. The resultant surfactant was used to emulsify a heavy crude oil from the western oil field of India. The oil-in-water emulsion developed with 60% oil content and 2 wt% surfactant exhibited a decrease in viscosity of 96% and a dramatic decrease in pour point where, the prepared emulsions were found to be flowing even at 1°C. The high stability of the emulsion has been attributed to the large reduction in the equilibrium interfacial tension (IFT) between the crude oil (diluted) and the aqueous phase of emulsions, which was almost nine times lower than that of no surfactant. These results suggested that the synthesized surfactants may be used to prepare a stable O/W emulsion for its transportation through offshore pipelines efficiently [94].

Furthermore, heavy oils in cold environments can be efficiently transported through a large diameter insulated pipelines at temperature below 273.15 K in the form of oil-in-brine emulsions containing 40–70% w/w of the dispersed oil with dissolving salts in the water phase at concentration sufficient to prevent freezing. These operating conditions permit the insulated pipeline to be buried in the ground without causing thawing of the permafrost, which in turn can cause damage to both the environment and the pipeline.

8. Conclusion

The plateau in conventional oil production and the corresponding increase in the demand for liquid fuels have motivated markets to respond with higher oil prices. And the current economics of oil are spurring the transformation of energy supplies, not away from oil, but toward new oils which are expected to be the dominant supplies of liquid-fuel through the twenty-first century based on the market forces. The amount of these new petroleum resources could continue to be revised upward as new technologies are developed. In this chapter, the mitigation technologies of low mobility of heavy crude oils used to facilitate their pipelining were explored. These methods have been divided into three categories based on reduction of viscosity and friction and in situ upgrading of these heavy crudes and bitumen. Each of these treatment techniques has special advantages and drawbacks. Therefore, there are many criteria to take into consideration on choice of each technique including the chemical and physical properties of the crude, regional logistics between the well-head and the refining site, operational and environmental concerns, distance of transportation, cost, and regulatory requirements. These criteria require that the petroleum industry tailor treatment strategies to fit the circumstances of its systems in accordance with the quality requirements of the pipeline operator as well as the content and quality demands of the refinery customer. Emulsification of crudes to form O/W emulsion was an appropriate method to reduce the viscosity of oils and enhance their flow under offshore conditions. Currently, in situ upgrading can be achieved during thermal enhanced oil recovery methods because this strategy characterized by the cost, environment and energy effectiveness. Also, the introduction of bacteria and bio-metal nanoparticles (Bio-NPs) in the reservoir to upgrade heavy oil, extra heavy oil, and bitumen in situ is an area of active research and hold great promise for improving flow of these crudes.

Author details

Emad Soliman
Polymer Materials Research Department, Advanced Technology and New Materials Research Institute (ATNMRI), City of Scientific Research and Technology Applications (SRAT-City), New Borg El-Arab City, Alexandria, Egypt

*Address all correspondence to: esoliman@srtacity.sci.eg

References

[1] Ali S. Heavy oil recovery-principles, practicality, potential, and problems. In: SPE Rocky Mountain Regional Meeting. Montana, USA: Society of Petroleum Engineers; 1974. pp. 1-10

[2] Thomas S. Enhanced oil recovery—An overview. Oil & Gas Science and Technology—Revue d'IFP. 2008;**63**:9-19. DOI: 10.2516/ogst:2007060

[3] Gordon D, Sperling D. Big oil can't get beyond petroleum. Cap Times. June 20, 2010. Available from: http://host.madison.com/ct/news/opinion/column/articleb1775726-9ef8-5e5a-85fb-ed56417475d8.html

[4] International Energy Agency (IEA), World Energy Outlook 2011. London; Press Launch. http://www.iea.org/weo/docs/weo2011/homepage/WEO2011_Press_Launch_London.pdf

[5] Speight JG. The Chemistry and Technology of Petroleum. New York: Marcel Dekker, Inc.; 1999. p. 11-14, 32, 306, 322-323

[6] Mullins OC, Sheu EY, Hammami A, Marshall AG. Asphaltenes, Heavy Oils and Petroleomics. New York: Springer; 2007. pp. 554-555, 621-622

[7] Merino-Garcia D. Calorimetric Investigations of Asphaltene Self-Association and Interaction with Resins. København: Technical University of Denmark; 2004

[8] Alboudwarej H, Felix JJ, Taylor S, et al. La importancia del petróleo pesado. Oilfield Review. 2006;**18**:38-39. Available from: https://www.slb.com/~/media/Files/resources/oilfield_review/spanish06/aut06/heavy_oil.pdf

[9] Curtis C, Kopper R, Decoster E, et al. Yacimientos de petróleo pesado. Oilfield Review, II. 2003;**23**:32-55

[10] Goodarzi N, Mai A, Bryan J, Kantzas A. Insights into non-thermal recovery of heavy oil. In: Presented at World Heavy Oil Conference. 2009. pp. 2006-2553

[11] Cesar Ovalles, Estrella, Rogel, John Segerstrom, Chevron ETC. Improvement of flow properties of heavy oils using asphaltene modifiers. In: SPE Annual Technical Conference and Exhibition. Denver, Colorado, USA. October 30 – November 2, 2011

[12] Ashrafi M, Souraki Y, Karimaie H, Torsaeter O. Experimental PVT property analyses for Athabasca bitumen. In: Presented at the Canadian Unconventional Resources Conference Held in Calgary, Alberta, Canada. November 15-17, 2011

[13] Werner A, Behar F, de Hemptinne JC, Behar E. Thermodynamic properties of petroleum fluids during expulsion and migration from source rocks. Organic Geochemistry. 1996;**24**(10-11): 1079-1095. DOI: 10.1016/S0146-6380(96)00082-4

[14] Zhang N, Zhao S, Sun X, Zhiming X, Chunming X. Storage stability of the visbreaking product from Venezuela heavy oil. Energy & Fuels. 2010;**24**:3970-3979. DOI: 10.1021/ef100272e

[15] Hart A. The catalytic upgrading of heavy crude oil in situ: The role of hydrogen. International Journal of Petroleum Science and Technology. 2012;**6**(2):79-96

[16] Ahmed NS, Nassar AM, Zaki NN, KhH G. Formation of fluid heavy oil-in-water emulsions for pipeline transportation. Fuel. 1999;**78**:593-600. DOI: 10.1016/S0016-2361(98)00157-4

[17] Martinez-Palou R, Maria de Lourdes M, Beatriz Z-R, Elizabeth

M-J, Cesar B-H, Juan de la Cruz C-L, et al. Transportation of heavy and extra-heavy crude oil by pipeline: A review. Journal of Petroleum Science and Engineering. 2011;**75**:274-282. DOI: 10.1016/j.petrol.2010.11.020

[18] Eskin D, Ratuloswski J, Akbarzadeh K, Pan S. Modelling asphaltene deposition in turbulent pipeline flows. Canadian Journal of Chemical Engineering. 2011;**89**:421-441. DOI: 10.1002/cjce.20507

[19] Fruman DH, Briant, J. Investigation of the Rheological Characteristics of Heavy Crude Oil-in-Water Emulsions. In: Proceedings of the International Conference on the Physical Modeling of Multi-Phase Flow. England: Coventry; 1983

[20] Nunez G, Briceno M, Mata C, Rivas H, Joseph D. Flow characteristics of concentrated emulsions of very viscous oil in water. Journal of Rheology. 1996;**40**:405-423. DOI: 10.1122/1.550751

[21] Schumacher MM. Enhanced Recovery of Residual and Heavy Oils. Park Ridge, NJ: Noyes Press; 1980

[22] Kessick MA, Denis CES. Pipeline Transportation of Heavy Crude Oil. U.S. Patent No. 4,343,323. 1982

[23] Saniere A, Hénaut I, Argillier JF. Pipeline transportation of heavy oil—A strategic, economic and technological challenge. Oil & Gas Science and Technology—Revue d'IFP. 2004;**59**(5):455-466

[24] Hart A, Shah A, Leeke G, Greaves M, Wood J. Optimization of the CAPRI process for heavy oil upgrading: Effect of hydrogen and guard bed. Industrial and Engineering Chemistry Research. 2013;**52**(44):15394-15406. DOI:10.1021/ie400661x

[25] Al-Roomi Y, George R, Elgibaly A, Elkamel A. Use of a novel surfactant for improving the transportability/transportation of heavy/viscous crude oil. Journal of Petroleum Science and Engineering. 2004;**42**:235-243. DOI: 10.1016/j.petrol.2003.12.014

[26] Raicar J, Procter RM. Economic considerations and potential of heavy oil supply from Lloydminster – Alberta, Canada. In: Proceedings of the Second International Conference on the Future of Heavy Oil and Tar Sands. Meyer RF, Wynn JC, Olson JC, eds., New York, McGraw- Hill. 1984. p. 212-219

[27] Guevara E, Gonzalez J, Nuñez G. Highly viscous oil transportation methods in the Venezuela oil industry. In: Proceedings of the 15th World Petroleum Congress. London: John Wiley and Sons. 1998. pp. 495-501

[28] Rimmer DP, Gregoli AA, Hamshar JA, Yildirim E. Pipeline emulsion transport for heavy oils. In: Schramm LL, editor. Emulsions: Fundamentals and Applications in the Petroleum Industry. Washington, DC, USA: American Chemical Society; 1992

[29] Speight JG. The Chemistry and Technology of Petroleum. 2nd ed. New York: Marcel Dekker, Inc; 1991. 760 p

[30] Szilas AP. Production and Transport of Oil and Gas. Gathering and Transportation. New York: Elsevier; 1986

[31] Jaimes MG et al., Efficiency of Artificial Lifting Systems in the Production of Heavy Crude, by the Use of Electrical Heating in Surface Facilities. In: the Proceedings of the World Heavy Oil Congress [WHOC13]. Aberdeen, Scotland: Instituto Colombiano del Petroleo; 2013

[32] Chang C, Nguyen QD, Ronningsen HP. Isothermal start-up of pipeline transporting waxy crude oil. Journal of Non-Newtonian Fluid Mechanics.

1999;**87**:127-154. DOI: 10.1016/S0377-0257(99)00059-2

[33] Anhorn JL, Badakhshan A. MTBE: A carrier for heavy oil transportation and viscosity mixing rule applicability. Journal of Canadian Petroleum Technology. 1994;**33**(4):17-21. DOI: 10.2118/94-04-02

[34] Gateau P, Henaut I, Barre L, Argillier JF. Heavy oil dilution. Oil & Gas Science and Technology—Revue de l'IFP. 2004;**59**(5):503-509. DOI: 10.2516/ogst:2004035

[35] Shigemoto N, Al-Maamari RS, Jibril BY, Hirayama A. A study of the effect of gas condensate on the viscosity and storage stability on Omani heavy crude oil. Energy & Fuels. 2006;**20**(6):2504-2508. DOI: 10.1021/ef060074h

[36] Urquhart RD. Heavy oil transportation: Present and future. Journal of Canadian Petroleum Technology. 1986;**25**(2):68-71. DOI: 10.2118/86-02-05

[37] Yaghi BM, Al-Bemani A. Heavy crude oil viscosity reduction for pipeline transportation. Energy Sources. 2002;**24**:93-102. DOI: 10.1080/00908310252774417

[38] Argillier J-F, Barre L, Brucy F, Dournaux J-L, Henaut I, Bouchard R. Influence of Asphaltenes content and dilution on heavy oil rheology. In: Proceedings of SPE International Thermal Operations and Heavy Oil Symposium. Vol. 69771. Venezuela: SPE; 2001. pp. 1-8

[39] Hasan SW, Ghannam MT, Esmail N. Heavy crude oil viscosity reduction and rheology for pipeline transportation. Fuel. 2010;**89**:1095-1100. DOI: 10.1016/j.fuel.2009.12.021

[40] Hu R, Crawshaw JP, Martin-Trusler JP, Boek ES. Rheology of diluted heavy crude oil saturated with carbon dioxide. Energy and Fuels. 2015;**29**(5):2785-2789. DOI: 10.1021/ef5020378

[41] Lederer EL. Viscosity of mixtures with and without diluents. In: Proceedings of the World Petroleum Congress. London. 1933;**2**:526-528

[42] Shu WR. A viscosity correlation for mixtures of heavy oil, bitumen and petroleum fractions, SPE 11280. SPE Journal. 1984;**24**(3):277-282

[43] Kirkwood KM, Foght JM, Gray MR. Prospects for biological upgrading of heavy oils and asphaltenes. In: Vazquez-Duhalt R, Quintero-Ramirez R (Eds.). Studies in Surface Science and Catalysis. 2004;**151**:113-143. DOI: 10.1016/S0167-2991(04)80145-9

[44] Speight JG. Chemical and physical studies of petroleum Asphaltenes. In: Yen TF, Chilingarian GV, editors. Asphaltenes and Asphalts. 1st ed. Amsterdam, the Netherlands: Elsevier Science B.V; 1994. pp. 7-65

[45] Rana MS, Sámano V, Ancheyta J, Diaz JAI. A review of recent advances on process technologies for upgrading of heavy oils and residua. Fuel. 2007;**86**:1216-1231

[46] Shah A, Fishwick R, Wood J, Leeke G, Rigby S, Greaves M. A review of novel techniques for heavy oil and bitumen extraction and upgrading. Energy & Environmental Science. 2010;**3**:700-714

[47] Yen TF. Filtration, Chapter 31. In: Environmental Chemistry: Chemical Principles for Environmental Processes. Vol. 4B. Upper Saddle River, NJ: Prentice Hall; 1998. pp. 1405-1439

[48] Joshi JB, Pandit AB, Kataria KL, Kulkarni RP, Sawarkar AN, Tandon D, et al. Petroleum residue upgradation via visbreaking: A review. Industrial and Engineering Chemistry Research. 2008;**47**:8960-8988

[49] Reynolds BE, Brown EC, Silverman MA. Clean gasoline via VRDS/RFCC. Hydrocarbon Process. Houston: Hydrocarbon Processing, International Edition. 1992;**71**(4): 43-51

[50] Ancheyta J, Sanchez S, Rodrıguez MA. Kinetic modeling of hydrocracking of heavy oil fractions: A review. Catalysis Today. 2005;**109**:76-92

[51] Charlos JC, Courdec JL, Page, JF. Heavy oil processing, a synthesis of the ASVAHL results. In: Proceeding E.C. 3rd European community symposium on new technologies for the exploration and exploitation of oil and gas resources. Luxembourg. March 22-24, 1988. Vol. 2, E. Millich et al. eds

[52] Greaves M, Xia TX. Downhole upgrading of Wolf Lake oil using THAI–CAPRI processes-tracer tests. Preprints of Papers-American Chemical Society, Division of Fuel Chemistry. 2004;**49**(1):69-72

[53] Beretta A, Ferrari P, Galbiati L, Andreini PA. Horizontal oil-water flow in small diameter tubes: Pressure drop. International Communications in Heat and Mass Transfer. 1997;**24**:231-239

[54] Joseph DD, Bai R, Chen KP, Renardy YY. Core annular flows. Annual Review of Fluid Mechanics. 1997;**29**:65-90

[55] Bannwart AC. Modeling aspects of oil-water core-annular flows. Journal of Petroleum Science and Engineering. 2001;**32**:127-143

[56] Isaacs JD, Speed JB. Method of Piping Fluids. US Patent 759374. 1904

[57] Bensakhria A, Peysson Y, Antonini G. Experimental study of the pipeline lubrication for heavy oil transport. Oil & Gas Science and Technology. 2004;**59**:523-533

[58] Peysson Y, Bensakhria A, Antonini G, Argillier JF. Pipeline lubrication of heavy oil: Experimental investigation of flow and restart problems. SPE Production & Operations. 2007;**22**:135-140

[59] Zagustin K, Guevara E, Nunez G. Core-annular Flow Process. US Patent 4753261. 1988

[60] Salager JL, Briceno MI, Brancho CL. Heavy hydrocarbon emulsions. In: Sjöblom J, editor. Encyclopedic Handbook of Emulsion Technology. New York: Marcel Dekker; 2001

[61] Ghosh S, Mandal TK, Das G, Das PK. Review of oil water core annular flow. Journal of Renewable and Sustainable Energy Reviews. 2009;**13**:1957-1965

[62] Wylde JJ, Leinweber D, Low D, Botthof G, Oliveira AP, Royle C, Kayser C. Heavy oil transportation: Advances in water continuous emulsion methods. In: Proceedings of the World Heavy Oil Congress. Aberdeen; Scotland. Canada Inc; 2012

[63] Bannwart AC. A simple model for pressure drop in core-annular flow. Revista Brasileira de Ciências Mecânicas. 1999;**21**:233-244

[64] Bai R, Chen K, Joseph DD. Lubricated pipelining: Stability of core-annular flow. Part 5 experiments and comparison with theory. Journal of Fluid Mechanics. 1992;**240**:97-132

[65] Visser RC. Offshore production of heavy oil. Journal of Petroleum Technology. 1989;**41**:67-70

[66] Arney MS, Ribeiro GS, Guevara E, Bai R, Joseph DD. Cement-line pipes for water lubricated transport of heavy oil. International Journal of Multiphase Flow. 1996;**22**:207-221

[67] Santos RG, Mohamed RS, Bannwart AC, Loh W. Contact angle measurements and wetting behavior

of inner surfaces of pipelines exposed to heavy crude oil and water. Journal of Petroleum Science and Engineering. 2006;**51**:9-16

[68] Ribeiro GAS. Topics in the transport and rheology of heavy crude oils. Minneapolis, MN; PhD dissertation, University of Minnesota. 1994

[69] Liu Y, Buckley JS. Evolution of wetting alteration by adsorption from crude oil. SPEFE (Mar. 1997) 5-11

[70] Langevin D, Poteau S, Hènaut I, Argillier JF. Crude oil emulsion properties and their application to heavy oil transportation. Oil & Gas Science and Technology. 2004;**59**(5):511-521

[71] Ahmed NS, Nassar AM, Zaki NN, Gharieb HK. Stability and rheology of heavy crude oil-in-water emulsion stabilized by an anionic–nonionic surfactant mixture. Petroleum Science and Technology. 1999;**17**:553-576

[72] Santos RG, Bannwart AC, Loh W. Physico-chemical properties of heavy crude oil-in water emulsions stabilized by mixtures of ionic and non-ionic ethoxylated nonylphenol surfactants and medium chain alcohols. Chemical Engineering Research and Design. 2011;**89**:957-967

[73] Ashrafizadeh SN, Kamran M. Emulsification of heavy crude oil in water for pipeline transportation. Journal of Petroleum Science and Engineering. 2010;**71**:205-211

[74] Lin CY, Chen LW. Emulsification characteristics of three and two phase emulsions prepared by ultrasonic emulsification method. Journal of Fuel Processing Technology. 2006;**87**:309-317

[75] Jiang J, Mei Z, Xu J, Sun D. Effect of inorganic electrolytes on the formation and the stability of water-in-oil (W/O) emulsions. Colloids and Surfaces A: Physicochemical and Engineering Aspects. 2013;**429**:82-90. DOI: 10.1016/j.colsurfa.2013.03.039

[76] McKibben MJ, Gillies RG, Shook CA. A laboratory investigation of horizontal well heavy oil-water flows. Canadian Journal of Chemical Engineering. 2000;**78**:743-751

[77] Khan MR. Rheological properties of heavy oils and heavy oil emulsions. Energy Sources. 1996;**18**:385-391

[78] Zaki NN. Surfactant stabilized crude oil-in-water emulsions for pipeline transportation of viscous crude oils. Colloids and Surfaces A: Physicochemical and Engineering Aspects. 1997;**125**(1):19-25. DOI: 10.1016/S0927-7757(96)03768-5

[79] Sun R, Shook CA. Inversion of heavy crude oil-in-brine emulsions. Journal of Petroleum Science and Engineering. 1996;**14**:169-182

[80] Hayes M, Hrebenar KR, Murphy PL, Futch LE, Deal JF, Bolden PL. Bioemulsifier-Stabilized Hydrocarbosols. US Patent 4793826. 1988

[81] Gutnick DL, Rosenberg E, Belsky I, Zinaida Z. Alpha Emulsans. US Patent 4395354. 1983

[82] Gutnick D, Bach HR. Compositions Containing Bioemulsifiers and a Method for their Preparation. US Patent 6512014. 2003

[83] Rivas H, Gutierrez X, Cardenas AE, Morales A. Natural Surfactant with Amines and Ethoxylated Alcohol. US Patent 5792223.1998

[84] Yan N, Masliyah JH. Demulsification of solids-stabilized oil in-water emulsions. Journal of Colloids and Surfaces A: Physicochemical and Engineering Aspects. 1998;**11**:15-20

[85] Lissant KJ. Emulsification and demulsification in oil recovery. In: Shah DO, Schechter RS, editors. Improved Oil Recovery by Surfactant and Polymer Flooding. UK: Academic Press; 1997

[86] Giuggioli A, De Ghetto G. Innovative technologies improve the profitability of offshore heavy oil marginal fields. In: International Meeting on Petroleum Engineering, Beijing, November 14-17. 1995

[87] Kokal SL, Maini B, Woo R. Flow of emulsions in porous-media. Advances in Chemistry. 1992;**231**:219-262. DOI: 10.1021/ba-1992-0231.ch006

[88] Simon R, Poynter WG. Pipeline Oil/water Mixtures. U.S. Patent. 1970. No. 3,5 19.006

[89] Hardy WA, Sit SP, Stockwell A. Field trials of transoil technology for emulsion pipelining of bitumen. In: Fourth UNITAR/UNDP Conference on Heavy Crude Tar Sands. Vol. 5. Edmonton, Alberta: UNITAR/UNDP; 1988. p. 222

[90] Urdahl O, Fredheim AO, LØken KP. Viscosity measurements of water-in-crude-oil emulsions under flowing conditions: A theoretical and practical approach. Colloids and Surfaces A: Physicochemical and Engineering Aspects. 1997;**123-124**:623-634

[91] Stockwell A, Sit SP, Hardy WA. Transoil technology for heavy oil transportation. Results of field trials at Wolf Lake. In: Paper SPE 18362 Presented at SPE European Petroleum Conference, October 16-19. Vol. 249. London, UK; 1988

[92] Bertero L, Dilullo A, Lentini A, Terzi L. An innovative way to produce and transport heavy oil through dispersion in water: Laboratory study and field test results. In: 69th Annual Technical Conference and Exhibition, New Orleans, LA, September 25-28. 1994

[93] Gregoli AA, Olah AM. Low-temperature Pipeline Emulsion Transportation Enhancement. US Patent 5156652. 1992

[94] Kumar S, Mahto V. Emulsification of Indian heavy crude oil using a novel surfactant for pipeline transportation. Petroleum Science. 2017;**14**:372-382. DOI: 10.1007/s12182-017-0153-6

Chapter 9

Challenges during Operation and Shutdown of Waxy Crude Pipelines

Mysara Eissa Mohyaldinn, Hazlina Husin, Nurul Hasan, Mohamed M.B. Elmubarak, Ahmed M.E. Genefid and Mahmood E.A. Dheeb

Abstract

Transportation of waxy crude oil faces great challenges due to shear and temperature dependency. At high temperatures, waxy crude exhibits low viscous Newtonian behavior where the resistance to flow due to friction is low, and hence low pumping pressure is required to transport it. At low temperatures, however, the crude exhibits shear thinning non-Newtonian behavior where its apparent viscosity becomes shear-dependent. In such cases, the operated pipeline needs to maintain a high pressure to guarantee a continuous flow. Moreover, due to heat transfer between the internal pipeline and surroundings, oil temperature declines along the pipeline. It follows that the crude viscosity and, hence, frictional resistance increase. If the flow is interrupted for any reason, i.e., emergency or planned shutdown, then the restartability of the pipeline becomes a challenge because of the nonexistence of heating generated from friction. In this chapter, the challenges normally facing transportation of waxy crude oil will be discussed. The chapter will introduce the rheological properties of waxy crude oil and explain and describe how these properties can affect the pressure losses inside the pipeline during its operation and shutdown. The measures that need to be considered when designing a waxy crude pipeline will be discussed.

Keywords: waxy crude, rheology, pipeline, oil transportation

1. Introduction

Flow assurance is one of the major technical challenges in oil and gas production operations. The formation of gas hydrate, wax, halite, asphaltenes, mineral scales, and naphthenic acids can all result in serious operational and economic concerns. The industry needs novel and innovative techniques to reduce the costs associated with flow assurance and to improve the overall operation performance. Nowadays, with the decrease in conventional oil reserves, oil companies are improving unconventional oil fields where the reserved hydrocarbons are heavy and possess complex properties. The most important consequence of these ultraheavy hydrocarbons is that they hinder the flow through different production and transportation systems. In particular, for oil transportation through pipelines, we have to consider the complexity of the fluid rheology as an important factor that needs to be fully controlled in order to generate a smooth transportation.

The crude oil that possesses a high amount of wax, known as waxy crude oil, can cause many operational obstacles during production, storage, and transportation. It is considered that the most important obstacle is the crystallization and sedimentation of wax in the crude oil in the pipes during transportation. While waxy crude oil is flowing through the pipeline, and due to continuous cooling of the oil caused by heat transfer toward surroundings, crystals of wax form and grow in size to form partial to total blockage of the pipe cross-sectional area. This process continues with shutdown time, and farther the distance from the pipeline inlet, the faster is the wax deposition on the pipe internal wall and the higher tendency of the wax crystals to enlarge in size. This results in a continuous decrease in the pipe internal diameter, and, as the frictional loss is inversely proportional to the pipe diameter, higher frictional pressure loss will occur. Because of that, the pumping pressure needs to be increased in order to preserve the desired flow rate. However, knowing the mobility of the flow is an important issue for effective and frugal process of a pipeline framework. The main concern of waxy crude oil is the stream restart after delay in planned or emergency shutting down of the flow process for any reason.

Wax precipitation and deposition also contribute significantly in changing the original rheology of the oil. While wax crystals are precipitated and deposited, viscosity of the crude oil rises to convert it to a high viscous fluid that adhered to the internal pipe wall to decrease the effective cross-sectional area of the pipe. A solution to this issue can be achieved by applying various methods to lower the viscosity and pour point. The most common methods extensively implemented are heating of the transported crude, dilution with lighter fluid, and adding chemical additives, known as flow improvers/modifiers, to the flow stream. Each one of the three methods has its own advantages and disadvantages; however, generally speaking, chemical additives stay the most commonly recognized solution as compared to the other choices.

Usually the waxy crude oil is pumped at temperatures above its wax appearing temperature (WAT) in order to prevent forming of wax on the wall of the pipes. The pressure pumped should be high enough to hold the stability of the fluid. When temperature gets lower due to heat transfer, the crude becomes more viscous. And if the temperature lowers enough to initiate wax deposition, then there will be a double effect that results in higher frictional pressure, viz., higher viscosity and smaller pipe diameter.

1.1 Definition of waxy crude oil

Waxy crude oil can be defined as the crude oil that contains high amount of long-chain paraffin wax (alkanes) components, making the crude to possess a high pour point and, may be, a low API gravity. A waxy crude oil is distinguished by the fact that it exhibits non-Newtonian rheological behavior at low range of temperature (i.e., about 20 F above the pour point). For such non-Newtonian behaved crude oils, the effective viscosity is not only temperature dependent but also a function of the effective rate of shear in the pipeline. At reservoir conditions, due to high pressure and temperature, wax molecules are normally dissolved in the crude oil to form a single continuous hydrocarbon liquid phase. While flowing through pipelines, especially in sub-sea environments, due to cooling down, wax crystals may start to precipitate from this continuous phase, and eventually the fluid may lose its liquidity behavior. This change of flow behavior normally starts to occur when the crude temperature becomes lower than its cloud point or WAT [1].

1.2 Wax chemical compound formation and properties

The wax existing in crude oil mostly contains paraffin hydrocarbon (C18-C36) recognized as paraffin wax and naphthenic hydrocarbon (C30-C60). The hydrocarbon element of wax is able to present in several phases, i.e., gas, liquid, and particles (solids), relying on the flow conditions, i.e., pressure and temperature. When the temperature of wax decreases, the agglomerates and the wax crystals from paraffin wax are recognized as microcrystalline or naphthenic hydrocarbon [2]. Chemical flow modifiers can be used to enhance the crude flowability at flow conditions where crystallization or gelling is expected. These flow modifiers can delay the crude crystallization by reducing its pour point and viscosity. According to [3], the temperature of the waxy crude oil at the time of injecting the chemical additives is an important factor to determine their efficiency. If the temperature is very low, surely some amount of wax will be formed, and this accumulated wax will affect the rendering of chemicals. It has been observed that a high molecular weight wax chemical flow modifier has better performance for lowering waxy crude pour point. The crystal growth development rate of the lower-molecular-weight wax inhibitor is much slower than that of the higher molecular weight wax inhibitor.

1.3 Wax appearance temperature or cloud point

Wax appearance temperature is the temperature below which wax starts to appear in a waxy crude liquid. When a heated waxy crude is cooled down to a temperature lower than WAT, the wax molecules form clusters of aligned chains. Once these nuclei reach a critical size, they become stable, and further attachment of molecules leads to growth of the crystal. Formation of these nuclei causes the fluid to take on a cloudy appearance, hence the name cloud point. This also is referred to as the wax crystallization temperature or WAT. If the WAT of a produced or transported waxy crude oil is found significantly higher than the temperatures expected to be encountered during production or transportation, then wax deposition problems should be expected, and precaution measures are to be taken to avoid the problem and lessen the consequences arisen from.

1.4 Factors controlling wax deposition

Wax deposition along the pipeline is affected by several factors that include temperature, pressure, wax molecular weight, and the other crude oil components. For a proper design, operation, and optimization of a waxy crude production and transportation system, the effect of each of these factors is to be investigated carefully following documented experimental procedures and standards.

Wax deposition onto the production system generally requires a nucleating agent, such as asphaltenes and inorganic solids. The wax deposits vary in consistency from a soft mush to a hard brittle material. Paraffin deposits will be harder if longer-chain n-paraffin is present. Paraffin deposits can also contain other materials such as asphaltenes, resins, gums, fine sand, silt, clays, salt, and water. High-molecular-weight waxes can deposit even in the higher-temperature sections of a well, while lower-molecular-weight fractions tend to deposit in lower-temperature regions. Prior to solidification, the solid wax crystals in the liquid oil change the flow properties from a Newtonian low viscosity fluid to a high viscous non-Newtonian shear-dependent fluid. With further temperature reduction, the oil may eventually turn into a very-complex-flow behavior gel with yield stress (i.e., becomes yield-pseudoplastic or yield-plastic non-Newtonian fluid). Regardless of the rheological behavior the oil is exhibiting, oil viscosity is always inversely proportional to the oil temperature. As the temperature of oil increases, the oil viscosity decreases and vice versa.

The wax solubility is also directly proportional to process temperature. According to Sadeghazad and Christiansen [4], when water is present in the crude, wax deposition tends to reduce. This is because the water decreases the oil's temperature drop keeping the solution above the pour point temperature. They stated that water is able to maintain the oil temperature because oil has only half the specific heat as compared to that of water.

The second factor that affects the wax deposition is the pressure. The wax present in oil has a positive divergence, i.e., the solubility of the wax present in the solution decreases with the increase in applied pressure. This is because of the intermolecular forces between molecules [4].

The third factor controlling wax deposition is the wax molecular weight. The higher the molecular weight of the wax, the lesser its solubility because of its melting point increasing with the increase in the molecular weight [4, 5]. Al-Shafey and his co-workers stated that the solution composition greatly affects the wax deposition as well [6]. A set of experiments conducted by Sadeghazad and Christiansen reveals that the cloud point decreases for a wax solution with a lighter composition, i.e., it would take longer time for the wax to deposit [4].

1.5 Wax deposition mechanism

The problem of paraffin wax may be described as a situation in which a predominantly organic deposit hampers the production of crude oil. The loss of the crude production from a well depends on the severity and location of the deposition. In a pioneering work, Burger and others [7] investigated four wax deposition mechanisms, namely, molecular diffusion of wax molecules, shear dispersion of wax crystallites, Brownian diffusion of wax crystallites, and gravity settling. Gravity settling of paraffin crystals in flow line conditions is negligible, because it's dominated by shear dispersion. These four wax deposition mechanisms are discussed in the following subsections:

1.5.1 Molecular diffusion

Molecular diffusion is the deposition mechanism prevalent in well tubing. To avoid the deposition in flowing well, the flowing oil temperature needs to be maintained above the cloud point throughout the flow journey until the oil reaches the wellhead [8, 9]. Deposition is enhanced as a result of radial heat transfer from the tubing core toward the surroundings. Due to wax deposition, a concentration gradient is formed in the oil as a result of temperature gradient profile, due to increasing solubility of waxes with increasing temperature. The concentration difference causes waxes in the solution to diffuse from the warmer oil, which has a greater concentration of dissolved waxes, to the colder oil, which has a lower concentration, resulting in molecular diffusion of the paraffin crystals toward the surface wall. The wax concentration gradient is triggered as the differential temperature at a cross section causes the particles near to the cold walls to start the deposition from the oil solution and develop an initial layer of deposit [10]. As the layer is deposited, a concentration gradient occurs between the bulk fluid and the wall causing more wax to be trapped and oil flowing through the wax as a porous media away from the wall, thus thickening the wax concentration. This mechanism is the common mechanism observed for wax deposition.

1.5.2 Shear dispersion

At low temperatures, shear dispersion is believed to be the most occurring mechanism. It deals most with particles that are settling on the surface of a cold pipe due to the grooved or rough surface as well as the intermolecular forces [11]. Yet it was concluded that the shear dispersion is not significant based on field operating experience as well as experimental investigations.

When wax particles are moving along while transporting oil through a pipeline on an average speed, shear dispersion occurs as a shearing effect near the wall. The speed of the flowing fluid is less near the wall due to the shearing and friction causing a shear dispersion. Thus the crystalized precipitates move toward the wall due to its higher weight and away from the turbulent flow of the crude. Once reaching the wall, such precipitates form an initial layer of deposition or get trapped in the matrix made by the molecular diffusion caused earlier [12]. Shear dispersion is most effective when the temperature of the turbulent flow is below the WAT, thus causing high wax precipitation. A shear dispersion coefficient expressed by Burger and his co-workers [7] is:

$$D_s = \frac{a^2 \gamma C_w^*}{10} \quad (1)$$

where D_s is the shear dispersion coefficient (m^2/s); a is the particle diameter, m; C_w is the wax volume fraction concentration excluding the wall, fraction; and γ is the oil shear rate on the wall, s^{-1}.

1.5.3 Brownian diffusion

When tiny solid crystals are suspended within the oil, they collide frequently with thermally vibrant molecules. Due to such collision, a Brownian movement is initiated. At a concentration differential of these particles, the motion will cause diffusion. Coefficient of the Brownian diffusion is expressed as:

$$D_b = \frac{RT}{6\pi\mu a N} \quad (2)$$

where R is the gas constant (J/mol.K); Ta is the absolute temperature (K); U is the viscosity (Pa.s); a is the particle diameter (μm); and N is the Avogadro's number (mol).

Yet as referred by Burger et al. [7], the Brownian diffusion can be ignored.

1.5.4 Gravity settling

As the waxy crystals are denser than the oil particles, they tend to settle down and deposit. However, according to some results, it is believed that the gravitational deposition is insignificant toward the wax deposition. The turbulent flow or the shear dispersion would disperse the settling particles thus eliminating the gravity settling.

1.6 Waxy crude rheology

The most commonly used parameter to describe fluids' rheology is the viscosity, defined as the amount of resistance exhibited by the fluid to start the deformation process once shear stress is applied. As the waxes are precipitating, the fluid changes into a non-Newtonian fluid. This generally occurs when the process temperature decreases below the WAT [13]. Moreover, the waxy crude oils possess high yield stresses when the temperature further declines to values below the pour-point temperature. When the temperature is lower than the pour-point temperature, the oil loses its total mobility and turns into a gel-like structure that would require critical value of stress to flow, known as yield stress value of the gel. At low temperatures (below WAT), waxy crude oils generally exhibit non-Newtonian flow behavior (namely, pseudoplastic fluids). The main feature of the pseudoplastic non-Newtonian fluid is the dependency of viscosity on shear rate (while it is constant for Newtonian fluids). **Figure 1** shows typical flow curves of fluids exhibiting different types of rheological behavior as proposed by Wardhaugh and others [3]. The upper curve (a) shows the variation of applied shear stress and shear rate as measured by a viscometer. The lower curve (b) shows the variation of fluid viscosity with shear rate. It is clear that the viscosity of Newtonian fluid is independent of shear rate, while the viscosity of non-Newtonian fluids decreases with increasing shear rate. Viscosity is the amount of resistance exhibited by the fluid to start the deformation process once shear stress is applied. Different correlations are used to derive viscosity as per the case. The value of viscosity of crude oil depends on many factors including the composition of oil, temperature, amount of gas dissolved, and pressure or stress. The viscosity can be calculated as the ratio of shear stress (Pa) to shear rate (s^{-1}); hence, the SI unit of viscosity is Pa·s.

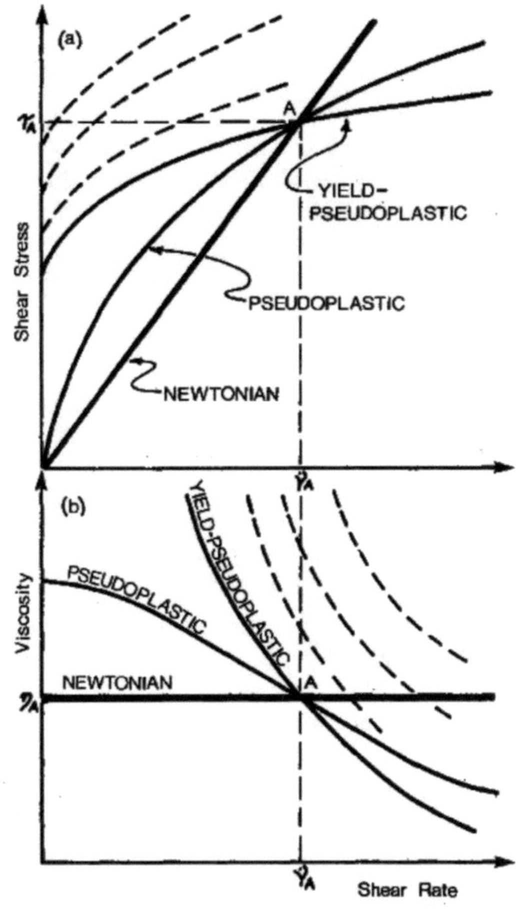

Figure 1.
Typical flow curves of fluids exhibiting different types of rheological behaviors [3].

As the fluid flows through a pipeline, it starts a deformation due to the slippage action occurring. Shear stress is a force that is acting parallel to the flow propagation of the fluid. The main cause of the shear stress is the friction exerted between the particles, which is proportionally related to the viscosity of the fluid. Mathematically, the shear stress is the force applied on a surface divided by the surface area, i.e., $\tau = \frac{F}{A}$, where τ is the shear stress in N/m^2 (or Pascal), F is the applied force in N, and A is the cross-sectional area of the fluid parallel to the applied force in m^2.

The shear rate is the rate at which the deformation caused by the shear stress is taking place. In Newtonian fluids, the shear rate increases as the flow rate increases. As in the case of flow through pipelines, the shear rate is the gradient of change in velocity across the radius of the pipe assuming that the top and bottom flows are not in similar velocities. Assuming two parallel plates with one moving while the other is stationary, the shear rate can be defined as $\frac{v1 - v2}{h}$ where $v1$ is the velocity at the top plane, $v2$ is the velocity at the bottom plane, and h is the gap between the plates. This radial velocity and shear distribution are similar to what happens during oil flow through pipelines. While flow velocity is constant along the pipeline (since the pipe diameter is constant), there is a radial velocity variation along the cross-sectional area due to the effect of shear. At the pipe wall, where the shear stress is maximum, velocity is zero, whereas the maximum velocity occurs at the center of the pipe. **Figure 2** shows an example of the radial velocity distribution at three values of flow rates [3], where A is the highest flow rate and C is the lowest flow rate as indicated by radial velocity distribution showing a turbulent flow and laminar flow, respectively.

1.7 Transportation of waxy crude through pipelines

Significant percentage of pipelines worldwide transport waxy crude oils with different amounts of wax contents and, hence, wide range of rheology complexity. In China, for instance, more than 80% of the produced oils are classified as waxy crude [14] with total annual production of over 100 million tons and total annual storage of more than 70 million tons [15]. Worldwide, half of the recoverable oil reserve is classified as heavy crude [16] with the waxy crude contributing about 20% to the total hydrocarbons reserve [17]. There was a noticeable increase in the production of waxy crude recently. The daily total production rate worldwide has increased from 1 million BOPD in 1960 to almost 24 folds (contributing to one third of the total world oil production) in 2009 [1].

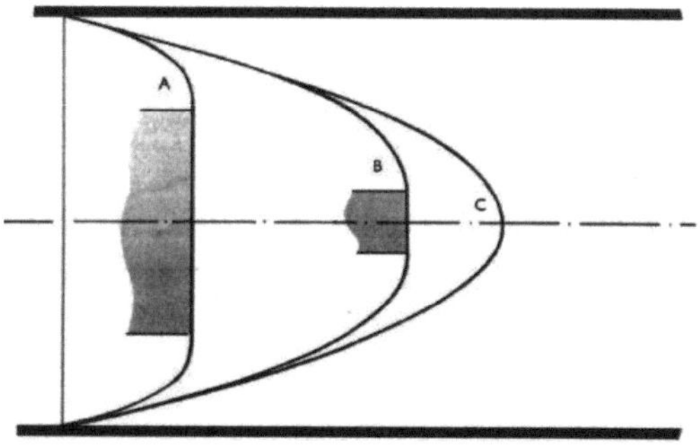

Figure 2.
Radial velocity distribution in the pipeline at three values of flow rate ($V_A > V_B > V_C$) [3].

Since pipelines are the main transportation methods of waxy crude oils, proper precautions and measures should be adopted to ensure a safe and smooth flow at various operating conditions. For long-distance pipelines, in particular, the temperature variation along the pipeline causes axial change of the rheological behavior which results in complexity of prediction of pressure losses at different segments along the pipeline.

Two cases need to be considered of a waxy crude pipeline, operating conditions and shutdown conditions.

2. Waxy crude pipeline operations

During operation, the flow of the transported waxy crude tends to facilitate the pipeline operation due to continuous shear of the fluid, from one hand, and the heat generation due to friction, from the other hand. The former causes reduction of the crude viscosity, which results in the decreasing of the generated frictional pressure losses, and hence, less pumping pressure is required. The latter adds a significant heating to the transported fluid, which can compensate part of the heat losses resulting from the heat transfer due to temperature difference between

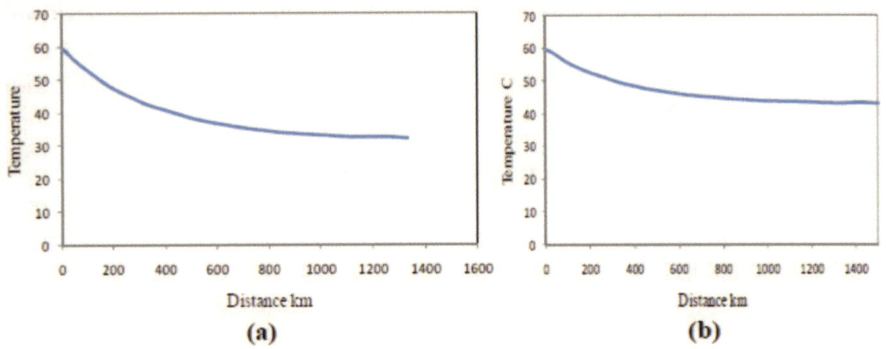

Figure 3.
Temperature distributions along Higleig-Portsudan pipeline (inlet temperature 60°C and flow rate 0.33 m^3/s) [18].

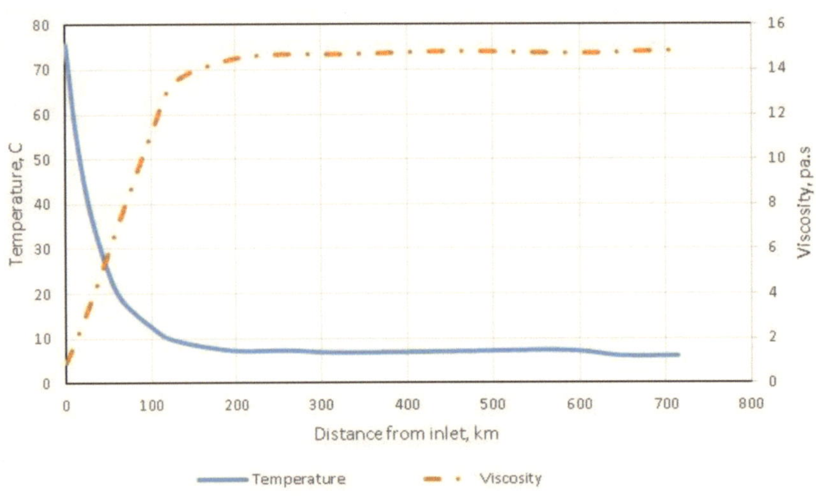

Figure 4.
Variation of temperature and viscosity along pipelines [19].

the fluid and surroundings. As a result, the temperature along the pipeline can be maintained to be always above the WAT. **Figure 3** is an example of temperature distribution along the Sudanese Higleig-Portsudan pipeline assuming two values of soil temperature [18]. The figure indicates that the shear action due to flow can maintain the temperature and delay the distance at which the temperature falls down to surrounding (soil) temperature. This distance is directly proportional to the fluid flow rate.

Figure 4 shows a typical variation of temperature and viscosity along a waxy crude pipeline [19]. As we go far from the inlet, temperature declines due to heat transfer between the fluid and the pipeline surroundings. The temperature

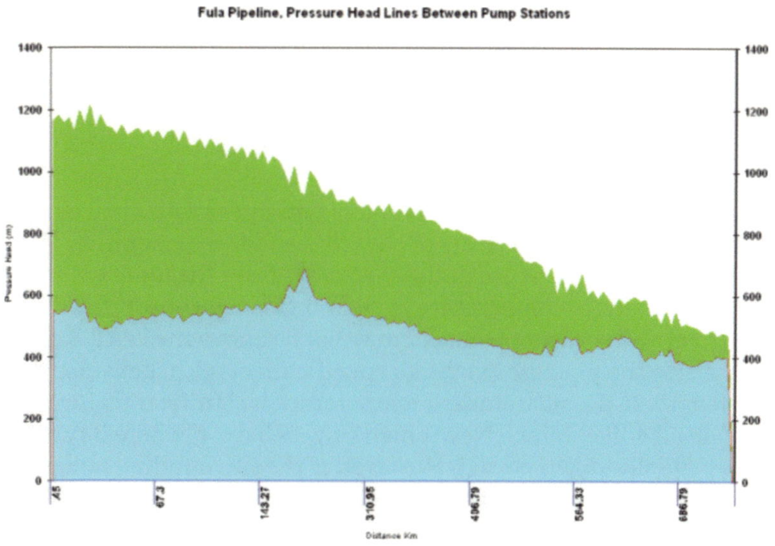

Figure 5.
Fula pipeline profile and pressure transverse between initial and terminal pump stations [20].

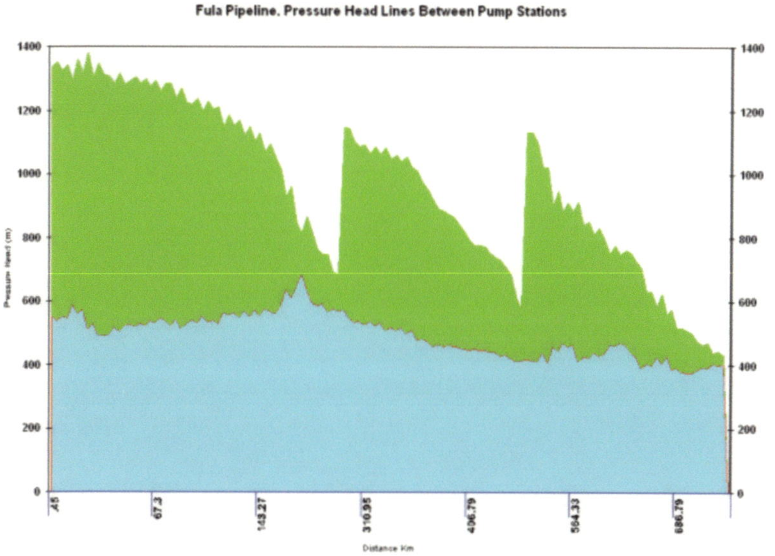

Figure 6.
Fula pipeline profile and pressure transverse between pump stations (3 pump stations) [20].

reduction leads to increase in viscosity, which eventually results in increase in frictional pressure loss per unit of length. The trend of the variation of pressure loss per kilometer length follows the same trend of viscosity variation.

To maintain a smooth operation of the pipeline, the total pressure losses due to elevation difference and friction (which is highly affected by the crude rheology) needs to be balanced by operating one or more pump stations. To facilitate the crude pumping, the crude rheology can be enhanced by heating or injection of chemical flow modifiers. **Figures 5** and **6** show the pressure traverse between pump stations of another Sudanese pipeline (Fula pipeline) assuming one and three pump stations, respectively [20]. The figures indicate that the inlet pumping pressure gradually declines along the pipeline to reach the next pump station at a specific terminal pressure, which may serve as a suction pressure to the successive pump station. This pressure profile is affected by flow rheology and wax precipitation inside the pipeline both of which are highly temperature dependent.

3. Waxy crude pipeline shutdown and restarting

Unlike operating pipelines, when a pipeline undergoes a planned or emergency shutdown, it loses the positive effects of shearing and heat generation due to continuous flow. The temperature declines steadily from the moment of shutdown until the time at which the temperature at all points along the pipeline reaches the surrounding temperature, which may or may not be above the WAT. Accordingly, the crude oil inside the pipeline exhibits a specific rheological behavior throughout the pipeline length. If the surrounding temperature is significantly higher than the WAT, the crude oil will exhibit a Newtonian flow behavior, where it can easily be restarted after the shutdown period. However, if the surrounding temperature is less than the WAT, then the fluid will exhibit a non-Newtonian behavior, and high restarting pressure is needed. **Figure 7** shows the temperature distribution along Higleig-Portsudan pipeline at the moment of shutdown and after every subsequent 12 h following the shutdown. The figure indicates that the temperature throughout

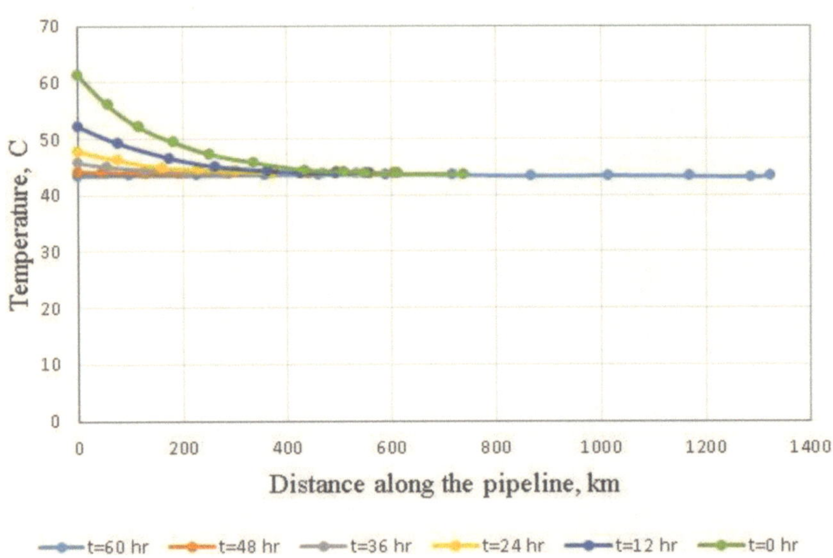

Figure 7.
Transient temperature distribution along the pipeline every 12 h after shutdown (inlet temperature 60°C, soil temperature 42°C, flow rate 0.33 m^3/s) [18].

the pipeline distance reaches the surrounding temperature after 2 days from the shutdown moment.

The most worse shutdown scenario is where the crude temperature drops below pour point and results in the gelling of the crude inside the pipeline. In such complex scenario, the restartability of the pipeline becomes a challenge, and the operator may become unable to restart it. It is, therefore, essential to avoid these complex scenarios by different measures such as heating, chemical additives, dilution of the crude with lighter fluids, and shortening the shutdown periods to ensure that the temperature is always at a value permitting a smooth restarting of the pipeline.

4. Conclusions

Transportation of waxy crude pipeline through pipelines can cause numerous problems that may impose safety, economical, and technical impacts on the pipeline operation. The severity of waxy crude-related problems is highly affected by the complexity of its rheological properties which in turn depends on the operating conditions (mainly temperature). During waxy crude pipeline operation, temperature declines along the axial length due to heat transfer caused by temperature difference between the transported crude and the surroundings. This temperature decline is concurrently encountered by a raise in temperature caused by heat generation due to friction which is proportionally related to the velocity gradients. This temperature variation along the pipeline causes axial variation in the crude rheological properties which results in variation in frictional pressure losses.

During planned and emergency shutdown of waxy crude pipelines, the problems will get worse due the absence of heat generation. The temperature declines steadily from the moment of shutdown until the time at which the temperature at all points along the pipeline reaches the surrounding temperature, which may or may not be above the WAT. Accordingly, the crude oil inside the pipeline exhibits a specific rheological behavior throughout the pipeline length. The problem may get worse when the surrounding temperature is below the WAT or even below pour point. In such case, the pipeline may need to be assisted by putting on some cost-effective measures to facilitate restarting up of the flow.

Author details

Mysara Eissa Mohyaldinn[1*], Hazlina Husin[1], Nurul Hasan[2], Mohamed M.B. Elmubarak[1], Ahmed M.E. Genefid[1] and Mahmood E.A. Dheeb[1]

1 Department of Petroleum Engineering, Universiti Teknologi PETRONAS (UTP), Perak, Malaysia

2 Petroleum and Chemical Engineering, Universiti Teknologi Brunei, Brunei

*Address all correspondence to: mysara.eissa@utp.edu.my

References

[1] Chala GT, Sulaiman SA, Japper-Jaafar A. Flow start-up and transportation of waxy crude oil in pipelines-a review. Journal of Non-Newtonian Fluid Mechanics. 2018;**251**:69-87

[2] Daungkaew S et al. Is there a better way to determine the viscosity in waxy crudes? In: SPE Asia Pacific Oil and Gas Conference and Exhibition. Australia: Perth; 2012

[3] Wardhaugh LT, Boger DV, Tonner SP. Rheology of waxy crude oils. International Meeting on Petroleum Engineering. Tianjin, China: Society of Petroleum Engineers. 1-4 November 1988

[4] Sadeghazad A, Christiansen RL. The Effect of Cloud Point Temperature on Wax Deposition in the 8 Abu Dhabi Internationat Petroleum Exhibition and Conference. Abu Dhabi: U.A.E; 1998

[5] Popoola CA et al. Triethanolamine (TEA) as flow improver for heavy crude oils. IOSR Journal of Applied Chemistry. 2015;**8**:34-38

[6] Al-Shafey H et al. Studies on the influence of long chain acrylic esters co-polymers grafted with vinyl acetate as flow improver additives of crude oils. Advances in Applied Science Research. 2011, 2011;**2**(5):476-489

[7] Burger ED, Perkins TK, Striegler JH. Studies of wax deposition in the trans Alaska pipeline. Journal of Petroleum Technology. 1981;**33**:1075-1086

[8] Bern PA, Withers WVR, Cairns RJR. Wax deposition in crude oil pipelines. In: European Offshore Technology Conference and Exhibition. London, United Kingdom: Society of Petroleum Engineers (SPE); 1980

[9] Leiroz AT, Azevedo LFA. Studies on the mechanisms of wax deposition in pipelines. In: Offshore Technology Conference. Houston, Texas: Society of Petroleum Engineers (SPE); 2005

[10] Zhang X, Queimada A, Szczepanski R, Moorwood T. Modelling the shearing effect of flowing fluid and wax ageing on wax deposition in pipelines. Asia, Kuala Lumpur, Malaysia: Offshore Technology Conference; 25-28 March 2014

[11] Rosvold K. Wax Deposition Models. Department of Petroleum Engineering and Applied Geophysics. M.Sc. Thesis. Trondheim: Norwegian University; 2008

[12] Bern PA, Withers VR, Cairns RJR. Wax deposition in crude oil pipelines. In: European Offshore Petroleum Conference and Exhibition. London, England: Society of Petroleum Engineers (SPE); 1980

[13] Venkatesan R, Creek JL. Wax deposition and rheology: Progress and problems from an operator. Houston Texas: Offshore Technology Conference; 3-6 May 2010

[14] Cheng Q et al. The study on temperature field variation and phase transition law after shutdown of buried waxy crude oil pipeline. Case Studies in Thermal Engineering. 2017;**10**:443-454

[15] Zhao J et al. Effect of physical properties on the heat transfer characteristics of waxycrude oil during its static cooling process. International Journal of Heat and Mass Transfer. 2019;**137**:242-262

[16] Ghannam MT et al. Rheological properties of heavy & light crude oil mixtures for improving flowability. Journal of Petroleum Science and Engineering. 2012;**81**:122-128

[17] Kumar L, Paso K, Sjöblom J. Numerical study of flow restart in the pipeline filled with weakly compressible waxy crude oil in non-isothermal condition. Journal of Non-Newtonian Fluid Mechanics. 2015;**223**:9-19

[18] Mohyaldinn ME. Hydraulic behaviour of Higleig-Portsudan pipeline at operation and shutdown conditions. Journal of Engineering Science and Technology. 2014;**9**(4):463-476

[19] Mohyaldinn ME. A predictive tool for thermal/hydraulic calculations of Fula pipeline. Journal of Petroleum and Gas Exploration Research. 2012;**2**(2):33-43

[20] Mohyaldinn ME. Analysis of pressure transverse between pump stations of fula pipeline under different operation scenarios. International Journal of Scientific & Technology Research. 2014;**3**(3):451-456

Chapter 10

Pre-Treatment of Heavy Crude Oils for Refining

Faith Uchenna Babalola and Alfred Akpoveta Susu

Abstract

The economic unattractiveness of heavy crude oils in the global oil market scene is fast changing due to the looming scarcity of light oils. There are huge reserves of heavy oils to which refiners are being reluctantly attracted. The fully developed conventionally available refineries are, however, incapable of processing these heavy crudes without some major modifications of both process and equipment, which will be capital intensive. This thrust promises to be very viable and yield great benefits since heavy oil is relatively cheap. Process and equipment modification begins with the required pre-treatments of these heavy crudes to facilitate efficient and cost-effective refining. Low API gravity, high amounts of impurities, and high viscosity are the major challenges addressed in the pre-treatment section. An option for a unit for the removal of some impurities for some special heavy oils is proposed. This should be followed by a multi-stage desalting unit that may be replaced with a centrifuge and a preliminary hydrotreating unit.

Keywords: heavy oil, pre-treatment, desalter, contaminants, impurities, efficient refining

1. Introduction

With the depletion in global reserves of light crude oil over the past few decades, the attention of refiners and researchers has been shifted to heavy and extra-heavy crude oils in recent times. It is estimated that global heavy oil reserves are over 3 trillion bbl, which can supply the world's energy needs for the next 100 years [1]. Refining processes for light crudes have been well developed and the products yield high economic value; this is in contrast with the more complex and more costly processing of refining heavy crudes which is less expensive but yields less attractive products. The trade-off, however, lies in the fact that these heavy crudes are currently becoming more abundant, and also with further development in their processing technologies, higher quality products will emerge and the return on investment will become quite attractive. It is estimated that there is twice the quantity of heavy crude in reserves than light crude and this ratio is most likely to increase in favor of heavier crudes.

The primary challenges in processing heavy crude oils are poor flowability due to high viscosity, low API gravity ($<20°$), and the presence of higher amounts of impurities than in light crudes. With treatments that increase flowability and remove impurities, the refining task is reduced to mainly cracking processes to yield desired products such as gasoline and middle distillates. Various technologies have been employed to improve heavy oil flowability in pipelines. Due to high viscosity and hence high pressure drop, pumping cost is extremely high and so energy intensive; alternative solutions to higher pumping power include pre-heating of the crude by heating of the pipes, dilution with light hydrocarbon fluids, oil-in-water emulsification, and partial upgrading [2, 3]. Flowability enhancement technologies for transporting heavy oil are outside the scope of this chapter but are available in literature [4, 5].

Upgrading of heavy oils is achieved by either a carbon-rejection process, a hydrogen-addition process, or a combination of both. These technologies produce light oil fractions, which can be processed in conventional refineries and, therefore, also fall outside the scope of this work but can be found in open literature [6]. To enhance refining, heavy crudes are first sent to pre-treatment units either on the drilling site or in the refinery to condition them for easy processing and render them equipment-friendly.

2. Unconventional crude oils

With global changes in technology and economy come concomitant changes in the definition of conventional and unconventional petroleum crudes. Generally, today, conventional oil includes light crude oil with natural gas and its condensates, while unconventional oil consists of a wider category, including heavy oil, extra-heavy oil, oil sands, gas to liquids, and other liquids. Conventional oils are cheaper and easier to produce and process than unconventional oils. These categorizations, however, may not remain fixed over time. With technological development, resources hitherto considered unconventional may migrate to the conventional category. According to the US Department of Energy (DOE), "unconventional oils" have not yet been strictly defined [7]. In fact, it has been argued that common classification of unconventional oils based on production techniques is imprecise and time-dependent. No universally accepted definition of "conventional" and "unconventional" oil has been accepted by the International Energy Agency [8].

2.1 Heavy and extra-heavy crude oils

Heavy oils are characterized as having boiling points above 650°F (343°C). These oils are termed "heavy" because of their relatively high density or specific gravity; they also have higher viscosity and heavier molecular composition. Heavy oils have an API gravity of less than 20° [9], and in 2010, the World Energy Council defined extra-heavy oils to have an API gravity of less than 10° and a reservoir viscosity of 10.000 centipoises or less [10]. Below an API of 10°, a heavy oil will sink in water rather than float and is termed extra heavy.

Heavy oils contain asphaltenes and resins; they are dense and viscous (heavy) due to the higher percentage of aromatics and naphthenes than alkanes. They also contain high amounts of nitrogen, sulfur, oxygen, and heavy metals. They have a higher proportion of compounds with over 60 carbon atoms and, hence, higher boiling points and higher molecular weights than light oils.

2.2 Transportation of heavy crude

Most existing pipelines are designed for light crude oil transportation and so cannot accommodate heavy oil unless they are modified. To transport heavy oil in these existing pipelines, diluting or drag reducing agents (DRAs) are added to the heavy oil. Sometimes, light oil is mixed with the heavy oil to enhance flowability, which, of course, reduces the value of the light oil [5]. With modifications of the pre-treatment unit proposed here to adequately handle water/oil emulsions, dilution with water may be a preferred method of improving flowability.

2.3 Environmental impact

Heavy oils have higher amounts of impurities and undesirable contaminants especially sulfur, metals, and nitrogen which impact negatively on the environment. Sulfur causes acid rain and combines with hydrogen to form hydrogen sulfide which is a major air pollutant. Sulfur also causes corrosion in pipelines and downstream equipment. Heavy metals, which are present in heavy oils, cause catalyst poisoning and fouling of process equipment and are toxic thereby posing disposal problems. At last, more carbon dioxide is generated both in the processing of heavy oils and in the consumption of products of heavy oil stock due to higher carbon-to-hydrogen ratio than in light oil [11].

3. Unwanted constituents and impurities

Some impurities in heavy oil occur as natural unwanted constituents of the oil, such as nitrogen, sulfur, oxygen, acids, and salts, while others are acquired by the oil during production; these include water and insoluble solids. These impurities and the processes for their removal are discussed in this section.

3.1 Effects of impurities

The impurities in heavy oil are responsible for its low economic value because they constitute a lot of problems to the pipelines, processing equipment, and the environment. Sulfur causes fouling in the equipment and environmental pollution in the form of hydrogen sulfide; nitrogen forms gums that lead to catalyst deactivation; oxygen can reduce the calorific value of the fuel; heavy metals such as nickel, vanadium, and lead are harmful to the environment and cause corrosion in process equipment and deactivate catalysts. Salts and acids also cause fouling of equipment, while insoluble solids cause plugging in pipes [1, 6, 7]. The list of the effects of impurities appears endless but suffices it to state here that the removal of these impurities is crucial to environmental preservation, process efficiency, and product quality.

3.2 Pre-treatments

Here, the various pre-treatment processes that ensure significant reduction or complete removal of impurities in heavy oils before they are taken to the refining process units are discussed. These pre-treatments mitigate downstream consequences which include corrosion and fouling of equipment, catalyst poisoning, more elaborate water treatment process, and higher energy cost. It is also worthy of note that prior to pre-treatment, the delivery system, handling and storage of the feed (heavy oil), goes a long way to affect the pre-treatment system requirements, efficiency, and cost [1].

3.2.1 Demulsifying

Crude oil emulsion is formed as dispersed water phase in an oil medium. The water globules are protected by an oil layer which prevents their coalescence. This protective oil layer increases in mechanical strength with time (aging) due to long storage, thus stabilizing the emulsion. Agitation during transportation in pipes has a yet more stabilizing effect on emulsions due to high shear, which leads to smaller sizes of water droplets. Smaller droplets are relatively more stable than larger ones; hence, it is more challenging to demulsify the crude oil emulsion at the refinery than at the production site [12]. Chemical demulsifiers or emulsion breakers are used to break the crude oil emulsion into oil and water phases. They destroy the interfacial film and enhance the coalescence of the water droplets. Some chemical demulsifiers include amines, polyhydric alcohols, acids, and polymers [13].

The quality of a demulsifier is determined by the rate of oil separation from water, amount of water left in the crude oil after separation, and quality of separated water for disposal. It is desirable to have a fast separation rate, a low amount of residual water in the crude oil, and a low percentage of oil in the separated waste water. Both the water phase and its dissolved salt contents need to be removed because they lead to corrosion in the refinery equipment. Dehydration in thermochemical steel tanks at temperatures as low as 50–55°C is effective for demulsifying heavy oil emulsions as well as stage-wise electrode-assisted desalting units [14]. Other demulsification techniques, such as biological, membrane, electrical, and microwave irradiation, have also been employed in the oil industry [15]. Low-temperature demulsifiers in delivery and storage facilities are used to improve feedstock quality and optimize desalter operations [16]. Some factors that need to be addressed in order to enhance demulsification and oil-water separation are discussed here.

a. *Agitation*: This should be minimized during transportation, pumping, and storage and avoided wherever possible to reduce the mechanical strength of the interfacial film between the water and the oil phases. Mechanical chokes and flow obstructions can be employed to reduce agitation, though a necessary degree of agitation is required to achieve an adequate mixing of the demulsifiers into the bulk of the oil-water emulsion.

b. *Heat*: When heat is applied to an oil-water emulsion, the viscosity of the oil is increased to enhance its flow, water expands and breaks the interfacial film, there is a higher rate of droplet collision leading to faster coalescence, and density difference between the oil and the water is increased. All these result in an increased rate of phase separation. There is, however, an optimum amount of heat requirement beyond which it can become disadvantageous and lead to loss of low-boiling hydrocarbons.

c. *Time*: The retention time, during which the emulsion is held at the treating temperature, affects the separation efficiency. Longer retention time results in greater separation efficiency. Conventional crudes are treated for about 10–30 minutes, while heavy oils would require a much longer retention time [12].

d. *Solids*: The presence of solids, whether oil-wet or water-wet, has a further stabilizing effect on emulsions and so needs to be removed, especially fine particles [17, 18].

3.2.2 Deasphalting

Heavy oils contain significant amounts of asphaltenes which, if not removed, will decrease the efficiency of the refining process and reduce product quality. Asphaltenes comprise a dark brown to black solid with no definite melting point but foams and swells when heated leaving a carbonaceous residue. Their molecular weight can span from 1000 to 100,000 [19]. Some common solids such as asphaltenes and waxes may be removed with the water after being water-wetted or dispersing them into the oil.

They contain hydrocarbons and heteroelements, namely oxygen, nitrogen, and sulfur. In asphaltenes, hydrogen-to-carbon ratio varies over a very narrow range of between 0.5 and 1.15%. Larger variations are seen in their heteroelement contents: oxygen varies from 0.3 to 4.9%, sulfur varies from 0.3 to 10.3%, and nitrogen varies from 0.6 to 3.3% [20].

The heteroatoms (sulfur, oxygen, and nitrogen) in asphaltenes cause other problems such as corrosion, catalyst deactivation, and environmental pollution. Solvent deasphalting processes have been used to significantly remove asphaltenes from heavy oils at mild temperature conditions [21]. In this process, an alkane (the solvent) is injected into the heavy oil feedstock to disrupt the dispersion of the oil components and, hence, cause the polar constituents to precipitate. Alkane solvents, most common of which is propane or a mixture of propane and butane, are used to obtain a good yield of deasphalted oil [22].

Practically, any conventional deasphalting process and equipment can be used. For example, the residual oil supercritical extraction (ROSE) process, which is currently used for residue fraction as feed, may be easily adapted and modified to accommodate the whole crude feedstocks. Some of the adaptations to address will include increasing the solvent/oil ratio, increasing the temperature, and the mixing effect, which may involve the use of pumps, mixing valves, or medium speed centrifuges. The deasphalted whole crude, after separation, may contain some of the solvents but has much less asphaltenes and metal content. The deasphalted crude stream is passed through a flash vaporization step to recover the solvent for reuse before it goes for refining [21].

The Solvahl process is a solvent deasphalting process suitable for application to heavy oils, extra-heavy oils, tar sands, and bitumen. It removes the asphaltene constituents, gives very high yields of deasphalted oil, and reduces metal content in the viscous feedstock. Various solvents used in this process include liquid propane, butane, pentane, hexane, and heptane depending on downstream process objectives and feedstock properties [23].

Solvent recovery for reuse is a crucial part of the deasphalting unit because the solvent is expensive and its ratio to the oil feed is high. Solvent recovery from the deasphalted oil is carried out at a high temperature of about 200°C. Carbon dioxide can be employed to effectively separate the solvent from the deasphalted oil at a much lower temperature of about 40°C. This is achievable due to the increase in the interaction between the CO_2 and the solvent [24].

Some classes of asphaltenes, which are insoluble in paraffin solvents, are removed using aromatic solvents. The drawback, however, is the high solvent-to-oil ratio required to remove most of the asphaltenes and the consequent high cost of solvent recovery. Asphaltene removal achieves sulfur content reduction, API gravity increase, viscosity reduction, and improvement in other oil properties. Some conventional deasphalting processes are done at the crude production site to upgrade heavy oil and bitumen but may also be carried out as a pre-treatment in the refinery. Deasphalting is achieved under high-temperature condition, which simultaneously achieves cracking and upgrading of the heavy oil. Subsurface deasphalting may be done for heavy oils using solvents but will not be addressed here.

3.2.3 Demetallization

Most crudes, especially heavy oils, contain various metals in different concentrations ranging from 1000 to a few million ppm depending on the origin of the crude. These metals are mainly sodium (Na), potassium (K), lithium (Li), calcium (Ca), strontium (Sr), iron (Fe), cupper (Cu), silver (Ag), manganese (Mn), tin (Sn), lead (Pb), cobalt (Co), titanium (Ti), gold (Au), chromium (Cr), vanadium (V), and nickel (Ni). These metals are present either as inorganic salts or as organometallic compounds such as porphyrins. They are accumulated in resins and asphaltenes and need to be removed to their barest minimum if not completely before refining because they lead to catalyst deactivation and corrosion in equipment and contribute to "acid rain." The water-soluble salts, such as sodium chloride and magnesium chloride, produce hydrochloric acid which is very corrosive. Calcium and magnesium combine with carboxylic acid to form soaps which act as emulsifiers and so are undesirable. Most of these metals combine with naphthenic acids to form complex organometallic compounds, while others are present as oil-soluble metalloporphyrins. These metallic complexes contain asphaltenic aggregates which have low reactivity; thus, most conventional metal removal methods require severe conditions [25]. The water-soluble salts are significantly removed during the desalting step before refining. The most abundant of trace metals in crude oils are nickel and vanadium which cause catalyst poisoning. It is, therefore, easier and preferable to demetallize heavy crudes before refining. The existing methods of demetallization can be broadly classified into three, namely:

i. thermal and chemical methods, where the nature of the organic matrix is modified;

ii. solvent and adsorption methods, where the metal-containing compounds are separated from the oil; and

iii. unconventional methods, where electric fields, magnetic fields, or radiation is employed [26].

Currently, demetallization is most commonly achieved during hydrotreating, deasphalting, and hydrocracking [27]. However, the efficiencies of these processes are greatly hampered by the presence of the metals that poison the catalysts. It has been reported that for greater effectiveness in metal removal, the usual hydrotreating catalysts should be dosed with very small amounts of active metals belonging to the sixth and seventh subgroups of the periodic table [28]. The chemical reaction, by which metals (especially the more common nickel and vanadium) are removed, is given in Eq. (1). By a sequence of two mechanisms, the porphyrins are first dehydrogenated to form precursor species which thereafter undergo ring-cleavage reactions leading to metal deposition on the catalyst surface.

$$(MP) + H_2 \rightleftharpoons (MPH_2) \longrightarrow M + PH_2 \tag{1}$$

where MP is the metalloporphyrin, MPH_2 is the intermediate hydrogenated metalloporphyrin, M is the deposited metal, and PH_2 is the resulting hydrocarbon.

Using chemical methods, metalloporphyrins react with an acid as given in Eq. (2):

$$MP + HX \rightleftharpoons MX + PH \tag{2}$$

where MP is the metalloporphyrin, HX is the acid, MX is the metal-acid complex, and PH is the resulting hydrocarbon [28].

It is, therefore, very necessary, especially for heavy metals, that a pre-treatment step be employed to first reduce or remove these harmful metals. There are newer technologies for metal removal, such as solvent extraction, by which up to 51% vanadium and over 65% nickel can be removed using cyclohexane and other organic solvents. Other methods include oxidation, electrochemical process, immiscible ionic liquid extraction, and adsorption [29]. Some natural and synthetic zeolites have been successfully employed to selectively remove nickel and vanadium [25, 30]. A special challenge encountered in the removal of calcium due to the presence of calcium naphthenate (CaNA) that is formed from calcium cation and the anion of naphthenic acids has been reported. An aqueous solution, containing a calcium-removal agent, was successfully employed to completely remove calcium from heavy crude oil [31].

3.2.4 Denitrogenation

Nitrogen is one of the major heteroatoms but is also present in crudes as nitrogen-containing compounds which need to be removed from crude oil to avoid downstream problems. Common nitrogen-containing compounds are mainly in form of complex structures such as porphyrins and quinolines as well as simple ones such as pyridine (C_5H_5N) and pyrrole (C_4H_5N), which are prone to free radical addition reactions to form gums and amines. These react with acid catalysts and cause deactivation [32]. Nitrogen content in most conventional oils is <0.3 wt%, but in heavy crudes, it can be as high as 1 wt% and can exceed that in some high nitrogenous crudes [33]. They are more thermally stable than sulfur compounds and so require higher heat to be removed. Though organonitrogen compounds occur at such low levels, they are undesirable because they cause catalyst poisoning in the refining process and generate a pollutant (NOx) upon combustion [34].

Hydrotreating units serve for effective denitrogenation and are the benchmark against which other methods are measured. Of the common heteroatoms, nitrogen is relatively more difficult to remove [35]. Hydrodenitrogenation removes nitrogen by forming other compounds such as ammonia, which are then separated from the oil. This method of nitrogen removal leaves the hydrocarbon part of the nitrogenous compound in the oil. A typical dehydrogenation reaction removes nitrogen in the form of ammonia as shown in Eq. (3). The common catalysts are cobalt, nickel, and molybdenum disulfide but are combined in appropriate ratios for the particular feedstock.

$$R_3N + 3H_2 \longrightarrow 3RH + NH_3 \tag{3}$$

Other methods, however, are also applicable: these remove the nitrogen-containing compounds completely from the oil thereby reducing the hydrocarbon content of the oil itself. One such method is liquid-liquid solvent extraction that exploits the difference in solubility parameters and employs polar solvents, acidic solvents, ionic liquids, and deep eutectic solvents [32]. Some others are liquid-solid solvent extraction, physical adsorption and chemisorption between the solid adsorbent and the nitrogen-containing compound [36]. Adsorption beds can be packed with activated carbon, silica gel, resins, polymers, and zeolites [37, 38]. Nitrogen limit is not directly given in fuel specification but is indirectly specified as gum content, storage stability, and thermal stability.

3.2.5 Deoxygenation

Oxygen is present as one of the heteroatoms and in other forms in oxygen-containing compounds that are natural contaminants in crude oil. A wide variety of oxygen-compounds have been identified in literature. More than 20 such compounds (containing both oxygen and nitrogen) in amounts significant enough to be regarded as above trace amounts have been reported [14]. Some of these compounds, being weakly acidic, are responsible for petroleum acidity and include carboxylic acids, cresylic acid, phenols, and naphthenic acids. Oxygen content in crude oils can vary from 0.05 to 1.5 wt%. Generally, heavy oils contain larger amounts of oxygen than light oils [39]. Naphthenic acids can be extracted using dilute caustic solutions; other oxygen-containing compounds, which are non-acidic, include esters, ketones, and amides but are quite negligible. These compounds along with sulfur are simultaneously removed during hydrotreating and other processes. Much of the unwanted oxygen is actually in the produced water, which is removed in the desalter. There is no special process to selectively remove oxygen alone from crude oil.

3.2.6 Desulfurization

Sulfur is a major drawback in the pricing of oil products because it contributes most significantly to equipment fouling and environmental pollution. Sulfur-containing compounds in crude oils vary in nature and concentration with boiling range; with the heaviest fraction, it containing the highest amount of sulfur. Compounds that contain aliphatic sulfur (sulfides and thiols) have been found easier to remove than compounds containing aromatic sulfur (thiophenic compounds), which are present in higher proportions in heavy crude oils [40]. High percentage removal of sulfur (above 70%) can be achieved using combined oxidants ($KMn0_4$-H_2O_2) in a formic acid catalyzed reaction system [41].

Conventional high-pressure hydrodesulfurization process is being replaced by more effective oxidative desulfurization (ODS) at mild conditions without hydrogen pressure but with various catalysts. ODS is sometimes assisted and further enhanced with ultrasound technology and the use of ionic liquids [42]. This has become necessary due to the shift in global oil supply from light to heavy oil, which usually has higher sulfur content. Microwave irradiation can also be used as a pre-treatment for heavy crude oil where the sulfur content is not relatively high though fragmentation and recombination reactions may not be avoidable [43]. Currently, however, most desulfurization processes are carried out during refining and not as a pre-treatment (except for crudes with very high sulfur content), and if as a pre-treatment, then it serves also to upgrade the feedstock.

3.2.7 Desalting

The desalting process derives its name from the fact that soluble salts in the crude oil enter the aqueous phase and are easily removed by phase separation in the desalter. The wash water to oil ratio is 3–4 for light oils of API > 30° and 7–10 for light oils of API < 30°; a pH of 6 is ideal for the wash water and may be adjusted using caustic soda or acid [44]. For the case of heavy oils, however, with API < 20° but >10°, much higher water volume ratios will be required.

The desalting unit is where virtually all the aforementioned pre-treatments take place to some extent apart from the deasphalting step. For heavy and extra-heavy crude oils, however, there may be need to include one or more of the above specific pre-treatment steps to be carried out before the general desalting process. This may be warranted if the heavy oil has a peculiar need for such a step; otherwise, a two or three stage desalting unit may very-well serve all the purposes needed to achieve about 90% removal of contaminants from the heavy crude. This is because heavy crudes have higher proportions and wider ranges of contaminants such that a one-stage desalting process alone will not be adequate to pre-treat them for efficient refining [1]. In a typical single-step desalting unit as shown in **Figure 1**, the crude oil is pumped from storage through a heater, and fresh water, dosed with demulsifying

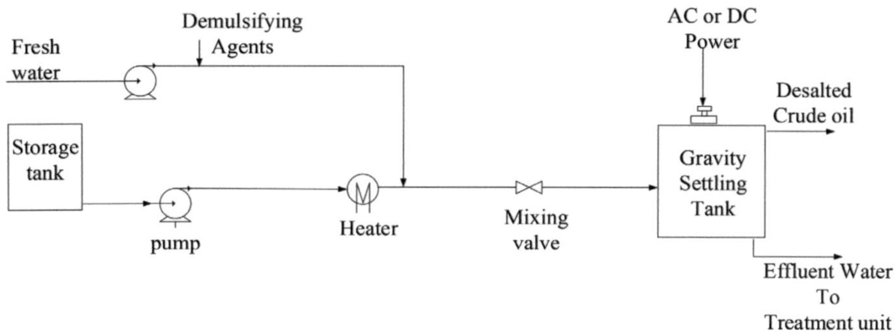

Figure 1.
Single-stage desalting unit.

agents, is added to the oil stream, which goes through a mixing valve for proper mixing. The stream then enters the gravity settling tank, where the water phase settles by gravity at the bottom of the tank, while the oil phase remains above it.

The separation of the oil from the water phase in the gravity settling tank is often aided by installing two electrodes between which an electrostatic AC or DC field is created with a potential of between 12,000 and 35,000 volts [45, 46]. These values are, however, most likely going to increase significantly in the desalting of heavy oils. The electrostatic field promotes coalescence of water globules and enhances separation. Effluent water is sent to water treatment unit, while the desalted crude is drawn from the top of the settling tank to the refinery for processing.

For heavy crude oils, however, except for crudes with exceptionally low amounts of contaminants, the desalting unit given in **Figure 2** is recommended, where the subunits in blue outlines are included for necessity. As the crude is pumped from storage, part or all of it is first diverted to a special treatment unit as the special need of the particular crude may require.

3.2.7.1 Special treatment unit

The traditional desalting and dehydration process was fully developed for light crudes of relatively high API gravity and low viscosity, but with the proportional increase in available heavier crudes, modifications have become necessary for effectiveness and efficiency in the desalting process. Heavy oil, especially with high contaminants, may be desalted after it is passed through its own special purification unit to significantly reduce the particular contaminant in question. For example, heavy oil with very high sulfur content can be first passed through a microwave irradiation unit before desalting [43]. Also, an adsorption packed bed with appropriate zeolite grades could be used to selectively reduce metal ions for a heavy crude, which is contaminated with a specific heavy metal [25]. Another possible case is a minor deasphalting unit using solvent extraction [21].

From the special treatment, the fresh water with demulsifying agents is added before the back-mixing by the mixing valve and then the stream flows to the first gravity settling tank. With the obvious need for a second gravity settling tank, the oil is sent to it for further separation, while the effluent water from the first settling tank is sent to water treatment unit. The desalted oil from the second tank goes to the final step in the pre-treatment section which is the hydrotreatment unit before it is sent to the refinery for processing, while the water from the bottom of the second tank is recycled as shown (**Figure 2**). This proposed arrangement is expected to achieve excellent results for the pre-treatment of

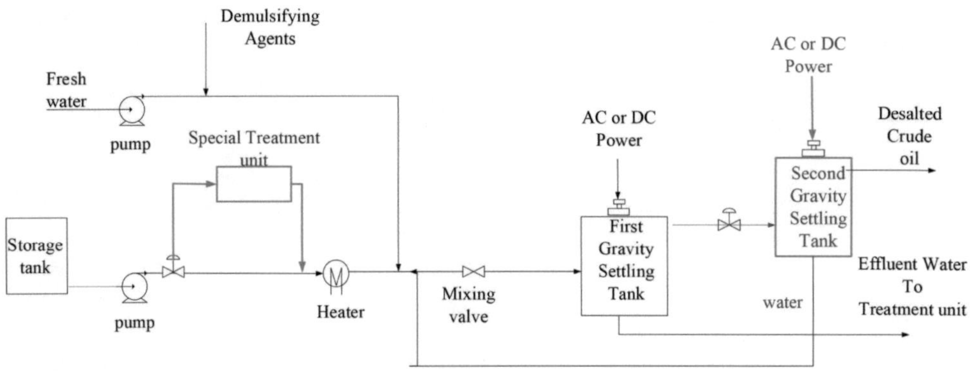

Figure 2.
Two-stage desalting unit with special treatment option.

heavy crudes. This special treatment unit may be bypassed if it is not required for a feedstock with minimal contaminants.

3.2.7.2 Multi-stage desalter

Heavy oils contain higher percentages of impurities and contaminants which cannot be removed by the conventional single-stage desalter. A two- or three-stage desalting unit will be required depending on the degree of contamination. A two-stage desalting unit is shown in **Figure 2**.

3.2.7.3 High-speed centrifuge

The limitations of the desalting process when treating heavy crude oil of API gravity of less than 20° and greater than 10° and of high viscosity can also be overcome by using a high-speed centrifuge. Well-designed high-speed centrifuges combine efficiency, reliability, product quality, and environmental friendliness. Gravity settling tanks are not very effective with heavy crudes even with the aid of an electrostatic field (AC or DC) due to their high viscosity. With a single-stage centrifuge technology, excellent separation between the oil and water phases is achievable [1]. The use of centrifuges leads to lower costs in energy consumption (in the electrostatic field), lower temperature and pressure in the operating conditions of the desalter, and less material and handling costs due to lower water-oil ratio and less amounts of demulsifying agents.

3.2.8 Hydrotreating

Hydrotreating is one major way of removing sulfur and other contaminants present in crude oil fractions. It can be applied at different stages of the refining process to different streams as required [47]. For the pre-treatment of heavy oils, however, hydrotreating needs to be employed primarily in its simplest form to significantly reduce or remove sulfur which can be as high as 8% by weight [19]. This unit is recommended to be placed immediately after the desalting tank and so serve as the last unit in the pre-treatment section before the refining process. Here, the desalted oil is mixed with hydrogen and heated to between 340 and 425°C, pressurized to about 14 MPa, and then passed over a suitable catalyst. Some commonly used catalysts are mainly cobalt, nickel, and molybdenum oxides on alumina. Biocatalysts have also been explored but have not yet become commercially attractive [32, 42, 44, 48].

In the hydrotreating unit, the primary target element is sulfur but many of the contaminants, which have been discussed earlier, are also removed simultaneously through several reactions as mentioned below:

- *Hydrodesulfurization* (HDS) reaction, which occurs between sulfur-containing compounds and hydrogen, leads to the removal of sulfur in the form of hydrogen sulfide. Eq. (4) shows such a reaction between ethanethiol (C_2H_5SH) and hydrogen (H_2). The hydrogen sulfide gas is captured and converted into elemental sulfur and sulfuric acid.

$$C_2H_5SH + H_2 \longrightarrow C_2H_6 + H_2S \tag{4}$$

- *hydrodenitrogenation* (HDN) by which nitrogen is removed mainly as ammonia;

- *hydrodemetallization* (HDM) by which metals are removed by being deposited on the catalysts;

- *hydrodeoxygenation* (HDO) by which oxygen is removed as water; and

- *hydrodearomatization* (HDA) by which olefins, naphthenes, and aromatics get saturated with hydrogen and become paraffins. Minor cracking also takes place to produce light alkanes.

The numerous products from the hydrotreating unit (HTU) are cooled using a series of heat exchangers then charged into a high-pressure gas separator, where hydrogen sulfide, ammonia, and the unreacted hydrogen are separated from the hydrotreated crude. The gases are sent to a scrubber with ethanol amine or other basic solutions to remove hydrogen sulfide for sulfur recovery, while hydrogen is recycled [49].

4. Modification requirements for heavy oil pre-treatment

Crude oil pre-treatment is a requirement before refining to remove contaminants which may be naturally part of the crude or acquired during production. A simple single-stage desalter can achieve sufficient purity when handing light crudes [16]. For heavy crudes, however, it is imperative to introduce some modifications since the conventional desalter will not be adequate. These modifications begin with the storage and pre-treatment process and extend all the way to the refining process and equipment. In the pre-treatment unit, a packed bed to achieve a significant degree of deasphalting, demetallization, deoxygenation, or denitrogenation should be introduced as an option at the beginning of the process depending on the particular need of the feedstock. The gravity settling tank (desalter) should be in two stages or may be replaced with a single high-speed centrifuge desalting tank. After desalting and the separation of the oil from the water phase, the desalted crude oil can then be partly hydrotreated in a mini hydrotreatment unit, particularly for heavy crudes with very high sulfur content. This applies to heavy crudes with >10–20° API gravity. **Figure 3** shows a schematic diagram of our proposed modification of the pre-treatment unit for efficient handling of heavy crude oil.

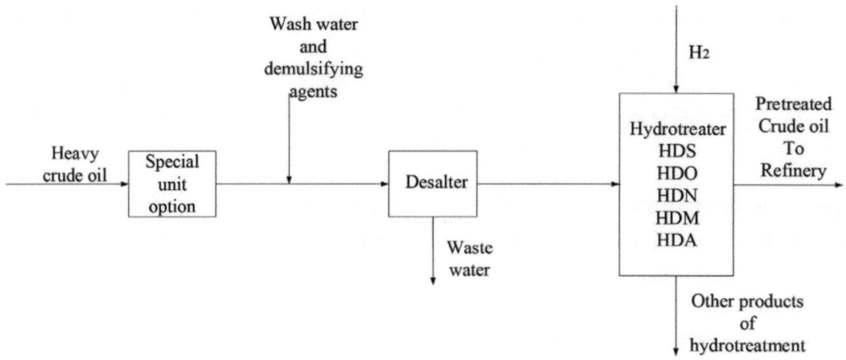

Figure 3.
Modified pre-treatment unit.

In the near future, extra-heavy crudes with an API gravity of <10° will be desalted in units with yet further modifications. This is because the API gravity of water is 10° and a crude with 10° API cannot be separated from the water phase by gravity. For extra-heavy oil, however (API < 10°), gravity settling tanks or centrifuges will still serve for phase separation; but in this case, waste water will be drawn from the top of the tank, while desalted crude will be drawn from the bottom.

5. Cost implications

Currently, heavy crude oil is more expensive to produce (upstream) but yields less economic returns to the producer because of the challenges encountered in its processing and the lower quality of its products; on the other hand, lighter crude is cheaper to produce and easier to process, and therefore, it is more expensive. With the economic turn of events lately, much of the global lighter crude reserves have been greatly depleted, while heavy crude reserve is in over-abundance, and refiners are now beginning to prefer to buy heavy crudes which are cheaper and modify their processes and equipment which is capital intensive. There is, however, a trade-off because the process modification is a one-time investment, while the long-term good product yield eventually compensates with a high return on investment. The development of new technologies for processing heavy oils needs to be rapid enough to forestall the effects of the dwindling supply of conventional crudes. It is also expected that in the course of process modifications, new technologies will serve to significantly reduce anticipated cost and favor heavy oil production. For example, in a modified desalting unit, the use of a high-speed centrifuge eventually results in significant savings in energy and material cost. This is due to lower temperature and pressure conditions, less amounts of required demulsifying agents, less quantity of wash water, and therefore less waste water generation. As refiners invest more in processing heavy oil, producers will more likely be motivated to invest in its production. The economic face of the oil market is likely to experience new paradigms as heavy oil processing technologies begin to compete with light oil refineries. For light oils, cost of refining, depending on oil configuration and refinery size, can vary widely from 4 to 15 USD/bbl, while for heavy oils, it is in the range of 20–22 USD/bbl [50, 51].

In basic refinery economics, the cost of crude oil and the price of refined products are highly volatile and nearly unpredictable because they are affected by economic, political, and environmental factors. Since refiners have no control over these factors, operational efficiency, which is measured by the ratio of output to inputs, is the basis for profit maximization and can be increased through continuous innovation [50]. As a determinant of profitability, facility expansion for complexity to handle a wider variety of feedstocks can impact on profitability up to the tune of 0.5 USD/bbl [51]. This implies that modifying the pre-treatment section of a refinery to handle heavy crude oils for efficient refining into "attractive" products is economically favorable. Since heavy oils are cheaper than light oils and the cost of crude oil is the largest input cost of a refinery, processing cheaper crudes into higher-value products through process modification usually improves profit margins. The increased global demand for lighter petroleum products made from heavier crudes (quality gap) drives downstream investors toward more complex refineries. A refinery's complexity is measured by its range of capacities to "add value" by its processes to a crude feedstock. The most common cost-based measure of complexity for refineries is the Nelson Complexity Index (NCI). It provides a relative measure of the construction cost of a refinery based upon its capacity to upgrade. The simplest topping refinery is assigned an index of 1.0 against which all others are measured. The higher the NCI number of a refinery, the more its complexity, cost, and capacity to add value to crude feedstocks. Since 2005, global trends have been toward refineries of higher complexities through newer technologies; hence, older refineries with low NCI numbers are fast fading out as lower-value crudes (such as heavy oils) increase in relative abundance. A refinery with a fluid catalytic cracker, alkylation, and hydrotreating units has an NCI of about 7.0, while an inclusion of hydrocracking, coking, and reforming units can raise its NCI number to about 14.1 [51].

For the pre-treatment section, therefore, the proposed necessary modification here will definitely cause a minor increase in the NCI number of the refinery due to the additional pieces of equipment. An overview of NCI numbers of complex refineries shows a wide range from 5 to 14 [52]; therefore, it is estimated that the pre-treatment section contributes only about 1–1.5% of the total NCI number and, hence, total cost of the refinery. The bulk of the refinery cost is in the refining processes and not in the pre-treatment; consequently, the pre-treatment equipment cost is certainly less than 2% of the total cost of the refining equipment with the modification proposed in this chapter even for pre-treating the most challenging heavy oil feedstocks.

6. Conclusion

It can be concluded from the issues discussed in this chapter that a modification of the conventional oil pre-treatment unit is required to ensure the adequate removal of impurities from heavy crude oils. First, a special optional unit is introduced at the beginning of the pre-treatment unit to treat crudes with any extremely high level of a particular contaminant such as heavy metals, asphalt, or sulfur. Second, the desalting stage is carried out in two stages aided by electrostatic fields, and third, the desalted crude is sent to a mini pre-hydrodesulfurization unit before it goes for proper refining. With high purity of pre-treated crude, refining processes will be enhanced, product quality will improve, and downstream problems will be reduced to their barest minimum or completely eliminated.

Conflict of interest

The authors declare no conflict of interest.

Appreciation

The authors thank Emmanuel O. Fajoye for assisting with the figures and Toluwaniyi S. Babalola for handling the IT matters.

Author details

Faith Uchenna Babalola* and Alfred Akpoveta Susu
Department of Chemical and Petroleum Engineering, University of Lagos, Lagos, Nigeria

*Address all correspondence to: fu_babalola@yahoo.com; fubabalola@unilag.edu.ng

References

[1] Perschke T. Desalting of heavy crude oil by using centrifugal technology. In: Society of Petroleum Engineers Kuwait Oil and Gas Show and Conference; 8-10 October 2013; Kuwait City, Kuwait. 2013. DOI: 10.2118/167365-MS. Available at: https://www.google.com/search?q=perschke+t%2C+Desalting+of+heavy+crude&oq=perschke+t%2C+Desalting+of+heavy+crude+&aqs=chrome..69i57j33.50899j1j9&client=ms-android-xiaomi-rev1&sourceid=chrome-mobile&ie=UTF-8

[2] Al-Roomi Y, George R, Elgibaly A, Elkamel A. Use of a novel surfactant for improving the transportability/transportation of heavy/viscous crude oil. Journal of Petroleum Science and Engineering. 2004;**42**:235-243

[3] Saniere A, Henaut I, Argillier JF. Pipeline transportation of heavy oils, a strategic economic and technological challenge. Oil and Gas Science and Technology. 2005;**59**(5):455-466

[4] Hart A. A review of technologies for transporting heavy crude oil and bitumen via pipelines. Journal of Petroleum Exploration and Production Technologies. 2014;**4**:327-336. DOI: 10.1007/s13202-013-0086-6

[5] Ghannam MT, Hasan SW, Abu-Jdayil B, Esmail N. Rheological properties of heavy and light crude oil mixtures for improving flowability. Journal of Petroleum Science and Engineering. 2012;**81**:122-128

[6] Castaneda LC, Munoz JAD, Ancheyta J. Current situation of emerging technologies for upgrading of heavy oils. Catalysis Today. 2014;**220-222**:248-273

[7] Gordon D. Understanding Unconventional Oil (PDF). Washington, DC: Carnegie Endowment for International Peace; 2013. [Accessed: December 28, 2013]

[8] Oil sands crude. In: The Global Range of Crude Oils. UK, Canada Crude Handout; 2013. [Accessed: December 28, 2013]. Available at: https://www.google.com/url?sa=t&source=web&rct=j&url=https://www.gov.uk/government/uploads/system/uploads/attachment_data/file/5005/canada-crude-handout.pdf&ved=2ahUKEwiE8JS8gu7kAhUTa8AKHWF6B5IQFjAOegQICRAB&usg=AOvVaw24AULma_wxSWdTK7ACnkTs

[9] Dusseault MB. Comparing Venezuelan and Canadian heavy oil and tar sands. In: Canadian International Petroleum Conference; 12-14 June 2001; Calgary, Canada. 2001 [Accessed: May 5, 2008]. Available at: https://www.google.com/url?sa=t&source=web&rct=j&url=https://search.spe.org/i2kweb/SPE/doc/onepetro:B3705AD4/&ved=2ahUKEwjZ-ov2g-7kAhXRTcAKHbwIACwQFjACegQIBBAB&usg=AOvVaw3t6Ouogf8OBorYe2MaDT9z

[10] Attanasi ED, Meyer RF. Natural bitumen and extra-heavy oil. In: Survey of Energy Resources. 22nd ed. London: World Energy Council; 2010. pp. 123-140. ISBN: 978-0-946121-26-7 [Accessed: August 24, 2013]. www.worlenergy.org

[11] Classification of Petroleum Products. 2019. Available from: http://www.petroleum.co.uk/classification

[12] Demulsification. 2019. Available from: https://petrowiki.org/Oil_demulsification

[13] Nour AH, Yunus RM, Jemaat Z. Chemical demulsification of water-in-crude oil emulsions. Journal of Applied Sciences. 2007;7(2):196-201. DOI: 10.3923/jas.2007.196.201

[14] Pinkovski YI. Pretreatment of crude petroleum before refining. Chemistry and Technology of Fuels and Oils. 1968;**4**(4):268-271

[15] Zolfaghari Z, Fakhru'l-Razi A, Chuah AL, Elnashaie SSEH, Pendashteh A. Demulsification techniques of water-in-oil and oil-in-water emulsions in petroleum industry. Separation and Purification Technology. 2016;**170**:377-407. DOI: 10.1016/j.seppur.2016.06.026

[16] Duggan G. Benefits of crude pretreatment for unconventional heavy crude processing. In: AIChE 12th Global Congress on Process Safety; 10-14 April 2016; Houston Texas. 2016. Available at: https://aiche.confex.com/aiche/s16/webprogram/Paper444224.html

[17] Menon VB, Wasan DT. Particle-fluid interactions with applications to solid-stabilized emulsions part III. Asphaltene adsorption in the presence of quinaldine and 1,2-dimethylindole. Colloids and Surfaces. 1987;**23**(4):353-362. DOI: 10.1016/0166-6622(87)80276-7

[18] Kokal S, Al-Juraid J. Reducing emulsion problems by controlling asphaltene solubility and precipitation. Presented at the SPE Annual Technical Conference and Exhibition; 27-30 September 1998; New Orleans, Louisiana; 1998. SPE-48995-MS. DOI: 10.2118/48995-MS

[19] Wauquier JP. Petroleum Refining 1 Crude Oil—Petroleum Products-Process Flow Sheets. Vol. 1. Paris: T-Editions Techip; 1995. Available from: http://www.editionstechnip.com/en/catalogue-detail/525

[20] Speight JG. The Chemistry and Technology of Petroleum. 4th ed. Boca Raton, Florida, USA: CRC Press, Taylor& Francis Group; 2006

[21] Salem SM, Abdelaleem GM, Elsayed NA, Saad WO. Improving the quality of petroleum crude oil by deasphalting. Journal of Engineering Sciences, Assiut University. 2011;**39**(4):885-896

[22] Speight JG. The Refinery of the Future. Norwich, New York: William Andrew; 2010. pp. 389-395. DOI: 10.1016/C2009-0-20064-X

[23] Speight JG. Heavy Oil Recovery and Upgrading. Houston, Texas, USA: Gulf Professional Publishing; 2019. pp. 807-821. DOI: 10.1016/C2016-0-04682-X

[24] Henning KD. Handbook of solvents—Solvent recycling, removal and degradation. In: Wypych G, editor. Use Health and Environment. 3rd ed. Toronto, Ontario, Canada: Chem Tech Publishing; 2019

[25] Ikyereve RE, Nwankwo C, Mohammed A. Selective removal of metal ions from crude oil using synthetic zeolites. International Journal of Scientific and Research Publications. 2014;**4**(5):1-3

[26] Mustafin IA, Shevchenko AA, Gilmanshina KA, Krasilnikova YV. The demetallization of west Siberian crude oil and its residues. International Journal of Applied Engineering Research. 2017;**12**(10):2338-2341

[27] Rana MS, Sámano V, Ancheyta J, Diaz JAI. A review of recent advances on process technologies for upgrading of heavy oils and residua. Fuel. 2007;**86**(9):1216-1231

[28] Ali MF, Abbas S. A review of methods for the demetallization of residual fuel oils. Fuel Processing Technology. 2006;**87**(7):573-584. ISSN: 0378-3820. DOI: 10.1016/j.fuproc.2006.03.001

[29] Kurbanova AN, Akhmetov NK, Yeshmuratov A, Zulkharnay RN, Sugurbekov YT, Demeuova G, et al. Removal of nickel and vanadium from crude oil by using solvent extraction and electrochemical process. Physical Sciences and Technology. 2017;**4**(1):74-80

[30] Ikyereve RE, Mohammed A, Nwankwo C. Performance of natural zeolites in the removal of metal ions from crude oil. ARPN Journal of Science and Technology. 2014;**4**(3):141-144. ISSN: 2225-7217

[31] Cho DW, Jung SJ, Beum HT, Jung TS, Yoon HC, Han SS, et al. Calcium removal from heavy crude oil. In: AIChE 12th Global Congress on Process Safety; 10-14 April 2016; Houston Texas. 2016. Available at: https://aiche.confex.com/aiche/s16/webprogram/Paper444224.html

[32] Prado GHC, Rao Y, de Klerk A. Nitrogen removal from oil: A review. Energy and Fuels. 2017;**31**(1):14-36. DOI: 10.1021/acs.energyfuels.6b02779

[33] Ball JS, Whisman ML, Wenger WJ. Nitrogen content of crude petroleum. Industrial and Engineering Chemistry. 1951;**43**(11):2577-2585. DOI: 10.1021/ie50503a047

[34] Hydronitrogenation. 2019. Available from: https://en.m.wikipedia.org/wiki/Hydrodenitrogenation

[35] Raje AP, Liaw SJ, Srinivasan R, Davis BH. Second row transition metal sulfides for the hydrotreatment of coal-derived naphtha. I. Catalyst preparation, characterization and comparison of rate of simultaneous removal of total sulfur, nitrogen and oxygen. Applied Catalysis, A: General. 1997;**150**:297-318. DOI: 10.1016/S0926-860X(96)00317-1

[36] Laredo GC, Vega-Merino PM, Trejo-Zárraga F, Castillo J. Denitrogenation of middle distillates using adsorbent materials towards ULSD production: A review. Fuel Processing Technology. 2013;**106**:21-32. DOI: 10.1016/j.fuproc.2012.09.057

[37] Hong X, Tang K. Adsorptive denitrogenation of diesel oil using a modified NaY molecular sieve. Petroleum Science and Technology. 2015;**33**:1471-1478. DOI: 10.1080/10916466.2015.1076844

[38] Hernández-Maldonado AJ, Yang RT. Denitrogenation of transportation fuels by zeolites at ambient temperature and pressure. Angewandte Chemie, International Edition. 2004;**43**:1004-1006. DOI: 10.1002/anie.200353162

[39] Crude Oil Composition. 2019. Available from: https://www.fkit.unizg.hr/_download/repository/PRPP_2013_Crude_oil_composition.pdf&ved=2ahUKEwi84tvs_ZDkAhXE2aQKHbCqD-IQFjABegQIDRAG&usg=AOvVaw0fVizvsKz2grs_-dUshVR9

[40] Papavinasam S. Corrosion Control in the Oil and Gas Industry. Oil and Gas Industry Network. 2014. Available from: https://www.sciencedirect.com/topics/engineering/hydrotreating

[41] Desalting. 2019. Available from: https://wwwnptel.ac.in/courses/103102022/8

[42] Gary JH, Handwerk GE, Kaiser MJ. Hydrotreatment. In: Petroleum Refining. 5th ed. New York: CRC Press; 2007. pp. 195-203

[43] Hydrotreatment Processes. 2019. Available from: https://www.e-education.psu.edu/fsc432/content/hydrotreatment-processes

[44] Gary J, Handwerck G, editors. Petroleum Refining: Technology and Economics. 4th ed. New York. USA: Marcel Dekker, Inc.; 2001. p. 49. ISBN: 0-8247-0482-7

[45] Pereira J, Velasquez I, Blanco R, Sanchez M, Pernalete C, Canelón C. Crude oil desalting process. In: Patel V,

editor. Advances in Petrochemicals. London, UK: IntechOpen; 2015. DOI: 10.5772/61274. Available from: https://www.intechopen.com/books/advances-in-petrochemicals/crude-oil-desalting-process [Accessed: July 22, 2016]

[46] Miadonye A., Snow S., Irwin DJG, Khan MR, Britten AJ. Desulfurization of heavy crude oil by microwave irradiation. Computational methods in multiphase flow.WIT Transactions on Engineering Sciences 2009; 63: 455-465 DOI: 10.2495/MPF090391

[47] Javadli B, de Klerk A. Desulfurization of heavy oil. Applied Petrochemical Research. 2012;**1**:3-19

[48] Haruna SY, Faruq UZ, Zubairu AY, Liman MG, Riskuwa ML. Comparative studies on reduction of sulphur content of heavy crude oil using $KMnO_4+H_2O_2/CH_3COOH$ and $KMnO_4+H_2O_2/HCOOH$ via oxidative desulphurization (ODS). American Journal of Applied Chemistry. 2018;**6**(1):15-24. DOI: 10.11648/j.ajac.20180601.13

[49] Houda S, Lancelot C, Blanchard P, Poinel L, Lamonier C. Oxidative desulfurization of heavy oils with high sulfur content: A review. Catalysts. 2018;**8**(9):344. DOI: 10.3390/catal8090344

[50] Farrell AE, Brabdt AR. Risks of the oil transition. Environmental Research Letters. Vol. 1. Bristol, UK: IOP Publishing Ltd; 2006. DOI: 10.1088/1748-9326/1/1/014004

[51] The Economics of Petroleum Refining. 2019. Available from: https://www.google.com/url?sa=t&source=web&rct=j&url=https://www.canadianfuels.ca/website/media/PDF/Publications/Economics-fundamentals-of-Refining-December-2013-Final-English.pdf&ved=

[52] Kaiser MJ. A review of refinery complexity applications. Petroleum Science. 2017;**14**(1):167-194. DOI: 10.1007/s12182-016-0137-y

Chapter 11

Recent Techniques for the Removal of Naphthenic Acid from Heavy Crude Oils

Feyisayo Victoria Adams, Chikaodili E. Chukwuneke and Bolade O. Agboola

Abstract

Naphthenic acid (NA) is one of the components of heavy crude oil. It is composed of a carboxylic acid functional group attached to a hydrocarbon molecule. Heavy crude oils with high concentration of NA are usually classified as poor quality oil and are sold at cheaper price. The presence of high concentration of naphthenic acid in crude oil often result in limited life span of equipment used in exploration and refining processes due to corrosion. In order to improve the quality of such crude oil and to save cost, it becomes very important to remove NA compounds from it. Researchers, as well as the oil and gas industries, have been working on how to overcome this challenge, and several techniques have been developed for the removal of naphthenic acid from the crude oil. This study thus envisages to present the various recent techniques available for the removal of NA from heavy crude oil. It has been seen that of all the recent physical and chemical methods that are available, catalytic-based methods are effective and their effectiveness depend on the temperature at which the catalyst is calcined as well as the surface area of the catalyst.

Keywords: catalysts, extraction, heavy crude oil, naphthenic acids, petroleum refining

1. Introduction

Naphthenic acid (NA), which is a mixture of alicyclic carboxylic acid and saturated aliphatic acid, is a component of a heavy crude oil and it is responsible for the heavy crude oil's acidity [1]. The presence of NA in heavy crude oil poses great challenge in the exploration of the crude oil; this is due to the corrosiveness of the acid. The removal of NA compounds from crude oil is known as the most important process in upgrading heavy oil. Caustic washing and blending of heavy acidic crude oil with conventional crude oil are two conventional methods used in the removal of NAs from heavy crude oil. The disadvantage of blending method is in its low efficiency with removal of about 2–3% [2] and the NAs are often not completely removed from the blend as a result, it is often used as a corrosion mitigation tool. Meanwhile, the limitation of caustic washing method is in the large amount of wastewater which is usually produced during the process [2].

Several deacidification removal techniques have been reported in the literature. These include catalytic destructive methods, non-catalytic destructive methods, and non-catalytic non-destructive methods [3, 4]. Thermal decomposition, which is an example of non-catalytic destructive methods, has shown to be ineffective in the reduction of acidity in heavy crude oil [5]. High temperature, which is higher than 400°C, is often used. This offers thermal cracking of the heavy crude oil, which then resulted in coke formation [1]. Catalytic decarboxylation process [6] has been used as an alternative to the non-catalytic destructive methods for naphthenic acids (NAs) removal from heavy crude oil. Polymeric compounds exhibiting basicity characteristics, on the other hand, can be used to neutralize NAs. Although, expensive neutralizing agents, which are difficult to recover from deacidified products, are needed in this method.

Extractive separation is another method used in the removal of NAs from heavy crude oils. Using this method, valuable hydrocarbons are usually removed and this makes the process expensive. Extractive separation method requires multistage wash using different types of solvents. A single NA compound or their derivatives such as salt and water-oil emulsion can also be used [3]. It has been reported that solutions of metal oxides, such as those of aluminum and magnesium, exhibit high adsorption capacity and are able to remove more than 95% of the NAs present in a liquid hydrocarbon feedstock at temperatures between 30 and 80°C [4, 7].

Adsorption process of NA removal from heavy crude oil can be used under low temperature condition for distillate fractions. Strong bases, such as alkali metal hydroxides (MOH), are often used as pre-treatment of the heavy oil prior to the adsorption process. The MOH, however, does not only provide an alkaline environment, but it neutralizes the acidic components of NAs in the heavy crude oil. The major challenge associated with this technique is its environmental challenge as disposal of MOHs into the environment creates nuisance to the environment [7].

Esterification process has been used to remove NAs from heavy crude oils. The NAs reacts with alcohol with or without the addition of a catalyst to form ester [8]. Oxides and carboxylates of metals have been successfully applied in NAs removal from heavy crude oils [9–11] using the esterification method at reaction temperatures between 250 and 350°C [12].

Organic bases such as monoethanolamine, 2-(Dibutylamino) ethanol [13], glycolic solutions [11], or imidazole derivatives [14] as alcoholic solutions have been used as alternative method to dilute caustic wash in the NAs removal from heavy crude oil. However, the amount of organic bases needed for a successful extraction is usually in excess. This results in amine been left in the petroleum phase, thus creating problems in the downstream catalyst units which is used for oil upgrading. On the other hand, lysine-derived amino acid ionic liquids (AAILs) have shown better extraction efficiency compared to other AAILs, because lysine-derived AAILs contain two amino groups which can react with carboxylic acid group to form zwitterionic species [3].

Zeolites [3] and metal oxides [15] have been successfully applied in NAs removal from heavy crude oil. The limitations associated with this process, however, are the moderately high operating temperatures of 300–400°C and the use of hydrogen [5] to increase deacidification rate, because hydrogen is often not available for the process. The use of caustic soda (NaOH) is a known industrial method for the removal of NAs from the kerosene/diesel fractions. The NAs are usually separated from the oil fraction as sodium naphthenates and is dissolved in water to form soluble fraction. Thereafter, acidifying sodium naphthenates with a mineral acid, will result in the recovery of water insoluble NAs [3]. However, due to poor solubility of NAs, their removal becomes incomplete [3].

Reactive extraction method has gained wide interest in the removal of NAs from oil. This came in response to the extreme economic pressure NAs removal poses on industries. Due to the decrease in efficiency and limitations of well-known processes and the demand for inexpensive and environmentally safe and high purity products, it is imperative to consider new processes. Thus, this work is aimed at compiling recent techniques used in the removal of NAs from heavy crude oil. These techniques include the physical and chemical methods which are reactive extraction and adsorption of naphthenic acids (physical methods) and esterification of naphthenic acids, decarboxylation of naphthenic acids, and supercritical fluids (SCF) (chemical methods).

2. Physical methods

2.1 Reactive extraction of naphthenic acids

Reactive extraction is a separation process in which reactions between materials to be extracted and the extractants take place. This extraction process depends on degree of extraction and different parameters such as the loading ratio, complexation equilibrium constant, distribution coefficient, type of complexes, etc. The distribution coefficient and the degree of extraction are used to analyze the extraction process. Equation (1) is used in the calculation of the distribution coefficient, K_d [16].

$$K_d = \frac{C_{org}}{C_{aqu}} \quad (1)$$

where, C_{org} and C_{aqu} are the total concentration of carboxylic acid in organic phase and the total acid concentration (dissociated and un-dissociated) aqueous phase at equilibrium, respectively.

The degree of extraction is given as the ratio of acid concentration in the extracted phase to the initial acid concentration in aqueous solutions at equilibrium as shown in Eq. (2) [16].

$$E = \frac{K_d}{(1 + K_d)} \times 100 \quad (2)$$

Kumar et al. [16] determined equilibrium complexation constant, extraction mechanism, degree of extraction, and the distribution coefficients using varied concentration of extractant (hydroxides). In this process, NA was dissolved in 0.125–0.650 mol/L of kerosene at temperature range of 301–333 K and atmospheric pressure. The distribution coefficient (K_d) of NA was found to be about—19.84 with degree of extraction of 56.48–95.16% at various temperature and concentration ranges. The highest degree of extraction (95.2%) was obtained at a concentration of 0.125 mol/L and at room temperature, while it decreased with increase in concentration. Meanwhile, the degree of extraction decreased as the temperature reached 60°C (333 K). It was observed that the degree of extraction increased with increase in the pH of the extractant [16].

Liquid-liquid extraction (LLE) has been found to be effective in the removal of NAs from heavy crude oils. However, the effectiveness of this process depends on the extractant and the experimental conditions such as temperature, contact time, and concentration. The use of ammonia or alcoholic alkaline solutions (such as potassium hydroxide (KOH) in alcohol) in the removal of NAs using LLE method is common and is effective. However, these extractants usually form stable emulsions [17]. Due to this limitation, other extractants have been proposed by several researchers [9, 18]. Liquid alkanols, ammonia in varied ratio with water are examples of the proposed extractants. Pre-treatment of heavy crude oil before extraction using alkoxylated amine and water has also showed better results in the removal of NAs [19]. Also, a mixture of NaOH and ethanol with heavy crude oil with the reagent/oil ratio being 0.4:1 (w/w) at room temperature showed an optimal amount of NaOH in crude oil after extraction to be 3000 μg/g under 5 min of extraction [14]. Liquid-liquid extraction (LLE) data were obtained for a ternary system of dodecane, naphthenic acid, and [DBU-Hex] [SCN] under atmospheric condition [20]. The solubility of NA was found to be higher in [DBU-Hex] [SCN] when compared to dodecane. Thus, [DBU-Hex] [SCN] was concluded to be better extractant for the removal of NA from heavy crude oil than dodecane.

Ionic liquids (ILs) have been considered as green solvents for extraction processes due to their unique properties [21, 22]. The physiochemical properties of ILs depend on the cations and anions of various ILs. The symmetry of the shape or/and alkyl chain in the cation could have effects on the viscosity, density, and surface tension of ILs, meanwhile ILs miscibility and their thermal stability could be affected by the anion [23]. The wide range of temperature for liquid phase and extremely low vapor pressure of ILs give them advantage over other solvents whether polar or non-polar organic solvents. Report has, however, showed that some ILs need large amount for effective extraction process for the removal of NAs. Also, the regeneration procedure for NA and ILs has been found to be complex [24, 25]. Low ILs/oil ratios of tetra-alkylphosphonium and tetra-alkylammonium hydroxide ILs were used in the removal of NA from a model acidic oil [26]. The results from the experiment showed tetra-alkylammonium-based ILs to be more efficient than tetra-alkylphosphonium-based ILs. It was observed that the longer the alkyl spacer length on the cation, the lower the extraction performance of the hydroxide-based ILs. The ILs used in the model acidic oil study showed a great reusability potential without losing its activity [26].

Neutralization method was used to remove NA from Doba crude oil using ILs ($[N_{2221}][MeCO_3]$ Triethylmethylammonium methylcarbonate) at an optimum temperature of 338 K and a removal efficiency of 73.1% was achieved [12]. Under the same condition, an efficiency of 80.4 and 69.6% was achieved using [N2221] [HCO3] Triethylmethylammonium hydrgoencarbonate and [N4441][MeCO3] Tributylmethylammonium methylcarbonate, respectively [12].

Crude oil containing dodecane and NA (TAN = 1.44 ± 0.01 mg KOH g^{-1}) was treated for NA removal using extraction method and 1-Decyl-3-ethylimidazolium phenolate.[C10mim][Phe] as extractant [27]. The optimum temperature, at which an efficiency removal of about 85.7% was achieved, was 320 K. On the other hand, when a crude oil having dodecane and NA (TAN = 4.74 ± 0.01 mg KOH g^{-1}) was treated, an efficiency of 83% was achieved at a temperature of 323.15 K [26]. Neutralization method was used in this case and Tetra methyl ammonium hydroxide, [N1111][OH] was used as the solvent. At a temperature of 323 K, a removal efficiency of 25% was obtained using extraction method with 1-ethyl-3-methylimidazolium ethyl sulfate [EMIM][EtSO4] as extractant [28].

Adsorption process was used for the removal of NA from crude oil containing white oil and NA (TAN = 1.20 mg KOH g^{-1}) [29], dodecane and NA (TAN = 3.46 ± 0.01 mg KOH g^{-1}) [24] and dodecane and NA (TAN = 2.93 ± 0.01 mg KOH g^{-1}) [21] at temperatures of 313 K, 360.15 K and room temperature, respectively. A removal efficiency of 97.61, ~71.5, and 100% was achieved in the process using 1-methylimidazole acetate [Bmim][CH3COO] [29],

1-decyl-1,8-diazobicycloundec-7-enethiocyanate [DBU-Dec] [SCN] [24], and Butyl methyl imidazolium functionalized silica SilPrBmim [21] as ILs, respectively.

2.2 Adsorption of naphthenic acids

Adsorption process has recently been used for the removal of NAs from heavy crude oils. Example of these is the removal of NAs from Sudanese crude oil using zeolite and clay [30]. A study by Zhang et al. [31] showed that the use of a naturally occurring clay gave adsorption capacity for NAs as high as about 70 mg-NA/g clay. Though this process is effective and efficient, its limitation is in the disposal of the used clay.

Lately, nanotechnology has been brought to light as the latest technique that can be applied in reducing TAN present in solutions. Simonsen [32] synthesized amino-functionalized iron oxide (Fe_3O_4) nanoparticles which has a high selectivity for NA. Harnessing the magnetic properties of the nanoparticles, a magnetic field is applied to the mixture for the sequestration of the NA. Among the nanoparticles synthesized: Fe_3O_4, Fe_3O_4/SiO_2, Fe_3O_4/SiO_2 NH_2/C_{18}, the amino-functionalized type showed a high affinity for NA and spectra from FTIR and UV-Visible Spectrophotometer confirmed that no other crude oil components were eliminated. This study achieved a remarkable feat of attaining a TAN reduction of 92% with this novelty.

3. Chemical methods

3.1 Esterification of naphthenic acids

Esters are a byproduct of reacting —COOH with —OH and this reaction is useful for NAs removal from solutions, since the major functional group for organic acids are carboxyl groups. The general form of an esterification reaction of NA takes the form below (**Figure 1**).

In the absence of any catalyst, Khan et al. [2] optimized the removal of NA from heavy acid crude oil by varying the pressure, reaction time, crude concentration, and temperature. A mixture of HAC Laguna and Kuwait Crude with TAN value of 2.57 mg KOH/g was esterified with methanol in a batch reactor and under the enhanced conditions of 250°C, 6.4 MPa, and 33.3 wt% to yield low TAN crude of 0.08 mg KOH/g (96.9% removal).

Another example of this application is observed in the esterification of Colombian crude by Quiroga-Becerra et al. [34]. Temperature was varied to see its effect on shrinking the TAN content and also FTIR analysis was done on products at each temperature profile. At 250°C, the outcome of the study was impressive as the TAN value fell from 7.33 to 0.47 mg KOH/g. FTIR data also support the post reaction TAN results, as a shrinking in the carboxyl group wave number between 1705 and 1710 cm^{-1} was observed. In addition, as a corollary to the esterification, a shift in the carbonyl group band to wave numbers 1730–1740 cm^{-1}, indicated a reduction in acidity.

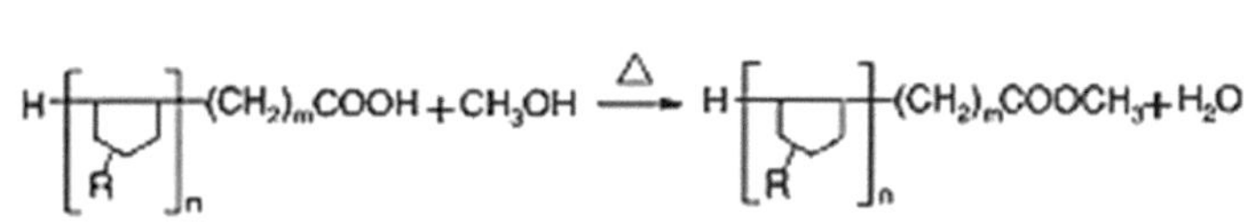

Figure 1.
Esterification reaction of NA [33].

3.2 Decarboxylation of naphthenic acids

In catalytic decarboxylation, commercially available chemicals or self-prepared materials are usually used as catalysts to determine their suitability and effectiveness in removing NAs from heavy crude oils with model compound before using them in a real situation [31]. Metal oxide catalysts have been found to be very effective in NAs catalytic decarboxylation. The beauty of catalysis lies in its ubiquitous application as it cuts across many chemical methods of NA removal. Every chemical reaction can be encouraged to favor NA removal by the use of active catalyst that provides an alternative pathway that ensures the lowering of the activation energy for the reaction to occur (**Figure 2**).

A study carried out by Shohaimi et al. [35] has shown the use of monometallic alkaline earth metal oxides (Ca/Al_2O_3 and Ce/Al_2O_3) as catalysts for the removal of NAs from acidic Petronas Penapisan Melaka refinery heavy crude oil with TAN of 2.43 mg KOH/g to be successful. It was observed that the Ca/Al_2O_3 catalyst was more effective (83.5%) in removing the NAs from crude oil compared to Ce/Al_2O_3 catalyst (71.2%). In another study, other types of catalysts ($Ni/Ca/Al_2O_3$, $Ni/Ba/Al_2O_3$, $Cu/Ca/Al_2O_3$, and $Cu/Ba/Al_2O_3$) were studied in TANX 2000 (concentration range of 10–50 mg/L) and ammonia solution in ethylene glycol with a concentration range of 100–1000 mg/L to remove NAs from Petronas Penapisan Melaka refinery Heavy Crude (TAN of 2.43 mg KOH/g), Petronas Penapisan Melaka refinery Light Crude (TAN of 2.52 mg KOH/g) and Korean Crude oils. The highest concentrations of TANX 2000 solution and ammonium solution in ethylene glycol exhibited the best experimental conditions. The catalyst ($Cu/Ca(10{:}90)/Al_2O_3$) with calcination temperature of 1000°C gave the most effective results for all the three types of crude oil. With the use of catalysts, there was improvement in the removal of NAs from all the three crude oils that were studied [36].

With the goal of removing NAs at low temperatures and on an economy budget, rare earth metals known to possess great catalytic potential were explored by Shukri et al. [37]. These researchers tested the use of alumina supported CeO-, ZnO-, and SnO-based catalyst coupled with Cu, Co, Mn, and Ni as dopants to deacidify Korean crude oil. Results showed that the calcination temperature, base catalyst type with large surface area, and the base chemical agent do play a role in optimizing the reduction of NA in Korean crude. The results proved CeO to be the best catalyst and Cu ($Cu/Ce(10{:}90)/Al_2O_3$) as the most effective dopant, with an optimum deacidification achieved when the aforementioned was calcined at a temperature of 1000°C. A concentration of 1000 mg/L for 4% NH3-PEG gave the best acidity cutback for the basic chemical agent used, but it was only able to cut down the TAN from 8.32 to 3.65 mg KOH/g using the Ce oxide-based catalyst. However, after it was augmented with 1% of 10 M KOH, the TAN was reduced to 0.56 mg KOH/g thereby meeting acceptable industrial standards of crude oil acidity [37].

The limitations involved in extant NA removal techniques such as emulsion formations via neutralization, high cost from physical adsorption processes, and cost of hydrogen from catalytic hydrogenation have hampered their industrial

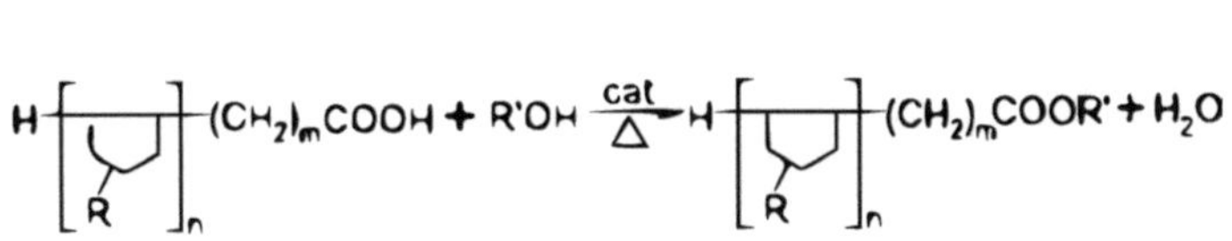

Figure 2.
A catalytic reaction for the removal of NA [10].

applications. In a study in which γ-Al2O3 supported 12-tungstophosphoric acid (TPA) catalyst was used in a fixed bed continuous flow reactor, factors such as increased methanol/crude oil ratio, temperature, and decreased liquid hourly space velocity (LHSV) downregulated the TAN value. Compared to other catalysts such as SnO/Al_2O_3 [38] and Mg-Al Hydrotalcite [39], TPA/γ-Al_2O_3 has proven to enhance the transesterification of NAs at lesser temperature (250°C). The TPA/γ-Al_2O_3 catalyst also decreased the Castilla crude oil TAN from 2 to <0.5 mg KOH/g with a methanol feedstock of 2 wt%. In addition, TPA/γ-Al_2O_3 can remain stable for up to 100 h of time-on-stream [40]. **Table 1** shows a compilation of scholarly works, where catalysts have been used for NA removal.

3.3 Supercritical fluids (SCF)

Improvements in chemical sciences and technology have impact in NA removal with the advent of employing SCF especially supercritical water (SCW) and supercritical methanol (SC-MeOH). The SCF by reason of definition are fluids that exist at temperatures and pressures beyond their critical points. They have been favored lately in NAs removal due to the global calls for greener processes that would mitigate environmentally toxic materials. The SC-MeOH has been reported to relieve refinery processes off the burden of corrosion issues associated with NAs in heavy crude oils. In conducted pilot experiments [45], various factors such as NA/SC-MeOH ratio, temperature, time, and pressure were studied to see their influence on TAN reduction in Laguna and Bachaquero-13 heavy crude oil. Among the process parameters studied, temperature was the key determinant on the TAN reduction as a rise from 300 to 400°C lead to a TAN decline from 56 to 96.9%. Other parameters such as pressure and NA/SC-MeOH ratio had little impact on TAN reduction. Hence, the use of the lowest test pressure of 10 MPa and highest NA/SC-MeOH ratio at 400°C. The aforementioned conditions are of a good report as it allows for the process designed to handle TAN reduction operation to be made on a low-cost budget.

Model oil	Catalyst	Reaction	Total acid conversion (%)	Reference
China crude	$SnO—Al_2O_3$	Esterification	82 (TAN before/after: 2.8/<0.5 mg KOH g^{-1})	[10]
Cyclo pentane carboxylic acid, cyclo hexane carboxylic acid, cyclo hexane propionic acid, and trans-4-pentyl cyclo hexane carboxylic acid dissolved in dodecane	MgO	Decarboxylation	~100	[41]
Bohai oil	NiAl-4-EG LDH	Esterification	~90	[42]
Bitumen derived heavy vacuum gas oil (HVGO)	ZnO	Decarboxylation	8.5	[43]
Heavy crude from petronas penapisan melaka	MgO	Neutralization	84.8	[44]

Table 1.
The model oil, catalyst type, reaction type, total acid conversion and the referenced work.

4. Conclusions

Selections of extractants and catalysts to be used for the removal of NAs from heavy crude oil whether they are to be used in the physical methods or the chemical methods are of utmost important. Calcination temperature, at which a catalyst is calcined, and the surface area of the catalyst were seen to have significant effects on its effectiveness in removing NAs from heavy crude oils. The higher the surface area of a catalyst, the more effective it is in its capacity to remove NAs from heavy crude oils. Also, catalysts, which are calcined at higher temperatures have been seen to exhibit high removal efficiency. A Cu/Ce (10:90)/Al_2O_3 catalyst calcined at 1000°C removed about 93% of NAs from Korean crude oil using catalytic deacidification method. Removal processes in which catalysts containing 10% Cu in them were used have been found to be effective in the removal of NAs from solutions. Metal oxides nanoparticles have also shown to be promising in the removal of NAs from solutions. Removal of NAs from acidic solutions dissolved in dodecane using decarboxylation process with calcined MgO as catalyst yielded about 100% removal of NAs from the acidic solutions. Although this was a modeled work, yet it looks promising if applied in real life system and heavy crude oils.

Author details

Feyisayo Victoria Adams[1,2]*, Chikaodili E. Chukwuneke[1] and Bolade O. Agboola[1]

1 Department of Petroleum Chemistry, American University of Nigeria, Yola, Nigeria

2 Faculty of Engineering and the Built Environment, Department of Metallurgy, School of Mining, Metallurgy and Chemical Engineering, University of Johannesburg, South Africa

*Address all correspondence to: feyikayo@gmail.com

References

[1] Mandal PC, Sasaki M. Total acid number reduction of naphthenic acids using supercritical fluid and ionic liquids. Recent Insights in Petroleum Science and Engineering. 2018:251-271. DOI: 10.5772/intechopen.71812. Available from: https://www.intechopen.com/books/recent-insights-in-petroleum-science-and-engineering/total-acid-number-reduction-of-naphthenic-acids-using-supercritical-fluid-and-ionic-liquids

[2] Khan M, Riaz A, Yi M, Kim J. Removal of naphthenic acids from high acid crude via esterification with methanol. Fuel Processing Technology. 2017;**165**:123-130

[3] Anderson K, Goodrich P, Hardacre C, Hussain A, Rooney DW. Removal of naphthenic acids from crude oil using amino acid ionic liquids. Fuel. 2013;**108**:715-722

[4] Duncum S, Osborne CG. Process for deacidifying a crude oil system. US Patent 6 464,859. 2002

[5] Blum S, Olmstead W, Bearden R. Thermal decomposition of naphthenic acids. US Patent 5 820,750. 1998

[6] Takemura Y, Nakamura A, Taguchi H, Ouchi K. Catalytic decarboxylation of benzoic acid. Industrial and Engineering Chemistry Product Research and Development. 1985;**24**(2):213-215

[7] Gillespie R, Arena BJ. Naphthenic acid removal as an adjunct to liquid hydrocarbon sweetening. US Patent 5 389,240. 1995

[8] Yan-zhao H, Jian-hua Z, Yan-yan W, Ben-Cheng W. Removal of naphthenic acids from high acid crude oil through catalytic esterification on zinc-aluminum hydrotalcite-like compounds. Journal of China University of Petroleum (Edition of Natural Science). 2010;**34**(6):147-156

[9] Sartori G, Savage DW, Blum SC, Dalrymple DC, Wales WE. Metal compounds as accelerators for petroleum acid esterification. US Patent. 1999

[10] Yan-zhen W, Jin-yun L, Xue-ying S, Hong-ling D, Chun-min S, Miao-miao Z. Removal of naphthenic acids from crude oils by fixed-bed catalytic esterification. Fuel. 2014;**116**:723-728

[11] Wang YZ, Liu YP, Liu CG. Removal of naphthenic acids of a second vacuum fraction by catalytic esterification. Petroleum Science and Technology. 2008;**26**(12):1424-1432

[12] Anderson K, Atkins MP, Goodrich P, Hardacre C, Hussain AS, Pilus R, et al. Naphthenic acid extraction and speciation from Doba crude oil using carbonate-based ionic liquids. Fuel. 2015;**146**:60-68

[13] Lu R, Xu X, Yang J, Gao J. Reduction of total acid number of crude oil and distillate. Energy Sources, Part A: Recovery, Utilization, and Environmental Effects. 2007;**29**(1):47-57

[14] Shi LJ, Shen BX, Wang GQ. Removal of naphthenic acids from Beijiang crude oil by forming ionic liquids. Energy & Fuels. 2008;**22**(6):4177-4181

[15] Zhang A, Ma Q, Wang K, Liu X, Shuler P, Tang Y. Naphthenic acid removal from crude oil through catalytic decarboxylation on magnesium oxide. Applied Catalysis A: General. 2006;**303**(1):103-109

[16] Kumar RB, Shinde SN, Gaikwad SG. Reactive extraction of naphthenic acid by using sodium hydroxide as an extractant.

International Journal of Advanced Engineering Technology. 2014;**5**: 103-106

[17] Gaikar VG, Maiti D. Adsorptive recovery of naphthenic acids using ion-exchange resins. Reactive and Functional Polymers. 1996;**31**(2):155-164

[18] Varadaraj R, Savage D, Wales WE. Removal of naphthenic acids in crude oils and distillates. US Patent 5 961,821. 1999

[19] Varadaraj R, Pugel T, Savage DW. Removal of naphthenic acids in crude oils and distillates. US Patent 6 096,196. 2000

[20] Najmuddin R, Mutalib M, Shah S, Suleman H, Lethesh K, Pilus R, et al. Liquid-liquid extraction of naphthenic acid using thiocyanate based ionic liquids. Procedia Engineering. 2016;**148**:662-670

[21] Shah SN, Pranesh M, Raj JJ, AbdulMutalib MI, Lethesh KC, Ghanem OB, et al. De-acidification of crude oil using supported ionic liquids phases. Separation and Purification Technology. 2018;**196**:96-105

[22] Ravilla UK, Banerjee T. Liquid-liquid equilibria of imidazolium based ionic liquid+ pyridine+ hydrocarbon at 298.15 K: Experiments and correlations. Fluid Phase Equilibria. 2012;**324**:17-27

[23] MacFarlane DR, Tachikawa N, Forsyth M, Pringle JM, Howlett PC, Elliott GD, Davis Jr. J.H, Watanabe M, Simon P, AngellC A. Energy applications of ionic liquids. Energy & Environmental Science 2014;**7**(1):232-250

[24] Shah SN, Ismail M, Ibrahim M, Mutalib A, Pilus RBM, Lethesh KC. Extraction and recovery of toxic acidic components from highly acidic oil using ionic liquids. Fuel. 2016;**181**:579-586

[25] Shah S, Lethesh KC, Mutalib MIA, Pilus RBM. Extraction and recovery of naphthenic acid from acidic oil using supported ionic liquid phases (SILPs). Chemical Product and Process Modeling. 2015;**10**(4):221-228

[26] Shah SN, Mutalib MIA, Pilus RBM, Lethesh KC. Extraction of naphthenic acid from highly acidic oil using hydroxide-based ionic liquids. Energy & Fuels. 2015;**29**(1):106-111

[27] Shah SN, Lethesh KC, Gonfa G, Ibrahim M, Mutalib A, Pilus RBM, et al. Extraction of naphthenic acid from highly acidic oil using phenolate based ionic liquids. Chemical Engineering Journal. 2016;**284**:487-493

[28] Coto B, Suárez I, Chirita M, Conde J, Giménez R, Rodriguez N, et al. Oil acidity reduction by extraction with [EMIM][EtSO4]: Experimental and model description. Separation and Purification Technology. 2019;**223**:234-242

[29] Shao X, Liu G, Yang J, Xu X. Research on the synthesis of ionic liquids/ layered double hydroxides intercalation composites and their application on removal of naphthenic acid from oil research on the synthesis of ionic liquids/ layered double hydroxides intercalation composites and their application on the removal of naphthenic acid from oil. Energy & Fuels. 2017;**31**(10):10718-10726

[30] Osman WSI, Dafaalla AA. Reduction total acid number (TAN) of some Sudanese crude oil by zeolite and clays. International Journal of Chemical and Petroleum Sciences (IJCPS). 2015;**1**(1):8-16

[31] Zhang A, Ma Q, Tang Y. Catalytic decarboxylation for naphthenic acid removal from crude oils: A theoretical and experimental study. In: Proceedings of the 227th ACS National Meeting; March 28–April 1 2004; Anaheim, CA. 2004. PETR 98

[32] Simonsen G, Strand M, Norrman J, Øye G. Amino-functionalized iron oxide nanoparticles designed for adsorption of naphthenic acids. Colloids and Surfaces A: Physicochemical and Engineering Aspects. 2019;**568**:147-156

[33] Wang Y, Liu Y, Liu C. Kinetics of the esterification of low-concentration naphthenic acids and methanol in oils with or without SnO as a catalyst. Energy & Fuels. 2008;**22**(4):2203-2206

[34] Quiroga-Becerra H, Mejía-Miranda C, Laverde-Cataño D, Hernández-López M, Gómez-Sánchez M. A kinetic study of esterification of naphthenic acids from a Colombian heavy crude oil. CT&F-Ciencia, Tecnología y Futuro. 2012;**4**(5):21-31

[35] Shohaimi NA, Azelee W, Bakar WA, Shukri NM, Shaidi AK. Extraction of naphthenic acids from acidic petroleum crude oil utilizing 2- methylimidazole with the aid of Ca/Al_2O_3 AND Ce/Al_2O_3 catalysts. e—Academia Special Issue TeMIC. 2019;**7**:47-57

[36] Shohaimi NAM, Jaafar J, Bakar WAWAA. Chemical technique for the treatment of naphthenic acid in various crude oils. In: Proceedings of the 24th Regional Symposium of Malaysia Analytical Sciences (SKAM 24); 21-23 November 2011; Langkawi, Kedah, Malaysia. p. 11

[37] Shukri N, Bakar W, Jaafar J, Majid ZM. Removal of naphthenic acids from high acidity Korean crude oil utilizing catalytic deacidification method. Journal of Industrial and Engineering Chemistry. 2015;**28**:110-116

[38] Wang YZ, Sun XY, Liu YP, Liu CG. Removal of naphthenic acids from a diesel fuel by esterification. Energy & Fuels. 2007;**21**(2):941-943

[39] Li X, Zhu J, Liu Q, Bencheng W. The removal of naphthenic acids from dewaxed VGO via esterification catalyzed by Mg–Al hydrotalcite. Fuel Processing Technology. 2013;**111**:68-77

[40] Rana B, Cho D, Cho K, Kim J-N. Total acid number (TAN) reduction of high acidic crude oil by catalytic esterification of naphthenic acids in fixed-bed continuous flow reactor. Fuel. 2018;**231**:271-280

[41] Oh H, Park J, Rhee Y, Kim J. Decarboxylation of naphthenic acid using alkaline earth metal oxide. Journal of Industrial and Engineering Chemistry. 2011;**17**(4):788-793

[42] Wang H, Liu X, Wu Y, Hou C, Qiu Y, Guo K. Microwave-assisted synthesis of ethylene glycol-intercalated NiAl LDHs and their application for intracrystalline catalytic esterIfication with naphthenic acids in crude oil. Energy & Fuels. 2017;**31**(9):9898-9904

[43] Ding L, Rahimi P, Hawkins R, Bhatt S, Shi Y. General naphthenic acid removal from heavy oils on alkaline earth-metal oxides and ZnO catalysts. Applied Catalysis A. 2009;**371**:121-130

[44] Akmar N, Shohaimi M, Azelee W, Abu W, Jaafar J, Shukri NM. Treatment of acidic petroleum crude oil utilizing catalytic neutralization technique of magnesium oxide catalyst. Modern Chemistry & Applications. 2013;**1**(3):1-5

[45] Khan MK, Insyani R, Lee J, Yi M, Lee JW, Kim J. Non-catalytic, supercritical methanol route for effective de-acidification of naphthenic acids. Fuel. 2016;**182**:650-659

Chapter 12

Nickel Tetraphenylporphine Extraction from Model Heavy Oil Using Ionic Liquids

Pradip Chandra Mandal

Abstract

Heavy metals containing compounds present in heavy oils create particular problems on their upgradation and refinement processes. The aim of this work is to extract nickel (II) tetraphenylporphine (NiTPP) from model heavy oils using ionic liquids (ILs). Exploring open literature, four different ILs such as 1-butyl-3-methylimidazolium bis(trifluoromethylsulfonyl)imide ([BMIM]TFSI), 1-ethyl-3-methylimidazolium octyl sulfate ([EMIM]OS), 1-butyl-3-methylimidazolium octyl sulfate ([BMIM]OS), and 1-methyl-3-octylimidazolium tetrafluoroborate ([OMIM] BF_4) were utilized in this study. Potential experiments were conducted in a batch-type extractor (design temperature of 250°C and pressure of 6 MPa) at temperatures of 0–180°C. The extracted samples were analyzed using UV-visible spectrophotometer to determine the concentration of metallic compounds. The results show that NiTPP extraction was increasing with increasing temperature and treatment time and approximately 63% NiTPP was extracted in [BMIM]OS at a temperature of 100°C and a treatment time of 40 min. It is also observed that NiTPP extraction was decreasing above temperature 100°C. The extraction kinetics followed first-order kinetics with respective activation energy and pre-exponential factor of 1.24 kcal per mol and 2.156×10^{-3} s^{-1}. Therefore, the IL, [BMIM]OS, is capable to extract NiTPP from model heavy oils at temperatures below 100°C indicating the effectiveness of this IL to extract NiTPP from heavy oils.

Keywords: heavy oil, heavy metal, ionic liquid, extraction kinetics, activation energy

1. Introduction

1.1 Heavy oil

Crude oil is a naturally occurring unrefined petroleum product that contains hydrocarbons 50–97%, nitrogen, oxygen, and sulfur-containing organic compounds which typically make up between 6 and 10%, while metals such as copper, nickel, vanadium, and iron account for less than 1% of the total composition. According to Koliander [1], some of the metals found originated from the crude itself, while others were added during production and transportation and migration of metals into the crude oil due to corrosion of pipelines, and tasks could also explain the existence of heavy metal in heavy oil. According to the API gravity (degree API = 141.5/specific gravity = 131.5), crude oil can be classified as follows:

Light crude oil: API gravity higher than 31.1.
Medium crude oil: API gravity from 31.1 to 22.3.
Heavy crude oil: API gravity from 22.3 to 10.
Extra-heavy crude oil: API gravity below 10.

Heavy oil is a crude oil having a viscosity generally greater than 10 cP and gravity below 22.3° API [2]. It contains high-molecular-weight hydrocarbons having more than 15 carbon atoms in its chain [3] and hence has low hydrogen-to-carbon ratios. In addition, it has elevated levels of heteroatom-containing compounds such as sulfur, nitrogen, oxygen, and metals. Moreover, it is highly acidic. The presence of heavy metal in heavy oil has been identified throughout the years, and although it only accounts for 1% of the total composition, accumulation of these metals over a long period of time could hinder the upgrading and the refinery process.

1.2 Heavy metal removal processes

Heavy oil generally needs to undergo several processes before it can be sold to the consumers to meet their specification. Metal removal process is one of the processes involved in the refinery stage where the hydrocarbons are treated to remove heavy metal before delivering into the downstream process to increase its market value. The presence of heavy metals, such as nickel and vanadium in the heavy oil, could lead to serious problems during any of the heavy oil refinement and upgradation processes. Hence, oil and gas companies of the world are interested in finding ways to remove the heavy metals from the heavy oil. Some of the problems caused by these metallic compounds are discussed below:

a. *Environmental effects*: Vanadium is the most common metal found in heavy oil. It produces vanadium pentoxide when it burns during the combustion process [4, 5]. Vanadium pentoxide is a toxic material and is harmful to the environment. In addition, less volatile metals and metal compounds can be found in great amount in the ashes of heating oil, which also disperse into the environment which could lead to a serious environmental problem [1]. Due to the hazardous and carcinogenic nature of heavy metal, it could have caused serious health issues, such as nervous system damage, cancer, and, in extreme cases, death.

b. *Industrial damages*: When heavy metal-containing heavy oil is burned during combustion process, the existing heavy metals would be stored in ashes which will result in the corrosion of the container. If heavy oil is used in power generation, it can lead to the corrosion of the turbine which will eventually lead to operational failure. Apart from this, once crude oil is produced, it will then be transported to a place where it will be processed. However, corrosion in pipelines could occur if there is a presence of heavy metal. In addition, the effective diameter of the pipeline could also be reduced due to the deposition of metal-containing compounds in the pipeline. In the oil industry, removal of the heavy metal is necessary to prolong the lives of catalysts in the fluid catalytic cracking [1].

c. *Cost*: As a result of decreasing amount of light hydrocarbon worldwide, oil and gas companies have taken interest in developing heavy oil into useful products. Generally, oil and gas companies invest a huge amount of money especially in terms of CAPEX (capital expenditure) and OPEX (operating expenditure). The presence of heavy metals in the heavy oil however causes spoilage to the equipment, hence resulting in further increment of the operating and maintenance costs. Thus, removal of heavy metal is a vital stage in upgrading the heavy oil.

Nowadays, there are various methods available in removing heavy metals from the heavy oils and its fractions which can be divided into several categories: physical, chemical, electrochemical, and catalytic hydro-treatment. Some of the methods in the category are chemical precipitation, adsorption process, membrane filtration, electrodialysis, and photocatalysis. These heavy metal removal processes come with its pros and cons with some of it being high in terms of operational costs, and other methods such as photocatalysis have a long processing time with very limited applications to the industry. The advantages and disadvantages of conventional heavy metal removal processes are summarized in **Table 1**.

Removal of heavy metals especially nickel (Ni) and vanadium (V) has been investigated in the recent years as more and more companies have taken the interest in upgrading heavy crude oil. Researchers all over the world have developed many metal removal methods and have published their findings in open literature. Although some of the methods have been proved to be efficient, unfortunately others have not shown a significant result within a specified amount of time. Therefore, further studies are essential to investigate the best way on removing heavy metal present in heavy oil. **Table 2** shows the experiments that have been done in removing heavy metal present in heavy oil under different sets of operating conditions.

Limited research works on the removal of heavy metal in heavy oil using ionic liquid (IL) are available in open literature. Most of the written papers are focusing on the removal of metal in wastewater or industrial effluent rather than in heavy oil.

1.3 History of IL

Ionic liquids (ILs) are chemical compounds of ion structure which have the melting point below 100°C or even at room temperature [16]. An IL contains positively

Treatment method	Advantages	Disadvantages	References
Photocatalysis	Less harmful and toxic by-products Simultaneous removal of both metals and organic pollutants	Limited applications Long removal time	[6, 7]
Chemical precipitation	Low initial cost Simple operation	Sludge generation requires additional operational cost for its disposal	[8]
Adsorption	Low cost Wide pH range Simple operating conditions	Production of waste products requires additional operation cost Low selectivity	[9, 10]
Membrane filtration	High separation selectivity Small space requirement	Requires frequent membrane replacement due to membrane fouling	[8]
Electrodialysis	High separation selectivity	Increased operational cost due to membrane fouling and high energy consumption	[11]

Table 1.
Advantages and disadvantages of conventional heavy metal removal processes.

Experiment	Operating conditions	Removal	References
Microwave-assisted nickel and vanadium removal from crude oil	Microwave power = 600–800 W Irradiation time = 3 min Dosage methane sulfonic acid = 2 wt%	Ni = 76.8% V = 83.2%	[12]
Using supercritical water (SCW) to remove nickel from Ni-EP	T = 450–490°C P = 25–45 MPa Time = 180 min	Ni = 90.64%	[13]
Mixed solvent (SCW) and toluene without the addition of any catalyst to remove vanadium	T = 410–490°C P = 25MPa Time = 180 min	V = 80.26%	[4]
Removing nickel from NiTPP using SCW in the presence of toluene	T = 450–490°C P = 25–35 MPa Time = 90 min	Ni = 65.65%	[14]
Effect of cationic starch on removal of Ni and V from crude oils under microwave irradiation	Microwave power = 300 W Irradiation time = 5 min T = 60°C	Iran crude oil Ni = 50%, V = 76% Shengli crude oil Ni = 60%, V = 79%	[15]

Table 2.
Methods in removing heavy metal from heavy oil.

charged cation and negatively charged anion. The commonly used cations are based on ammonium, sulfonium, phosphonium, imidazolium, pyridinium, picolinium, pyrrolidium, etc. with different substitutes. The generally used anions are halogen (first-generation ILs), organic ([CH3COO]$^-$, $[CF_3CO_2]^-$, [PhCOO]$^-$, etc.) or inorganic ($[BF_4]^-$, $[PF_6]^-$, $[AlCl_4]^-$, etc.). Though the properties of ILs depend on mutual fit of cation and anion, size, geometry, and charge distribution, they well fit in the assumptions of green chemistry. In 1998, Anastas and Warner have formulated 12 principles of green chemistry [17]. ILs meet three pillars of these principles—safer solvents (principle no. 5), provision of energy efficiency (principle no. 6), and usage in catalytic reactions (principle no. 9) [16]. **Figure 1** represents a simplified model of an irregular system of ions present in the IL structure.

IL precursors are quaternary ammonium halide salts that are known from the 1890s. But these salts are widely used and tested in the twentieth century for different applications such as synthesis, disinfection, fabric softener, wood preservation, and so on. These salts are used as IL from 1996.

A low-melting organic salt has been observed as red oil in the mid-nineteenth century for the first time. This oil was produced as a by-product in the Friedel-Crafts reaction of benzene alkylation with aluminum chloride as a catalyst. The red oil known as chloroaluminate, σ-complex, was structured after the popularization of nuclear magnetic resonance technique as shown in **Figure 2**.

Gabriel and Weiner produced ethanolammonium nitrate with the melting point (MP) of 52°C in 1888. Paul Walden [18–20] in 1914 produced ethylammonium nitrate with the MP of 12°C. This compound is considered as the first IL by many researchers [20], and Walden is known as the father of ILs. But this compound has no use due to its high reactivity.

In 1934, US patent no. 1943176 has claimed that the cellulose has dissolved in 1-benzylpyridinium chloride, 1-ethylpyridinium chloride, and so on at the temperature above 100°C. In 1948, just after World War II, 1-ethylpyridinium bromide has appeared in another US patent [19] for the electrodeposition of aluminum

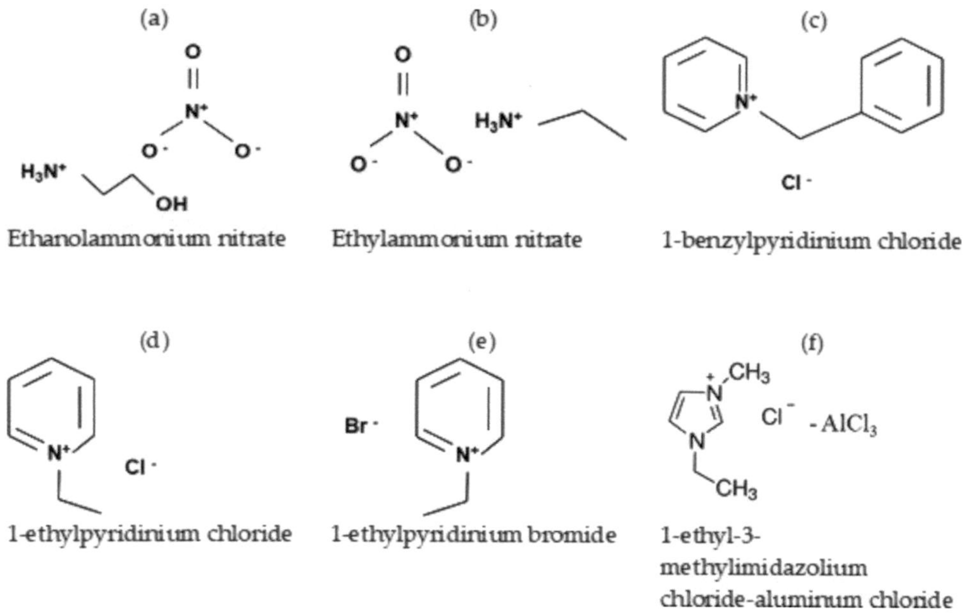

Figure 1.
Structure of historic ILs (ILs developed in the 20th century).

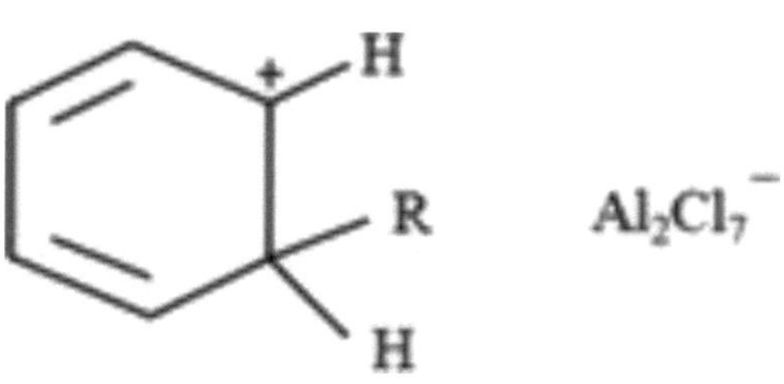

Figure 2.
Chloroaluminate, σ-complex, in the Friedel-Crafts reaction.

from aluminum (III) chloride. At that time, they were unable to investigate the IL because of the complex mixture of bromide and chloride salts. Later, in 1975, the physical and chemical properties of this IL were investigated by Osteryoung group aided by Bernard Gilbert [19]. The IL attracted the international researchers in the 1970s as aluminum chloride-based molten salts were used in the 1940s for the formation of nuclear warhead batteries. Wilkes in 1970 has tried to develop better batteries for nuclear warheads and space probes as they needed molten salts to operate [18]. Wilkes and Hussey have discovered an IL system which is liquid at room temperature known as 1-ethyl-3-methylimidazolium chloride-aluminum chloride ([EMIM]Cl-$AlCl_3$). This was the first appropriate example of IL system [19]. ILs emerged as solvents in the late 1990s.

ILs are interesting compounds due to their variable properties such as they are not all noncorrosive, but they can be designed to be noncorrosive [21]. Generally, ILs have negligible vapor pressure, high thermal and chemical stability, nonflammability, and numerous potential applications, such as microwave-assisted organic synthesis, catalysis, biocatalysis, separation, extraction, electrochemistry, nanomaterial synthesis, polymerization reactions, and corrosion inhibitors [22]. They have potential efforts to replace and optimize the existing petroleum-based technologies and processes.

This chapter explores the capability of ILs to extract NiTPP from model heavy oil that is not disclosed previously in open literatures.

2. Materials and methods

2.1 Materials

Based on literature survey, four specific ILs as stated in **Table 3** were utilized to extract nickel tetraphenylporphyrin (NiTPP) from model heavy oil. These ILs were purchased from Sigma-Aldrich and used without any modification. The heavy metal-containing compound, NiTPP, and the solvent, toluene, were also purchased from Sigma-Aldrich and used without further treatment.

2.2 Model heavy oil preparation

Model heavy oil was prepared by dissolving NiTPP in toluene. For producing 250 mL 0.06 M model heavy oil solution, exactly 0.015 g of NiTPP was taken in a beaker. Then required volume of toluene was added in the beaker at least three occasions at a volume of 50 mL in each occasion to ensure complete and stable mixing. The beaker was then placed on a hot plate, and the solution was stirred using a magnetic stirrer at a temperature of 50°C and a speed of 200 rpm. After the completion of mixing, the resultant solution was transferred into 250 mL volumetric flask, and toluene was added up to the mark to make a 250 mL solution.

(a)	(b)
Name: 1-butyl-3-methylimidazolium bis(trifluoromethylsulfonyl)imide Synonym: [BMIM]TFSI Empirical Formula: $C_{10}H_{15}F_6N_3O_4S_2$ Molecular weight: 419.36 g/mol Structural Formula: 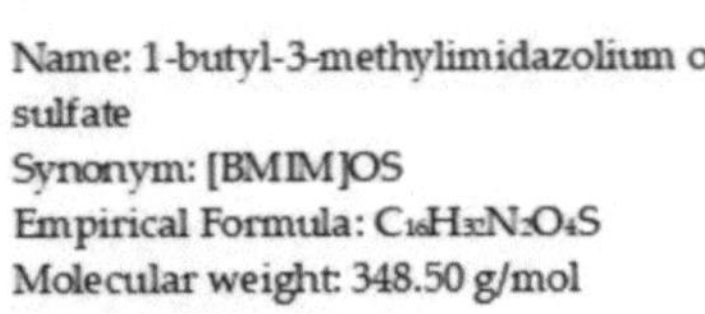 Signal word: Danger Target organ: Nervous system	Name: 1-ethyl-3-methylimidazolium octyl sulfate Synonym: [EMIM]OS Empirical Formula: $C_{14}H_{28}N_2O_4S$ Molecular weight: 320.45 g/mol Structural Formula: 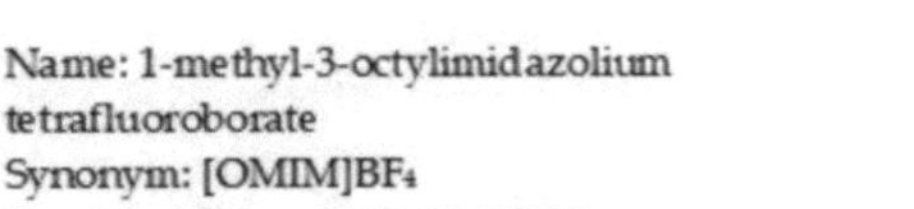 Melting point: -9 degC Density: 1.095 g/cc (20 degC)
(c)	**(d)**
Name: 1-butyl-3-methylimidazolium octyl sulfate Synonym: [BMIM]OS Empirical Formula: $C_{16}H_{32}N_2O_4S$ Molecular weight: 348.50 g/mol Structural Formula: 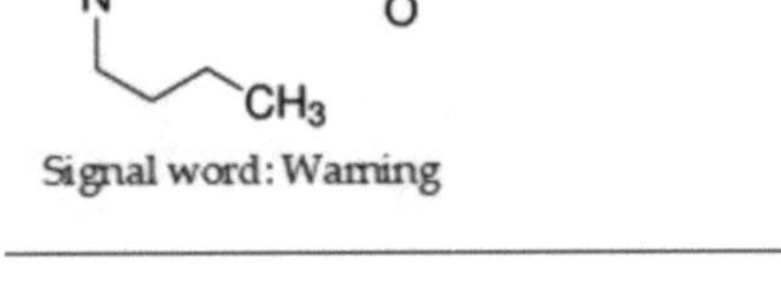 Signal word: Warning	Name: 1-methyl-3-octylimidazolium tetrafluoroborate Synonym: [OMIM]BF_4 Empirical Formula: $C_{12}H_{23}BF_4N_2$ Molecular weight: 282.13 g/mol Structural Fomula: 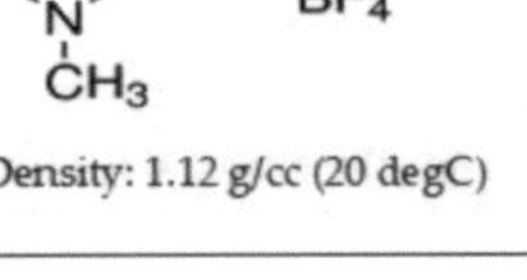 Density: 1.12 g/cc (20 degC)

Table 3.
General information of four used ILs.

2.3 Experimental method

The ILs and model heavy oil were taken in an autoclave reactor at a ratio of 1:10 (v/v). Then the reactor was loaded into an oven which was preheated to the planned temperatures of 0–180°C. The reactor was removed from the oven after a specific treatment time of 10–40 min. and kept under a water bath to cool the reactor up to room temperature. The samples were then collected into the separating funnel by washing the reactor interior at least three times with toluene. Two layers were formed after allowing sufficient time—the top is NiTPP solution portion, and the bottom is NiTPP-extracted IL portion. The top portion was separated and made ready for analysis.

2.4 Analytical method

Out of two layers, the separated top layer was analyzed using UV-visible spectrophotometer (UV-VIS), Perkin Elmer Lambda 26 model, to find out the concentration of NiTPP. A cuvette, size of 12.5 × 12.5 × 45 mm, was used to place the samples in the UV-VIS. Toluene is used as the reference solvent.

A calibration curve, a plot of the concentration of NiTPP vs. absorbance, was plotted using five known samples of NiTPP at the beginning of experiments. The best-fit curve was a straight line almost passing through the origin. After analyzing the treated samples, the concentration of NiTPP in the samples was calculated using the aforesaid calibration curve. The following term is used in this study.

NiTPP extraction was calculated using Eq. (1):

$$\text{NiTPP extraction (\%)} = \frac{C_0 - C_t}{C_0} \times 100 \quad (1)$$

where C_0 is the initial concentration of NiTPP in the toluene solution and C_t is the concentration of NiTPP in the toluene solution at treatment time, t.

The experiments were carried out three times. The error margin of the achieved data was less than 5 with 96% confidence level.

3. NiTPP extraction using ILs

3.1 Effect of ILs on NiTPP extraction

Four toluene (boiling point 110.6°C at 0.01 MPa and critical point 318.64°C at 4.109 MPa) insoluble imidazolium-based ILs such as [BMIM]TFSI, [EMIM]OS, [BMIM]OS, and [OMIM]BF_4 were utilized in this study to extract NiTPP from model heavy oil as they have a capability to extract some metals. Hildebrand's solubility parameter (HSP) of different solvents and ILs are displayed in **Table 4**. NiTPP is a purple solid crystal. Mandal et al. in 2011 [14] have disclosed that this compound is soluble in toluene (HSP = 18.3) and xylene (HSP = 18.20) but insoluble in water (HSP = 48.00) indicating that the HSP of this compound is close to 18.00. The toluene-soluble NiTPP was purple in color and had intensive absorption peaks in the visible region.

The IL, [BMIM]TFSI was not dissolved in toluene and water at room temperature. It was not capable of extracting NiTPP at this temperature. Hamidova et al. in 2015 [25] have disclosed that [BMIM]TFSI IL has a density of 1436.76 and 1385.21 kg/m^3 at 25 and 80°C. When NiTPP solution was treated with [BMIM]TFSI at a temperature of 75°C and a treatment time of 30 min, the color of [BMIM]TFSI did not change (**Figure 3a**) at all, indicating that it was not able to extract NiTPP

Name of solvent or ILs	Hildebrand's solubility parameter (HSP) ($MPa^{0.5}$) [23, 24]
[BMIM]TFSI	26.70 at 25°C
[EMIM]OS	30.20 at 25°C
[BMIM]OS	22.83 at 25°C, 23.25 at 55°C
[OMIM]BF_4	22.50 at 25°C
Toluene	18.30 at 25°C
Xylene	18.20 at 25°C
Water	48.00 at 25°C

Table 4.
HSP of different solvents and ILs.

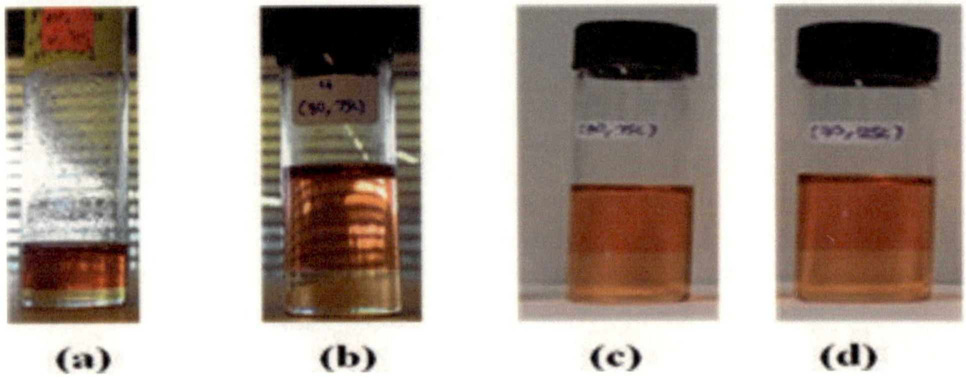

Figure 3.
Effect of ILs on NiTPP extraction at a temperature of 75°C and a treatment time of 30 min: NiTPP extraction using (a) [BMIM]TFSI, (b) [EMIM]OS, (c) [BMIM]OS and (d) [OMIM]BF4.

though the density of this IL was reduced with the increase of temperature. This was probably due to the unequal HSP of NiTPP and [BMIM]TFSI.

Similarly, the IL, [EMIM]OS (density = 1095 kg/m^3) was not dissolved in toluene at room temperature. It is lighter than [BMIM]TFSI. When NiTPP solution was treated with [EMIM]OS at a temperature of 75°C and a treatment time of 30 min., the color of [EMIM]OS slightly changed (**Figure 3b**) due to some of the NiTPP being extracted into this IL. The IL, [BMIM]OS, is insoluble in toluene. When the NiTPP solution was treated with [BMIM]OS at a temperature of 75°C and a treatment time of 30 min., the color of [BMIM]OS was darker than the color of [EMIM]OS (**Figure 3c**). The IL, [OMIM]BF_4 (density = 1120 kg/m^3), is insoluble in toluene at room temperature. When the NiTPP solution was treated with [OMIM]BF_4 at a temperature of 75°C and a treatment time of 30 min, the color of [OMIM]BF_4 was darker than the color of [EMIM]OS (**Figure 3d**) but very close to [BMIM]OS. This means that [BMIM]OS and [OMIM]BF_4 are more efficient in order to extract NiTPP from the NiTPP solution. For the next studies, [BMIM]OS is used due to its availability and cheapness.

3.2 Effect of temperature of [BMIM]OS on NiTPP extraction

Variation of NiTPP extraction in [BMIM]OS as a function of temperature is shown in **Figure 4**. It is observed that NiTPP extraction was increasing with temperature up to 100°C, and after that, NiTPP extraction was decreasing with increasing

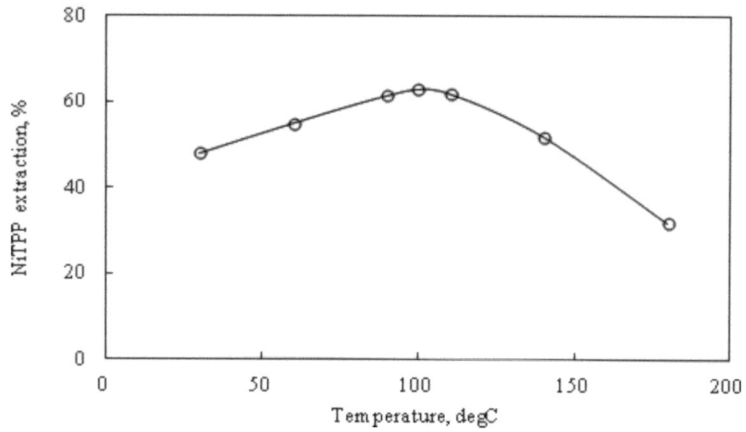

Figure 4.
Variation of NiTPP extraction in [BMIM]OS at a function of temperature.

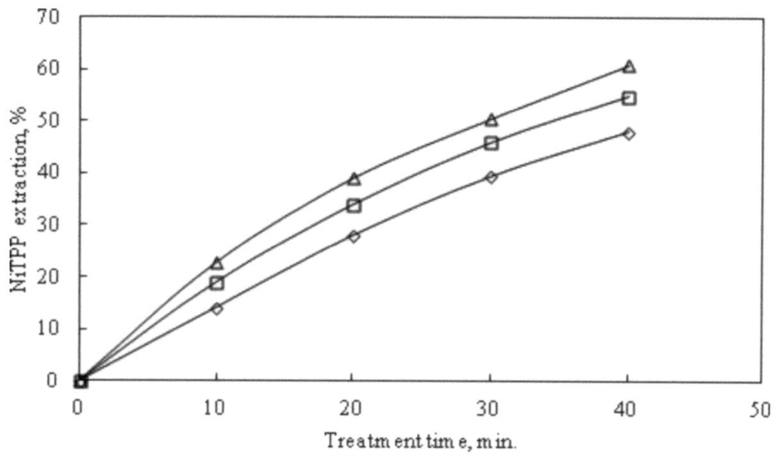

Figure 5.
Variation of NiTPP extraction at a function of temperature and treatment time (symbol: ◊, at temperature of 30°C; □, at temperature of 60°C; and Δ, at temperature of 90°C).

temperature. Maximum 63% NiTPP was extracted at a temperature of 100°C and a reaction time of 40 min. Thus, [BMIM]OS was effective to extract NiTPP at temperatures below 100°C. The reason is that viscosity of IL normally reduces gradually with the increase of temperature. Hence, NiTPP mixed well with IL by increasing the collisions between them. Such viscosity-temperature criteria no longer exist after a temperature exceeding 100°C for the case of IL [BMIM]OS because this IL decomposed at a temperature exceeding 100°C. Therefore, low temperature was favorable to extract NiTPP using this IL.

3.3 NiTPP extraction at temperatures below 100°C

Figure 5 shows that NiTPP extraction increased with increasing reaction temperature and time. Approximately 61% NiTPP was extracted at a temperature of 90°C and a treatment time of 40 min. One reason for this is that NiTPP and IL mix properly with increasing temperature, increasing their respective collisions events and thus increasing treatment rate. Mandal et al. in 2015 [26] have disclosed that IL molecules themselves could form cage structure through specific chemical bonds with the increase of treatment time. In this cage, the objective molecules would be captured through the formation of liquid clathrate due to the π-π interaction between ILs and objective molecules. This was the reason of increasing NiTPP extraction with the increase of treatment time.

3.4 NiTPP extraction kinetics

Extraction kinetics is important for good extractor design. To determine the extraction order, a kinetic plot of *ln(1-X)*, where *X* denotes fraction of NiTPP extraction, versus reaction time (*t*, min) is constructed (**Figure 6**). The NiTPP extraction was close to zero at zero treatment time. Each set of data gives a straight line obtained with the least square method that passes almost exactly through the origin indicating first-order kinetics with respect to NiTPP extraction, and it follows the kinetics equation [Eq. (2)]:

$$ln(1 - X) = kt \tag{2}$$

where k is rate constant in time^{-1}.

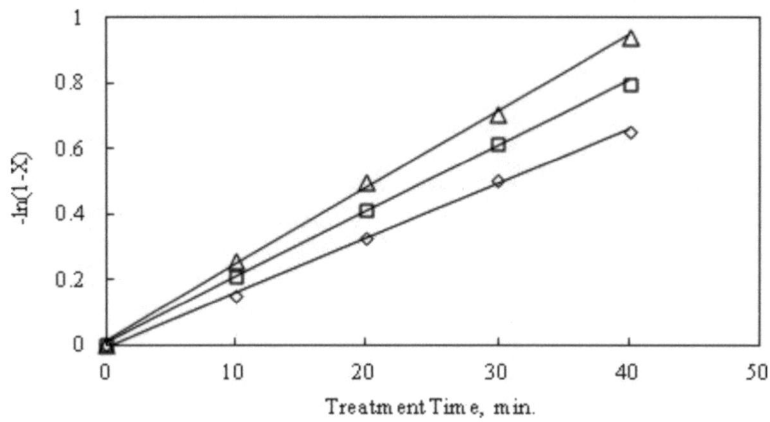

Figure 6.
NiTPP extraction kinetics plot (symbol: ◊, at temperature of 30°C; ◊, at temperature of 60°C; and Δ, at temperature of 90°C).

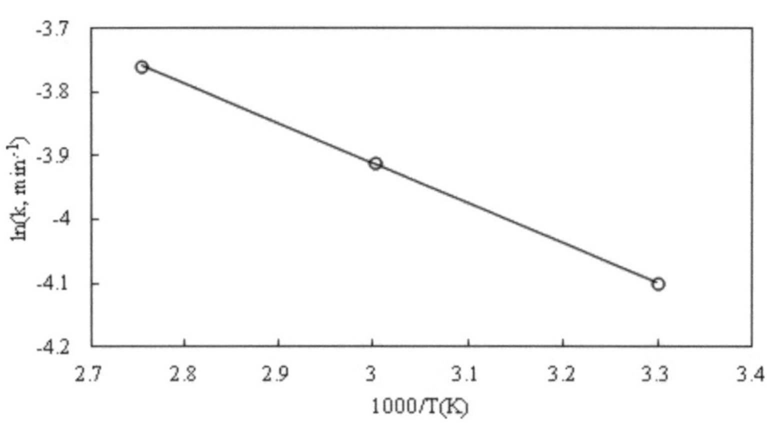

Figure 7.
Arrhenius plot.

Reaction kinetics analyses of this study were tested and evaluated to a maximum of 61% NiTPP extraction as kinetics data were more compatible with first-order kinetics at this extent [27]. From the slope of the straight lines of **Figure 6**, the respective rate constant was determined, and an Arrhenius-type temperature dependency acidity removal rate constant plot was then plotted (**Figure 7**). By exploring this plot, activation energy and pre-exponential factor of the said extraction process were estimated with respective values of 1.24 kcal per mol and $2.156 \times 10^{-3}\ s^{-1}$. Thus, the Arrhenius equation for NiTPP extraction can be rewritten as shown in Eq. (3):

$$k_{NiTPP\ extraction} = 2.156 \times 10^{-3} e^{\frac{-1.24}{RT}}, \text{temperature below } 100^{\circ}\text{C} \tag{3}$$

4. Conclusions

The IL-based metal extraction from heavy oil process is presented in this study. The IL, [BMIM]OS, is capable to extract NiTPP from model heavy oil, and approximately 63% NiTPP was extracted at a temperature of 100°C and a treatment time of 40 min. The extraction is dependent on temperature and treatment time at temperatures below 100°C. The time dependency extraction kinetics follows first-order kinetics with activation energy of 1.24 kcal per mol and pre-exponential factor of $2.156 \times 10^{-3}\ s^{-1}$. Therefore, IL-based metal extraction processes can attract the researcher's attention due to their greenness and separability tendency after using.

Acknowledgements

This work was supported by the Yayasan Universiti Teknologi Petronas (YUTP), Universiti Teknologi PETRONAS (UTP), Malaysia.

Author details

Pradip Chandra Mandal
Department of Pipeline Construction, Titas Gas T and D Co. Ltd, Dhaka, Bangladesh

*Address all correspondence to: pradipbd2002@yahoo.com

References

[1] Koliander W. The metal content of crude oils and its influence on crude oil processing. In: 16th World Petroleum Congress; 11-15 June 2000; Calgary, Canada. Vol. 2000. p. 1

[2] Faergestad IM. Heavy oil. In: Oilfield Review 2016; Schlumberger. 2016. Available from: https://www.slb.com/resources/oilfield_review/or_en_intro_article.aspx

[3] Trevisan OV, Lisboa ACL, França FA, Trindade W. Oil production in offshore fields: An overview of the Brazilian technology development program. In: World Heavy Oil Conference; 1-7, Pequim, China. 2006

[4] Mandal PC, Wahyu D, Sasaki M, Goto M. Non-catalytic vanadium removal from vanadyl etioporphyrin (VO-EP) using a mixed solvent of supercritical water and toluene: A kinetic study. Fuel. 2012;**92**(1):288-294. DOI: 10.1016/j.fuel.2011.07.002

[5] Mandal PC, Goto M, Sasaki M. Removal of nickel and vanadium from heavy oils using supercritical water. Journal of the Japan Petroleum Institute. 2014;**57**(1):18-28. DOI: 10.1627/jpi.57.18

[6] Barakat M. Removal of toxic cyanide and Cu(II) Ions from water by illuminated TiO_2 catalyst. Applied Catalysis B: Environmental. 2004;**53**(1):13-20. DOI: 10.1016/j.apcatb.2004.05.003

[7] Kajitvichyanukul P, Ananpattarachai J, Pongpom S. Sol–gel preparation and properties study of TiO_2 thin film for photocatalytic reduction of chromium(VI) in photocatalysis process. Science and Technology of Advanced Materials. 2005;**6**(3-4):352-358. DOI: 10.1016/j.stam.2005.02.014

[8] Kurniawan TA, Chan GY, Lo W, Babel S. Physico–chemical treatment techniques for wastewater laden with heavy metals. Chemical Engineering Journal. 2006;**118**(1-2):83-98. DOI: 10.1016/j. cej.2006.01.015

[9] Aklil A, Mouflih M, Sebti S. Removal of heavy metal ions from water by using calcined phosphate as a new adsorbent. Journal of Hazardous Materials. 2004;**112**(3):183-190. DOI: 10.1016/j.jhazmat.2004.05.018

[10] Babel S, Kurniawan TA. Various treatment technologies to remove arsenic and mercury from contaminated groundwater: An overview. In: Proceedings of the First International Symposium on Southeast Asian Water Environment. Bangkok, Thailand; 24-25 October, 2003. pp. 433-440

[11] Mohammadi T, Razmi A, Sadrzadeh M. Effect of operating parameters on Pb^{2+} separation from wastewater using electrodialysis. Desalination. 2004;**167**:379-385. Available from: http://www.desline.com/articoli/5618.pdf

[12] Shang H, Liu Y, Shi JC, Shi Q, Zhang WH. Microwave-assisted nickel and vanadium removal from crude oil. Fuel Processing Technology. 2016;**142**:250-257. DOI: 10.1016/j. fuproc.2015.09.033

[13] Mandal PC, Wahyu D, Sasaki M, Goto M. Nickel removal from nickel etioporphyrin (Ni-EP) using supercritical water in the absence of catalyst. Fuel Processing Technology. 2012;**104**:67-72. DOI: 10.1016/j.fuproc.2011.07.004

[14] Mandal PC, Wahyu D, Sasaki M, Goto M. Nickel removal from nickel-5,10,15,20-tetraphenylporphine using supercritical water in absence of catalyst: A basic study. Journal of Hazardous Materials. 2011;**187**(1-3):

600-603. DOI: 10.1016/j. jhazmat.2011.01.059

[15] Wang S, Yang J, Xu X. Effect of the cationic starch on removal of Ni and V from crude oils under microwave irradiation. Fuel. 2011;**90**:987-991. DOI: 10.1016/j. fuel.2010.11.036

[16] Pernak J, Rzemieniecki T, Materna K. Ionic liquids "in a nutshell" (history, properties and development). Chemik. 2016;**70**(9):471480. Available from: http://www.chemikinternational.com/wp-content/uploads/2016/10/9_16_Pernak_GB.pdf

[17] Anastas PT, Warner JC. Green Chemistry: Theory and Practice. Oxford, New York: Oxford University Press; 1998

[18] Keskin S, Kayrak-Talay D, Akman U, Hortacsu O. A review of ionic liquids towards supercritical fluid applications. Journal of Supercritical Fluids. 2007;**43**:150-180. DOI: 10.1016/j. supflu.2007.05.013

[19] Plechkova NV, Seddon KR. Application of ionic liquid in the chemical industry. Chemical Society Reviews. 2008;**37**:123-150. DOI: 10.1039/ B006677J

[20] Koel M. Ionic Liquid in Chemical Analysis. CRC Press, Tailor & Francis Group. Boca Raton, London, New York. ISBN: 1-42004646-2; 2008. p. XXVII

[21] Mandal PC, Sasaki M. Total acid number reduction of naphthenic acids using supercritical fluid and ionic liquids. In: Recent Insights in Petroleum Science and Engineering. Rijeka, Croatia: Intech; 2018. pp. 251-272. DOI: 10.5772/intechopen.71812

[22] Martinez-Palou R, Luque R. Applications of ionic liquids in the removal of contaminants from refinery feedstocks: an industrial perspective. Energy & Environmental Science. 2014;7:2414-2447. DOI: 10.1039/C3EE43837F

[23] Marciniak A. The solubility parameters of ionic liquids. International Journal of Molecular Sciences. 2010;**11**:1973-1990. DOI: 10.3390/ijms11051973

[24] Sistla YS, Jain I, Khanna A. Validation and prediction of solubility parameters of ionic liquids for CO2 capture. Separation and Purification Technology. 2012;**97**:51-64. DOI: 10.1016/j. seppur.2012.01.050

[25] Hamidova R, Kul I, Safarov J, Shahverdiyev A, Hassel E. Thermophysical properties of 1-butyl-3-methylimidazolium bis(trifluoromethylsulfonyl)imide at high temperatures and pressures. Brazilian Journal of Chemical Engineering. 2015;**32**(01):303316. Available from: http://www.scielo.br/scielo.php?pid=S010466322015000100303&script=sci_abstract

[26] Mandal PC, Abdalla MA, Moniruzzaman MD. Acidity reduction of naphthenic acid using imidazolium chloride based ionic liquids. International Journal of Applied Engineering Research. 2015;**10**(89):69-74. Available from: http://www.ripublication.com/ijaer.htm

[27] Mandal PC, Shiraishi T, Wahyu D, Sasaki M, Goto M. Kinetics and reaction pathways of heptylbenzene decomposition in supercritical water. Journal of Chemical Engineering of Japan. 2011;**44**(7):486-493. DOI: 10.1252/jcej.10we296

Chapter 13

Catalysts for Hydroprocessing of Heavy Oils and Petroleum Residues

Ching Thian Tye

Abstract

With the increasing demand of petroleum-derived products due to the world population and development, upgrading of crude oil with heavier quality and petroleum residues is unavoidable. Hydroprocessing is a preferable process for heavy oil upgrading. The process is operated with the presence of a catalyst, and catalysis plays an important role in the process. An overview regarding the catalyst design such as the catalyst active metal, active phase, support properties, and catalyst structure for heavy oil hydroprocessing is provided. There also include some recent advancements related to catalytic hydroprocessing of heavy oils and residue processes. Further catalyst performance improvement will likely come from catalyst optimization and better catalyst deactivation resistance resulting from metal poisoning and coke formation.

Keywords: heavy oil, residue, hydrocracking, hydroprocessing, mesoporous, macroporous

1. Introduction

The increasing world population and economic growths cause significant increases in oil demand. With the finite conventional fossil fuel reserves, the production based on alternative renewable sources and nonconventional oils such as heavy crude oils and tar sand bitumen is expected to increase. Heavy oil, as the general term for nonconventional heavy feedstock, is considered as an alternate suitable source for transportation fuels, energy, and petrochemicals to fulfill the requirements of modern civilization [1]. This indicates that processing of heavy oils in the petroleum refining industries is growing.

In general, heavy oil is defined as any liquid petroleum with an API gravity less than 20°. API gravity is an expanded density scale used widely in petroleum industry. Heavy oil with API gravity below 10.0°API sinks under the water as it is heavier compared with water (10.0°API), and it is termed extra-heavy oil [2]. Compared with conventional light oil, heavy oils have higher viscosity, higher density, and lower API gravity. The properties of heavy oils vary according to its origin and synthetic route.

The main issue with heavy crude oil is that it produces a lower percentage of gasoline and diesel fuels and a higher percentage of residue fraction when distilled by an oil refinery. In addition, heavy crude oil has more negative impacts to the environment than its light counterpart, and its refining requires the use of more advanced techniques. Petroleum residues are the heavy fraction remaining after distilling petroleum crudes at atmospheric pressure (atmospheric residue) or at reduced pressure of 25–100 mmHg (vacuum residue; boiling point >525°C). Petroleum residue has high average molecular weight (> 1000 Da number average for vacuum residue) and higher asphaltene content [3]. Asphaltenes are insoluble in n-alkanes such as n-pentane and n-heptane but soluble in benzene or toluene [4]. In addition, residues have high heteroatom content such as sulfur (~ 5 wt%), nitrogen, and heavy metal (i.e., vanadium and nickel > 100 ppm). These heteroatoms tend to poison catalysts and cause problems during processing and, therefore, need special processing in the refinery to remove them.

Petroleum residues can be converted into lighter oil or more value-added products using bottom-of-the-barrel conversion processes or residue upgrading processes [1]. No matter which type of process is used, a substantial fraction of residue molecules can be cracked off as fragments to become liquids in the transportation fuel and vacuum gas oil boiling ranges. However, due to the high asphaltene content, overly converted residues can push the selectivity toward the thermodynamically favored but lower valued products, namely, coke and hydrocarbon gases.

Hydrogen addition and carbon rejection are the two major approaches to upgrade petroleum residues into lighter oils in petroleum industry. Carbon rejection, such as coking process, produces a lot of cokes during the heavy oil upgrading processes which have low or no market value. In the past few decades, hydrogen addition approach especially hydroprocessing has gained prominence in heavy oil or vacuum residue upgrading processes [1]. This approach upgrades heavy feeds to distillate fractions and generates less coke.

Various hydroprocessing reactor technologies such as fixed-bed, ebullated-bed, or slurry-phase reactors are used to upgrade heavy residues [5]. The principles of these reactor operations are almost the same but differing with respect to some technical minutiae and tolerance of impurities [1, 6]. Typically, hydroprocessing of heavy oils is an upgrading process carried out at temperature around 380–480°C and hydrogen pressure at 80–270 bar with the presence of a catalyst [7]. H-Oil™ and LC-Fining are the processes designed to convert heavy residue to lighter fractions by hydroprocessing using an ebullated-bed reactor. Eni slurry technology (EST) and CanMet by Energy Research Laboratories, Canada, are examples that are using slurry-phase reactors.

In general, hydroprocessing involves hydrocracking and hydrotreating. The main objective in heavy feedstock upgrading is to convert heavier residue into lighter distillates in such a way to increase hydrogen-to-carbon (H/C) ratio of the product. Thus, hydrocracking reaction is the main concern. Hence, sometimes hydroprocessing of heavy oils is also called hydrocracking of heavy oils. During the process, larger molecules such as asphaltenes and resins are cracked and being converted to lighter liquid hydrocarbons such as gasoline, diesel, and gas oil. Meantime, various hydrotreating reactions such as hydrodesulfurization (HDS), hydrodenitrogenation (HDN), and hydrodemetalation (HDM) are also occurring during the process under the operating conditions. This leads to better product oil quality and contributes toward hydrotreating process in the later stage.

Hydroprocessing is the most versatile of the modern heavy feed conversion processes. The flexibility of the operating conditions with respect to both the feedstock and product separation has provided the most economical refinery balance relative to supply and demand [1]. Catalyst plays an important role attributed to this flexibility in hydroprocessing, in which upgrading the most problematic feeds such as vacuum residue is derived from heavy crudes, as well as the direct upgrading of the latter. During reactions, catalyst helps to reduce the activation

Reactor type	Catalyst size, mm [8, 9]	Catalyst in reactor, particle/cm^3 [9]
Fixed bed	~1.2–3.0	120
Ebullated bed	~0.8	250
Slurry system	~0.002	2.4×10^9

Table 1.
Typical catalyst properties for residue hydrocracking.

energies leading to higher yield and selectivity for the products in favor. Various factors influence the product selectivity; these include reaction operating conditions and catalyst properties such as active sites, shape, size, chemical compositions, etc.

Reaction operating conditions associate directly with the productivity. Heavy oil hydrocracking is typically carried out at a relatively higher temperature. Since the reaction operating conditions for different technologies are different, to ensure the desirable yield and selectivity, it is important to consider the characteristics of the feed, appropriate reactor system, and catalysts for the hydrocracking of vacuum residue [1].

For different types of reactor operation, the catalyst properties especially the catalyst particle sizes deployed also vary. The typical catalyst properties used in different types of reactors are given in **Table 1**.

Generally, supported sulfide catalysts containing group VIB and VIII metals especially cobalt, molybdenum, tungsten, or nickel are used in heavy oil upgrading processes. However, impurities present such as heavy metals in the residues decrease the catalytic activities [1, 10].

Establishing an understanding of the key catalyst properties that influence catalyst performance in hydrocracking of heavy residue is important. This provides a basis to develop a better-performance catalyst for heavy oil hydrocracking.

2. Catalyst design for hydroprocessing of residues

Hydrocracking reactions require a bifunction catalyst with high cracking and hydrogenation activities. The high cracking activity is provided by an acidic support, whereas the hydrogenation activity is provided by metals on the support. The acidic catalyst support, such as acid-treated clay, alumina, or silica-alumina, is used to promote cracking and support the metals such as nickel, tungsten, platinum, and palladium that supply the hydrogenation function. These highly acidic catalysts are sensitive to the feed with high-nitrogen compound, which easily neutralizes the acid sites. Therefore, catalyst used for heavy oil hydrocracking process is different from the normal distillate hydrocracking process. Further, heavy oil that is derived from different geographical areas possess different properties. Therefore, it is important to have a specific designed catalyst that is suitable to the targeted heavy oil hydrocracking. Here, a key challenge is to design and synthesize catalysts that have high activity for hydroprocessing. Main components in catalyst design for hydroprocessing of heavy oils are discussed in the following sections.

2.1 Transition metals for hydrocracking catalyst

Cracking of C—C bonds require a high amount of energy; therefore, hydrocracking reactions occur significantly only at high temperature (~400°C and above). The main function of catalysts is to provide hydrogenation-dehydrogenation reactions that lead to higher liquid hydrocarbon yield and suppress coke formation. Platinum-group precious metals such as platinum and palladium are the representative catalysts with excellent hydrogenation capacity, and they are also commonly used in many other reactions. However, these platinum-group precious metals showed weak resistance especially to sulfur [11]. Sulfur escapes and reacts with the catalyst during upgrading process as extra-heavy oil containing high concentration of sulfur [12]. Therefore, in order to cope with heavy residue, which is high in sulfur content and other impurities such as nitrogen, and heavy metals that can easily deactivate or poison the catalyst, platinum-group precious metals have not been used as catalysts for the upgrading of extra-heavy oil.

Mo, Ni, Co, and W are the common established transition metals used as the active ingredient in catalyst for hydrocracking of heavy residue. Sulfides of Mo and W are usually used, and Ni and Co are used in promoting species components in the petroleum industry. These metal sulfides not only promote relatively high hydrogenation but are also good in hydrotreating activities such as HDN, HDS, and HDM during heavy residue upgrading process. The abovementioned metal sulfides are favored as well in terms of their availability and the cost comparing to the noble metals such as platinum and palladium. There are also studies on using other promoters such as phosphorus (P), fluorine (F), and boron (B) on the catalysts [1].

The reactive metals are usually deposited on catalyst support by wet or incipient wetness impregnation methods. The impregnated catalysts are then calcined, reduced, and used for the hydrocracking reactions.

2.2 State of active phase

Sulfides of metals, e.g., MoS_2 and WS_2, have been used as the active phase in hydrocracking catalyst. NiMo and CoMo sulfides were identified to exhibit superior HDS activity [13, 14].

Sulfidation of supported oxidative metal catalysts is normally carried out by thermally reacting with sulfur compound prior to hydroprocessing in order to have the best catalyst performance. For dispersed catalyst, sulfidation of the catalyst can be done either ex situ or in situ. Most oil-soluble catalyst metal precursors, which are available in the oxide, can be sulfided in situ by thermally decomposing sulfur compound in the heavy residue upon the reaction severity and transformed from inactive oxides to sulfides during the upgrading process [15]. However, the complete sulfidation of active components can be challenging [16], and incomplete sulfidation reduces the catalyst performance to a certain degree. Dispersed metal oxide catalyst can also be presulfided ex situ by using H_2S as the sulfiding agent followed by dispersion in oils [14].

Sulfidation is a complex reaction. Strong interaction between metal and support can cause poor reducibility of the metal species, which eventually makes the sulfidation of the catalyst difficult and leads to poor activity [17]. In spite of similar activities for the hydrocracking reaction, W-based catalysts have received less attention than Mo catalysts. This is due to their lower sulfidation ability, on which the catalytic activation is strongly dependent [18].

Carbide catalysts are also being tested in heavy residue upgrading [19, 20]. Studies have shown that the electron distribution of the *d* orbitals and consequent catalytic activity of WC are similar to those of platinum, while the resistance to sulfur is higher than platinum [21]. Density functional theory (DFT) studies have shown that WC can act as a catalyst with a strong hydrogenation capability without being poisoned by sulfur [22]. However, there are problems with carbide catalysts such as nano-sized particles of WC are difficult to obtain because of the high temperature in the carburization process, and metal contamination occurs when WC in bulk is obtained through a ball milling process [20, 23].

2.3 Properties of support material—acidity and porous structure

Catalyst support plays an important role in a catalyst. Support provides space for the active metal to stay upon as well as acid sites for cracking activity. Large surface area of support is generally favorable for catalyst as it can be interpreted directly into more active sites available on the catalyst surface which normally lead to better catalyst performance. Porous structure support exhibits large surface area. Hence, other than the chemical composition of catalyst, textural properties of support such as size, shape, surface area, and porosity have a great influence especially for heterogeneous catalyst system.

2.3.1 Acidic properties of the support

Acid sites of the catalyst promote cracking activity. Materials such as alumina, crystalline zeolite, amorphous silica-alumina, and mixture of crystalline zeolite and amorphous oxides with high acidity (Lewis and Brønsted acid sites) are used as catalyst base or support in order to promote more cracking activity during reactions. These supports usually have a structure with numerous micropores (pore diameter < 2 nm), which provide high surface area and acidic properties [24].

In terms of chemistry at the molecular level, zeolites and amorphous silica-alumina share some similarities. However, zeolites, which have the crystalline structure, offer higher activities and better controlled selectivity than the amorphous materials. Typically, using zeolite-based catalysts in hydrocracking results in greater cracking activity owing to its greater acidity. In addition, the use of zeolite-based catalyst improves thermal or hydrothermal stability and resistance to nitrogen/sulfur compounds.

However, catalyst deactivation occurs quickly when these conventional supported catalysts are used in heavy residue hydroprocessing. This is due to pore blocking caused by relatively high concentrations of heteroatoms, metal poisoners, and asphaltenes commonly found in heavy fractions. This significantly reduces the availability and accessibility of the active sites at the catalyst surface in the pores by the reactant molecules. Hence, conventional supports are not suitable for upgrading of extra-heavy oils and vacuum residues. It is also believed that the catalyst support's intrinsic acid sites (e.g., γ-Al_2O_3) participate in sediment or coke formation during hydrocracking reaction [25]. To avoid pore blocking problem, this leads to the introduction of meso (2–50 nm)/macropores and high pore volume into the catalyst support for hydrocracking catalyst.

Therefore, greater attentions are paid to the properties of catalyst especially support materials such as the size of the particles, pore volume and pore size distribution, and the shape of the particles to maximize utilization of the catalyst.

2.3.2 Support material porosity and pore size distribution

High asphaltene content in heavy oils leads to a higher tendency to form coke during processing compared to conventional light crude oils. This is due to the accumulation of the coke precursors on both external and internal configurations of the catalyst. In addition, rapid deposition of metallic poisons narrows down the catalyst pore size, restricting the accessibility by reactant species and products. These may cause complete pore blockage. As a result, the catalyst activities and performance reduce markedly. Hence, typical good catalyst design for heavy oils upgrading processing must offer (1) accessibility to large asphaltene agglomerates in order to increase their diffusion and transformation and (2) catalytic active sites or support adsorption sites that are less favorable to sediment or coke formation [3].

Also, it is clear that the pore structure of the catalyst has a close relationship with the conversion, selectivity, and stability properties of the catalysts during hydrocracking reaction [26].

In order to overcome the difficulty of asphaltene diffusion through small pores, the development of the porous shaped support or catalyst structure with macropores (>100 nm) is rationalized. However, too many macropores (ca. above 30% of total pore volume) should not be created, since shaped particles could turn fragile [3]. Mesoporous alumina is a rigid porous material with a mutually interconnected or isolated network structure which has not only the characteristics of a crystalline phase of alumina but also the characteristics of a porous material [27]. Catalysts having pores between 7 and 20 nm diameter showed higher activities than those catalysts having pores between 3 and 7 nm which was reported for hydrocracking of Athabasca oil sand bitumen [28]. In order to retain the mechanical strength, stability, and accessibility to large asphaltene agglomerates of the catalyst during heavy residue hydrocracking reaction, the multimodal porous structures or hierarchical pore system is developed in hydrocracking catalyst. Hierarchical pore system in catalyst not only integrates different functionalities in the catalyst but also allows the realization of a multiple step reaction in one single catalyst in a cascade way without any separation processes.

Hydrotreating effect especially HDS of atmospheric residue is a well-established residuum upgrading process which requires catalysts designed to remove and accumulate metals and to desulfurize the feed [8]. A wide-pore NiMo/γ-alumina was tested for its performance in a vacuum residue hydroprocessing. The catalyst showed a remarkably high activity for HDM and asphaltene conversion reactions in the vacuum residue hydroprocessing [29].

Other than active metal and phase, pore size distribution has been found to affect the catalyst activity in hydroprocessing of vacuum residue. A hydroprocessing study using Kuwait vacuum residue as feedstock has shown that the effect of catalyst pore size was significantly different for different hydrotreating reactions [30]. Sulfided NiMo catalysts with unimodal and bimodal distributions with different proportions of meso- and macropores were prepared. For sulfur removal (HDS), a unimodal pore catalyst with maximum pore volume in the medium mesopore range (10–25 nm diameter) showed the highest activity. For HDM and HDN reactions, large-pore catalysts, having a major proportion of pore volume in 100–300 nm diameter pores, were found to be more effective.

2.4 Methods to create meso-/macro- and hierarchical pore system

γ-Al_2O_3 is the most commonly used support in catalysis for hydroprocessing. There are methods developed to prepare mesoporous and hierarchical texture support for hydrocracking catalyst. Two most important techniques for preparation of the material with the desired hierarchical texture are by hydrothermal and template methods.

2.4.1 Hydrothermal method

Hydrothermal method is an approach where the mixed solution (aluminum precursor, solvent, etc.) is poured into a sealed reactor. In the reactor, relatively high-temperature and high-pressure conditions promote the dissolution and recrystallization of poorly soluble or insoluble material. γ-Al_2O_3 is then produced by calcination of precipitated hydroxides [27, 31]. During the hydrothermal process, under the nonrestricted conditions, the crystal grows to its largest possible size, and its various characteristics such as shapes, high degree of crystallinity, uniform distribution, and lighter particle agglomeration are developed [32]. The hydrothermal conditions such as water temperature, pressure, and viscosity of the solution are closely related to the development of the crystal morphology. The same type of crystal can be produced with different morphologies under different hydrothermal conditions [27, 33].

Stanislaus et al. [29] studied the mechanism of pore widening in γ-alumina under hydrothermal conditions in the presence and absence of additives such as P, F, phenol, and acetic acid. The formation and growth of boehmite into large crystallites by rehydration of γ-alumina were found responsible for pore enlargement. Li et al. [34] synthesized alumina from aluminum ammonium sulfate, urea, and polyglycol 2000 by hydrothermal method. It was reported that mesopore alumina of different morphologies can be fabricated merely by adjusting hydrothermal temperature.

2.4.2 Template method

Template method is a technology which is widely applied in recent years. It is a synthetic method for controlling the structure, particle size, and morphology of materials through utilization of a template [27].

Mesoporous alumina is normally prepared via surfactant templates. Su et al. [35] used $Al_2(SO_4)_3$ and $NaAlO_2$ as aluminum precursors for preparing lamellate structure mesoporous alumina with crystalline framework walls in the presence of nonionic surfactant PEG6000. Well-crystallized mesoporous γ-alumina was formed after the samples were calcined at 600°C for 3 h. Mesoporous alumina prepared in this way exhibited a large surface area (279 m^2/g) and a very rich porosity with large mesopores, and both the pore volumes and the pore sizes increased with the addition of the surfactant in the precipitation process [35].

Zhou et al. [36] used a promising strategy to prepare well-defined and uniform pore structure (WDUPS) Al_2O_3 by coating specially treated SiO_2 opals with Al_2O_3. The pore size of the WDUPS Al_2O_3 was tailored in the range of meso- to macro-size by using microspheres of different sizes to fabricate the SiO_2 opal. When the SiO_2 opals were coated with optimal amount of Al_2O_3, the obtained WDUPS Al_2O_3 showed a uniform acid density.

Recently, another hard template method, which uses polystyrene (PS) microspheres, was used to prepare a range of hierarchically porous alumina samples. The alumina precursor was mixed either with dry powder of PS microspheres (dry) or PS water suspension (wet) during preparation before pelletizing and calcination. The "wet" method gives pellets with a narrower distribution of macropores in size and the greater material mechanical strength. The approach allows tuning textural parameters of hierarchically porous alumina samples in different ranges of specific surface, mesopore volume, and total pore volume [37].

In a recent review by Galadima and Muraza [38] on hierarchical zeolites in hydrocracking catalysts, in general, the generation of the mesoporous zeolite systems could be achieved either during the initial crystallization process or through post-synthesis treatment processes. In either case, the resulting materials could have hierarchical arrangement of pore system or unimodal system. Another strategy adopted is the application of hard templates for synthesizing hierarchical zeolite. For achieving an efficient mass transport and high catalytic performance, the design of hierarchically porous catalytic supports is an interesting and most effective strategy.

3. Catalyst structure—activity relations

The catalytic activity is related to the structure of the catalyst. Daage and Chianelli [39] proposed the "rim-edge" model of MoS_2 for hydrotreating, in which

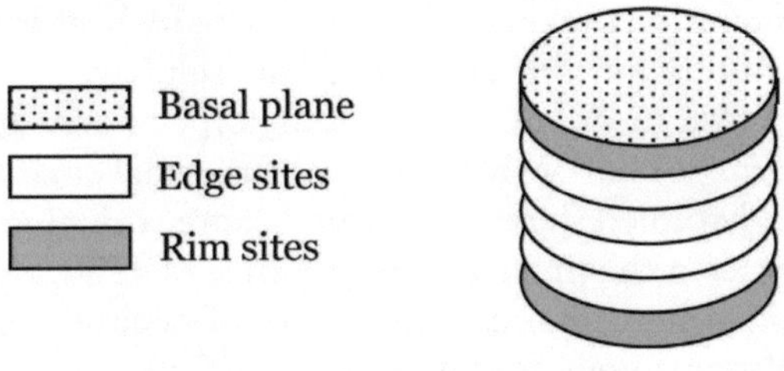

Figure 1.
Rim-edge model [39].

two kinds of exterior sites are responsible for hydrogenation (HYD) and hydrodesulfurization (HDS) (**Figure 1**). MoS_2 forms hexagonal or triangular slabs, and the rim sites around the exposed basal planes contribute to both HYD and HDS, and the exposed edge sites around the interior layers of MoS_2 stacks are accepted as active sites only in HDS.

The number of stacking layers of MoS_2 was then reported to increase the hydrogenation activity by promoting planar adsorption of aromatic rings on the edge sites [40]. Later, Lauritsen et al. [41] proposed the "brim site" model as the active site of HDS, based on observations of MoS_2 nanoparticles using scanning tunneling microscopy. The brim sites associated to Mo edges that are fully sulfided exhibit electron exchange capability as in typical metal catalysts.

In 2014, Chen et al. [42] compared the structure of hexagonal and triangular MoS_2 catalysts with their activities in hydrotreating. The hexagonal MoS_2 with more Mo edge sites was reported to have higher activity in HYD, but the triangular MoS_2 with more S-edge sites exhibited higher activity in HDS.

Similar catalyst structure-activity relations have also been found for W-based catalyst. In 2016, a DFT study that seeks to understand the active sites of "W-edge" or "S-edge" of NiWS phases showed that the surface concentration of "W-edge" and "S-edge" sites is closely related to the intrinsic hydrogenation activity of NiWS catalyst [43].

4. Recent advancement for heavy residue hydroprocessing

Low-quality heavy feedstock leads to quick catalyst deactivation, high coke formation, and large pressure drop in the operation of a fixed-bed reactor. In such a case, ebullated-bed or slurry-phase reactors can be an effective choice. These are systems designed with continuous addition and withdrawal of catalyst which allow one to control the catalyst deactivation due to coke formation as well as thermal exchanges in the reactor.

Supported catalysts are usually used in a fixed-bed or ebullated-bed reactor system for heavy oil hydroprocessing. On the other hand, nano-size unsupported dispersed catalysts are used in a slurry-phase reactor. In this case, the catalyst is in dispersed form such as unsupported metal particles, oil-soluble metals, or metal precursors that form in situ active phase during the process. Slurry-phase hydrocracking processes exhibit high selectivity and yield toward the product. For slurry-phase hydrocracking process of vacuum residue, both heterogeneous and homogeneous catalysts, depending upon the physical properties of the catalyst, are used. Catalyst development for heavy residue hydroprocessing is always ongoing for a better-performance catalyst and a longer catalyst run-length.

4.1 Supported solid catalyst

Carbon or carbon black has been tested as catalyst support for hydroprocessing. It is cheap and has a low tendency to form coke-related compounds (due to very weak acid sites). It has an affinity for V and Ni porphyrine-like compounds, which could be refrained from poisoning the catalyst by attacking the catalyst active sites. Meantime, it also has high HDS and HDN activities [44] and has no micropores below 3 nm which could be easily occluded, therefore turning active Ni and Mo metals into inactive ones [3]. However, carbon or carbon black supports are known for their low mechanical resistance since the structure of shaped catalyst particles easily collapses especially when used as catalyst in the ebullated-bed reactor [3]. To overcome the aforementioned mechanical resistance issue, a bimodal alumina modified with (8–18 wt%) carbon black composite support, containing 11–20% of total pore volume as macropores (>100 nm), was developed, and these prospective supports were impregnated with Ni and Mo. The addition of carbon black to alumina, after inert atmosphere pyrolysis, has produced strong shaped particles, suitable for high-colliding attrition applications, as commonly found in the ebullated-bed reactors. HDS and HDN can be enhanced when combining carbon black and macropores. It is also noted that catalysts containing macropores were reported more efficient in HDM than catalysts containing no macropores [3].

A series of mesoporous alumina catalysts with different textural properties were prepared by using the sol-gel method with the varied aluminum precursor, chain length of the surfactant, and calcination temperature. Mesoporous alumina with large surface area (270–380 m^2/g) and pore size (3–19 nm) was successfully synthesized. The activity of mesoporous alumina-supported molybdenum catalysts with different properties in hydrocracking of residual oil was determined at 400°C. Higher conversion and higher liquid yield were achieved by using catalysts with larger mesopore size than catalysts with smaller pore size and similar acidity [45].

Dong et al. [46] synthesized a nest-like hollow γ-AlOOH microspheres constructed from numerous hierarchically organized nanowires via a template-free simple hydrothermal approach, from which hierarchically porous γ-Al_2O_3 microspheres with average macropores of 900 nm, mesopores of 20 nm, and a pore volume of 0.93 cm^3/g were obtained readily. The nest-like γ-Al_2O_3 microspheres were tested as catalyst supports. The prepared MoNi/γ-Al_2O_3 (hierarchically pore texture)-supported catalyst was applied for HDM catalysis. The catalysts exhibited superior catalytic performance and longer life due to enhanced diffusion of the reactants, compared to the catalysts prepared with the commercial γ-Al_2O_3.

A theoretical estimation of catalyst performance using geometrical characteristics of the porous media based on Monte Carlo methods and the graph theory was done on evolution of alumina catalyst texture during macromolecule conversion in heavy oil hydroprocessing [47]. In this study, a unimodal mesoporous structure of conventional catalyst and bimodal meso-/macroporous structure of the catalyst were modeled. Deactivation was modeled by the monotonic increase of alumina grain radius, which represented deposition of coke and metal species onto the surface of grains. There was a correlation observed for the hierarchical texture model and the experiment. Both the hierarchical texture in model and experiment were with prolonged catalyst lifetime.

A series of NiMoS/γ-Al_2O_3 catalysts were prepared with varying Mo loading amounts, Ni addition, P addition, and porosity of γ-Al_2O_3 support, in order to design a suitable catalyst for hydrocracking of vacuum residue. Activity tests were conducted in an autoclave batch reactor at 420°C and 10 MPa H_2 in the presence of 0.26 wt% catalyst and 33.0 wt% tetralin. The amount of Mo loadings on γ-Al_2O_3 was observed most favorable at 8.0 wt% with the asphaltene conversion of 62.9%. The addition of Ni had minimal effect on the vacuum residue hydrocracking but significant effect on the HDS activity. The increase in porosity of the γ-Al_2O_3 support reported to play an important role in enhancing the asphaltene conversion to 68.5% [48].

These metal-supported catalysts exhibit good performance in HDM, HDS, and hydrocracking reactions. However, when upgrading extra-heavy oil using these metal-supported catalysts, a critical problem of a short lifetime can result from deactivation of the active sites due to coke or sulfur deposition [14]. Hydrocracking processes that use heterogeneous catalyst of fixed-bed, ebullated-bed, or slurry reactors are always with the problems of feed diffusion, pressure drop, and mass transfer. The intraparticle mass transfer between liquid and solid phases, particle size, and mixing matter which must be taken into consideration [49]. All these concerns make limitations to feedstock used for the commercial application. Application of dispersed catalyst in slurry-phase hydrocracking is an alternate option to resolve these issues.

4.2 Unsupported dispersed catalyst

Coke formation resistance is vital to a hydrocracking catalyst since it determines the durability and lifetime of the catalyst. It is known that coke formation is markedly inhibited when the active species are applied as dispersed nanoparticle catalysts [14]. The application of the highly dispersed catalysts in the slurry-phase hydroprocessing of heavy residues favors the rapid uptake of hydrogen and deactivates the intermediate free radical moieties in the liquid phase, thereby suppressing coke formation, increasing total conversion, and enhancing quality of liquid product [50].

Dispersed catalyst can be classified into either water-soluble or oil-soluble [50]. Oil-soluble dispersed catalyst is generally preferred because it has a better catalyst activity as it can disperse uniformly in oil. To prepare oil-soluble dispersed catalysts, metals are usually introduced into the oil-soluble precursors to form an organometallic compound [1]. The metal precursors are then homogeneously dispersed in the reactor containing heavy oil or residue. The catalyst precursor will be activated (sulfided) in situ by reacting with the sulfur compound in the oil. Typical oil-soluble dispersed catalysts are such as molybdenum naphthenate and ammonium heptamolybdate [51]. For a water-soluble catalyst, pretreatments such as dispersion, emulsion, and dehydration are needed [1].

Catalyst or catalyst precursor and its dispersibility in the oil during the hydroprocessing of vacuum residue have direct correlation to the system performance. Among the transition metals, Mo- and W-based dispersed catalysts have been the focus of the study in hydrocracking of extra-heavy oils or vacuum residue. For instance, exfoliated MoS_2 was applied as a dispersed catalyst for Cold Lake bitumen upgrading [52].

In [52], the exfoliated MoS_2, a two-dimensional MoS_2 material, was prepared via chemical exfoliation of MoS_2 particles that were dispersed in water. The results were compared with MoS_2 prepared in situ by the decomposition of molybdenum naphthenate. Although liquid yield and coke suppression were similar among those mentioned catalysts, a better hydrogenation activity, especially HDN, and asphaltene and microcarbon residue (MCR) removal were obtained with the exfoliated MoS_2 [52, 53]. The improved hydrogenation is believed to be a consequence of increased rim-edge sites [39] associated with the exfoliated MoS_2.

In the case of using tungsten, unsupported nickel tungsten sulfide (NiWS(x)) particles, where x is the actual molar ratio of Ni/W (x = 0, 0.005, 0.01, 0.02), were applied as a dispersed catalyst for hydroprocessing of a vacuum residue (2.3 °API) at 400°C with an initial H_2 pressure of 70 bar. Increasing the Ni content was found to increase the degree of sulfidation of tungsten, which promotes formation of Ni-W-S phases and enhances the overall catalytic activity. Among the NiWS(x) dispersed catalysts, the NiWS(0.02) catalyst showed the highest performance in total liquid product yield (87.0 wt.%), commercial fuel fraction yield (51.9 wt.%), and sulfur removal (86.5%) with coke formation (4.0 wt.%) suppressed efficiently [14].

Nano-sized tungsten carbide (2.8 nm) particles were synthesized using removable ceramic coating method and then were applied as dispersed catalysts to hydrocracking of vacuum residue, which was carried out at 400°C for 4 h with initial hydrogen pressure of 70 bar. The temperature programmed desorption analysis results showed that nano-sized tungsten particles are capable of superior hydrogen adsorption (compared to bulk tungsten carbide) as well as high catalytic performance, not only in commercial liquid yield (naphtha, middle distillate, and gas oil; 46.7 wt.%) but also in coke formation (5.9 wt.%) [20].

The effects of Co, Ni, or Mo precursors that have different oil solubility and oxidation state on the slurry-phase hydrocracking of vacuum residue at 400°C and 9.5 MPa H_2 were reported in [54]. The metal precursors were found to form nanoscaled dispersed particles of MoS_2, Co_9S_8, and Ni_3S_2, with the following vacuum hydrocracking performances in the order of $MoS_2 > Co_9S_8 \gg Ni_3S_2$, based on the same metal loading of 0.113 mmol. Among the oil-soluble Mo precursors used, Mo-hexacarbonyl, Mo-octoate, and Mo-naphthenate, Mo-octoate has an intermediate oxidation state forms the smallest particles of 5.8 nm in size and exhibits the best activity in the vacuum hydrocracking.

Shin et al. [55] tested six newly synthesized metal precursors as catalysts for hydrocracking of vacuum residue. New group VI transition metal complexes, (LM $(CO)_3$) (M = Mo or W, L1 = 3-phenyl-1-propyne, L2 = 4-phenyl-1-butyne, and L3 = 5-phenyl-1-pentyne) were prepared by simply stirring $M(CO)_3(RCN)_3$ (R = Me, Et) in tetrahydrofuran solution. The catalytic activity was compared to the commercially available material, Mo-octoate. $LMo(CO)_3$ showed similar activities to Mo-octoate. Among all, $L3Mo(CO)_3$ which had the longest alkyl chain showed a comparable result of low yield for coke and gas products [55].

Different morphologies of oil-dispersed MoS_2 catalysts were obtained by a ligand stabilization method using $Mo(CO)_6$ as a Mo precursor and trioctylphosphine oxide as a coordinating agent to identify the active site of MoS_2 in the hydrocracking of vacuum residue. It was observed that the MoS_2 forms a nanoscaled monolayer from 5 to 10 nm in size. The effect of the oil-dispersed MoS_2 catalysts with different morphology on the slurry-phase vacuum residue hydrocracking process was investigated at 400°C and 9.5 MPa H_2. The turnover frequency (TOF), activity based upon the H_2 consumption rate per lateral metal number, of the dispersed MoS_2 catalysts in hydrocracking shows a good correlation with the rim-site Mo dispersion of the MoS_2 slabs based on the same metal loading of 0.113 mmol [56].

An oil-soluble W precursor, $W(CO)_6$, was used to prepare WS_2 in situ in the hydrocracking of vacuum residue, and its activity was compared to that of oil-dispersed MoS_2 prepared using the Mo precursor. Vacuum residue was treated in a batch reactor at 419°C at 9.5 MPa H_2 with the metal loading of 0.113 mmol. The TOF shows a higher value for WS_2 (0.709 s^{-1}) than MoS_2 (0.573 s^{-1}) catalyst. A higher asphaltene conversion for WS_2 (39.2%) than MoS_2 (35.8%) was obtained. These results suggest that the WS_2 catalyst exhibits a higher intrinsic activity than the MoS_2 catalyst in the residue hydrocracking process. It was noted that the dispersibility of WS_2 (58.0%) was found better than MoS_2 (3.3%) [17].

Liu et al. [57] proposed a simple method for the preparation of presulfided oil-soluble MoS_2 catalysts. The catalyst precursor, cetyltrimethylammonium heptamolybdate (CTATTM), was synthesized by dissolving ammonium tetrathiomolybdate in an aqueous solution containing excess cetyltrimethylammonium bromide, which could be directly dissolved in a vacuum residue. The obtained catalyst was evaluated in slurry-phase hydrocracking of vacuum residue at 410°C with an initial H_2 pressure of 10 MPa for 1 h. Remarkable hydrocracking activity was observed, with a reduction of the resin content from 25.21 to 3.54 wt% and of the C7-asphaltene content from 6.82 wt.% to almost zero. With using the catalyst, the liquid yield increased from 75.03 to 96.43 wt.%, coke yield and gas yield were 0.19, and 3.38 wt.%, respectively, significantly lower than the values of 14.05 and 10.92 wt.%, respectively, obtained without using the catalyst. Additionally, with increasing Mo content, HDM conversion increased from 42.5 to 83.4 wt%, while HDS conversion increased from 21.6 to 59.2 wt%. The CTATTM precursor shows good solubility and stability in heavy oil, leading to easy formation of small MoS_2 particles with stacking numbers of 1 and 2 and slab lengths of 5–11 nm. This presulfided oil-soluble MoS_2 catalyst shows a great potential for further industrial applications [57].

5. Catalyst deactivation

Hydrocracking of heavy oils differs markedly from that of light feeds owing to the fact that the catalysts used for this process are deactivated fast due to the presence of asphaltenes and metal containing molecules [58]. Restraining the deactivation of the catalysts is one of the main challenges in development of hydrocracking catalysts for heavy oil upgrading regardless of the reactor system chosen. A better understanding of the catalyst deactivation is one of the most important aspects to improve the catalytic performance in heavy oil and petroleum residue refining processes. A good commercial catalyst is known by its activity, selectivity, and stability.

The main causes of loss of activity in the catalyst are due to the formation of coke and deposition of undesirable carbonaceous products, metallic compounds, and asphaltenes on the catalyst surface. In addition, the structural changes including thermally or attrition of catalyst are also responsible for the deactivation [1].

In industrial catalytic processes, the loss of catalytic activity and/or selectivity over time is a major and continuing concern. The degree of catalyst deactivation is greatly influenced by the feed properties and the associated operating conditions. Generally, performance of catalysts decreases with time. In order to maintain constant product yields and/or quality, the loss of catalytic activity is compensated by periodic increases of reaction temperature [7]. Though significant costs, in the magnitude of billions of dollars annually, to industry are due to catalyst replacement and process shutdown, catalysts activity decay over time is unavoidable [59].

Catalyst deactivation is a complex process. Various model equations are developed in order to model the deactivation mechanisms such as coke deposition and metal accumulation. In the model, agents affecting the catalyst activity including metal content and coke precursors need to be taken into consideration. To achieve that, detailed characterizations of spent catalysts obtained at different reaction conditions, time on stream, and reactor position need to be carried out through various experiments. With the aforementioned information, a better deactivation model for heavy oil hydrocracking catalyst can be developed and further applied in reactor design, simulation, and optimization [59].

6. Conclusion

The key developments in better catalyst understanding in association with particle size, pore size distribution, and activity grading have greatly improved the levels of conversion, increased run-length, and enhanced product qualities while maintaining residue fuel stability. Further catalyst performance improvement will likely come from catalyst optimization and better catalyst deactivation resistance resulting from metal poisoning and coke formation. The future challenge for a refiner will be to ensure a more efficient conversion of heavier feedstock to warrant a cleaner transportation fuel and sustainable environment for the generations to come.

Acknowledgements

Special thanks to Dr. Guan Tien Tan for his helpful comments, suggestions, and support while preparing the document.

Author details

Ching Thian Tye
School of Chemical Engineering, Universiti Sains Malaysia, Nibong Tebal, Pulau Pinang, Malaysia

*Address all correspondence to: chcttye@usm.my

References

[1] Sahu R, Song BJ, Im JS, Jeon Y-P, Lee CW. A review of recent advances in catalytic hydrocracking of heavy residues. Journal of Industrial and Engineering Chemistry. 2015;**27**:12-24. DOI: 10.1016/j.jiec.2015.01.011

[2] Speight JG. The Desulfurization of Heavy Oils and Residua. New York: Marcel Dekker Inc; 2000

[3] López-Salinas E, Espinosa JG, Hernández-Cortez JG, Sánchez-Valente J, Nagira J. Long-term evaluation of NiMo/alumina–carbon black composite catalysts in hydroconversion of Mexican 538°C+ vacuum residue. Catalysis Today. 2005;**109**:69-75. DOI: 10.1016/j.cattod.2005.08.011

[4] Gray MR. Upgrading Petroleum Residues and Heavy Oils. Vol. 22. New York: Marcel Dekker, Inc; 1994

[5] Cherzer J, Gruia AJ. Hydrocracking Science and Technology. New York: Marcel Dekker; 1996

[6] Furimsky E. Selection of catalysts and reactors for hydroprocessing. Applied Catalysis A: General. 1998; **171**(2):177-206. DOI: 10.1016/S0926-860X(98)00086-6

[7] Gary JH, Handwerk GE, Kaiser MJ. Petroleum Refining Technology and Economics. 5th ed. New York: CRC Press, Taylor & Francis Group; 2007

[8] Morel F, Kressmann S, Harle V, Kasztelan S. Processes and catalysts for hydrocracking of heavy oil and residues. In: Froment GF, Delmon B, Grange P, editors. Hydrotreatment and Hydrocracking of Oil Fractions. Amsterdam: Elsevier Science B.V; 1997

[9] Eser S. Uses of Hydrocracking. FSC 432 Petroleum Processing. 2018. Available from: https://www.e-education.psu.edu/fsc432/content/uses-hydrocracking [Accessed: August 1, 2019]

[10] Speight JG. New approaches to hydroprocessing. Catalysis Today. 2004;**98** (1-2):55-60. DOI: 10.1016/j.cattod.2004.07.053

[11] Dhandapani B, St Clair T, Oyama ST. Simultaneous hydrodesulfurization, hydrodeoxygenation, and hydrogenation with molybdenum carbide. Applied Catalysis A: General. 1998;**168**(2):219-228

[12] Li K, Akeredolu BA, Renehan AM, Yang Y, Batzle M, Evans RJ, et al. Correlation of chemical and physical properties of an Alaska heavy oil from the Ugnu formation. Fuel. 2013;**103**: 843-849. DOI: 10.1016/j.fuel.2012.07.037

[13] Hur YG, Kim M-S, Lee D-W, Kim S, Eom H-J, Jeong G, et al. Hydrocracking of vacuum residue into lighter fuel oils using nanosheet-structured WS_2 catalyst. Fuel. 2014;**137**:237-244. DOI: 10.1016/j.fuel.2014.07.094

[14] Hur YG, Lee D-W, Lee K-Y. Hydrocracking of vacuum residue using NiWS(x) dispersed catalysts. Fuel. 2016;**185**:794-803. DOI: 10.1016/j.fuel.2016.08.027

[15] Bellussi G, Rispoli G, Landoni A, Millini R, Molinari D, Montanari E, et al. Hydroconversion of heavy residues in slurry reactors: Developments and perspectives. Journal of Catalysis. 2013; **308**:189-200. DOI: 10.1016/j.jcat.2013.07.002

[16] Okamoto Y, Hioka K, Arakawa K, Fujikawa T, Ebihara T, Kubota T. Effect of sulfidation atmosphere on the hydrodesulfurization activity of SiO_2-supported Co- Mo sulfide catalysts: Local structure and intrinsic activity of the active sites. Journal of Catalysis. 2009;**268**:49-59. DOI: 10.1016/j.jcat.2009.08.017

[17] Jeong H-R, Lee Y-K. Comparison of unsupported WS_2 and MoS_2 catalysts for slurry phase hydrocracking of vacuum residue. Applied Catalysis A: General. 2019;**572**:90-96. DOI: 10.1016/j.apcata.2018.12.019

[18] Reinhoudt HR, Troost R, van Langeveld AD, van Veen JAR, Sie ST, Moulijn JA. The nature of the active phase in sulfided NiW/γ-Al_2O_3 in relation to its catalytic performance in hydrodesulfurization reactions. Journal of Catalysis. 2001;**203**:509-515. DOI: 10.1006/jcat.2001.3363

[19] Levy RB, Boudart M. Platinum-like behavior of tungsten carbide in surface catalysis. Science. 1973;**181**(4099): 547-549. DOI: 10.1126/science.181.4099.547

[20] Kim CH, Hur YG, Lee SH, Lee K-Y. Hydrocracking of vacuum residue using nano-dispersed tungsten carbide catalyst. Fuel. 2018a;**233**:200-206. DOI: 10.1016/j.fuel.2018.05.091

[21] Cheng Z, Zha S, Liu M. Stability of materials as candidates for sulfur-resistant anodes of solid oxide fuel cells. Journal of the Electrochemical Society. 2006;**153**(7): A1302. DOI: 10.1149/1.2198107

[22] Xi Y, Huang L, Forrey RC, Cheng H. Interactions between hydrogen and tungsten carbide: A first principles study. RSC Advances. 2014;**4**(75): 39912-39919. DOI: 10.1039/C4RA06225F

[23] Hunt ST, Nimmanwudipong T, Roman-Leshkov Y. Engineering non-sintered, metal-terminated tungsten carbide nanoparticles for catalysis. Angewandte Chemie International Edition in English. 2014;**53**(20): 5131-5136. DOI: 10.1002/anie.201400294

[24] Rana MS, Sámano V, Ancheyta J, Diaz JAI. A review of recent advances on process technologies for upgrading of heavy oils and residua. Fuel. 2007; **86**(9):1216-1231. DOI: 10.1016/j.fuel.2006.08.004

[25] Cornelius EB, Milliken TH, Mills GA, Oblad AG. Surface strain In oxide catalysts-alumina. The Journal of Physical Chemistry. 1955;**59**(9):809-813

[26] Nasser GA, Kurniawan T, Tago T, Bakare IA, Taniguchi T, Nakasaka Y, et al. Cracking of n-hexane over hierarchical MOR zeolites derived from natural minerals. Journal of Taiwan Institute of Chemical Engineering. 2016; **61**:20-25. DOI: 10.1016/j.jtice.2015.11.025

[27] Xie Y, Kocaefe D, Kocaefe Y, Cheng J, Liu W. The effect of novel synthetic methods and parameters control on morphology of nano-alumina particles. Nanoscale Research Letters. 2016;**11**:259. DOI: 10.1186/s11671-016-1472-z

[28] Sekhar MVC. Pore structure engineered catalysts for hydrocracking heavy feeds. Studies in Surface Science and Catalysis. 1988;**38**:383-392. DOI: 10.1016/S0167-2991(09)60671-6

[29] Stanislaus AM, Al-Dolama K, Absi-Halabi M. Preparation of a large pore alumina-based HDM catalyst by hydrothermal treatment and studies on pore enlargement mechanism. Molecular Catalysis. 2002;**181**(1):33-39. DOI: 10.1016/S1381-1169(01)00353-3

[30] Absi-Halabi M, Stanislaus A, Al-Mughni T, Khan S, Qamra A. Hydroprocessing of vacuum residues: Relation between catalyst activity, deactivation and pore size distribution. Fuel. 1995;**74**(8):1211-1215. DOI: 10.1016/0016-2361(94)00042-P

[31] Busca G. Catalytic materials based on silica and alumina: Structural features and generation of surface acidity. Progress in Materials Science. 2019;**104**:215-249. DOI: 10.1016/j.pmatsci.2019.04.003

[32] Zhang Y, Wang YF, Yan YH. Development and application of hydrothermal method in growing low-dimensional artificial crystal. Bulletin of the Chinese Ceramic Society. 2002;**3**: 22-26

[33] Hu CC, Wu YT, Chang KH. Low-temperature hydrothermal synthesis of Mn3O4 and MnOOH single crystals: Determinant influence of oxidants. Chemistry of Materials. 2008;**20**(9): 2890-2894. DOI: 10.1021/cm703245k

[34] Li YH, Peng C, Zhao W, Bai MM, Rao PG. Morphology evolution in hydrothermal synthesis of mesoporous alumina. Journal of Inorganic Materials. 2014;**29**(10):1115-1120

[35] A-p S, Zhou Y, Yao Y-h, C-m Y, Du H. A facile rout to synthesis lamellate structure mesoporous alumina using polyethylene glycol 6000 (PEG, molecular weight = 6000) as structure directing agent. Microporous and Mesoporous Materials. 2012;**159**:36-41. DOI: 10.1016/j.micromeso.2012.04.002

[36] Zhou Z, Chen S-L, Hua D, Wang Z-G, Chen A-C, Wang W-H. Tailored ordered porous alumina with well-defined and uniform pore-structure. Chemical Engineering Journal. 2013;**223**:670-677. DOI: 10.1016/j.cej.2013.01.024

[37] Parkhomchuk EV, Fedotov KV, Semeykina VS, Lysikov AI. Polystyrene microsphere-template method for textural design of alumina – an effective catalyst support for macromolecule conversion. Catalysis Today. 2019. DOI: 10.1016/j.cattod.2019.07.027. (In Press)

[38] Galadima A, Muraza O. Hydrocracking catalysts based on hierarchical zeolites: A recent progress. Journal of Industrial and Engineering Chemistry. 2018;**61**:265-280. DOI: 10.1016/j.jiec.2017.12.024

[39] Daage M, Chianelli RR. Structure-function relations in molybdenum sulfide catalysts: The "rim-edge" model. Journal of Catalysis. 1994;**149**:414-427. DOI: 10.1006/jcat.1994.1308

[40] Hensen EJM, Kooyman PJ, Van der Meer Y, Van der Kraan AM, De Beer VHJ, Van Veen JAR, et al. The relation between morphology and hydrotreating activity for supported MoS_2 particles. Journal of Catalysis. 2001;**199**(2):224-235. DOI: 10.1006/jcat.2000.3158

[41] Lauritsen JV, Nyberg M, Nørskov JK, Clausen BS, Topsøe H. Hydrodesulfurization reaction pathways on MoS_2 nanoclusters revealed by scanning tunneling microscopy. Journal of Catalysis. 2004;**224**:94-106. DOI: 10.1016/j.jcat.2004.02.009

[42] Chen J, Maugé F, El Fallah J, Oliviero L. IR spectroscopy evidence of MoS_2 morphology change by citric acid addition on MoS_2/Al_2O_3 catalysts – A step forward to differentiate the reactivity of M-edge and S-edge. Journal of Catalysis. 2014;**320**:170-179. DOI: 10.1016/j.jcat.2014.10.005

[43] Alphazan T, Bonduelle-Skrzypczak A, Legens C, Boudene Z, Taleb A-L, Gay A-S, et al. Improved promoter effect in NiWS catalysts through a molecular approach and an optimized Ni edge decoration. Journal of Catalysis. 2016;**340**:60-65. DOI: 10.1016/j.jcat.2016.05.004

[44] Vít Z. Comparison of carbon- and a lumina-supported Mo, CoMo and NiMo catalysts in parallel hydrodenitrogena tion and hydrodesulphurization. Fuel. 1993;**72**(1):105-107. DOI: 10.1016/0016-2361(93)90383-D

[45] Looi PY, Mohamed AR, Tye CT. Hydrocracking of residual oil using molybdenum supported over mesoporous alumina as a catalyst. Chemical Engineering Journal. 2012;**181-182**: 717-724. DOI: 10.1016/j.cej.2011.12.080

[46] Dong Y, Chen Z, Xu Y, Yang L, Fang W, Yi X. Template-free synthesis

of hierarchical meso-macroporous γ-Al2O3 support: Superior hydrodemetallization performance. Fuel Processing Technology. 2017;**168**:65-73. DOI: 10.1016/j.fuproc.2017.08.034

[47] Semeykina VS, Malkovich EG, Bazaikin YV, Lysikov AI, Parkhomchuk EV. Optimal catalyst texture in macromolecule conversion: A computational and experimental study. Chemical Engineering Science. 2018; **188**:1-10. DOI: 10.1016/j.ces.2018.05.005

[48] Park H-B, Lee Y-K. Designing supported NiMoS2 catalysts for hydrocracking of vacuum residue. Fuel. 2019;**239**:1265-1273. DOI: 10.1016/j.fuel.2018.10.154

[49] Chen Y-W, Hsu W-C, Lin C-S, Kang B-C, Wu S-T, Leu L-J, et al. Hydrodesulfurization reactions of residual oils over CoMo/alumina-aluminum phosphate catalysts in a trickle bed reactor. Industrial and Engineering Chemistry Research. 1990; **29**:1830-1940

[50] Nguyen MT, Nguyen NT, Cho J, Park C, Park S, Jung J, et al. A review on the oil-soluble dispersed catalyst for slurry-phase hydrocracking of heavy oil. Journal of Industrial and Engineering Chemistry. 2016;**43**:1-12. DOI: 10.1016/j.jiec.2016. 07.057

[51] Tye CT, Smith KJ. Hydrodesulfurization of dibenzothiophene over exfoliated MoS_2 catalyst. Catalysis Today. 2006;**116**(4): 461-468. DOI: 10.1016/j.cattod.2006.06.028

[52] Tye CT, Smith KJ. Cold Lake bitumen upgrading using exfoliated MoS_2. Catalysis Letters. 2004;**95**(3-4): 203-209. DOI: 10.1023/B:CATL.0000027295. 76634.ac

[53] Tye CT, Smith KJ. Catalytic activity of exfoliated MoS2 in hydroprocessing reactions. Topics in Catalysis. 2006;**37** (2-4):129-135. DOI: 10.1007/s11244-006-0014-9

[54] Kim S-H, Kim K-D, Lee D, Lee Y-K. Structure and activity of dispersed Co, Ni, or Mo sulfides for slurry phase hydrocracking of vacuum residue. Journal of Catalysis. 2018b;**364**:131-140. DOI: 10.1016/j.jcat.2018.05.002

[55] Shin S, Lee JH, Jo YK, Nguyen MT, Park BK, Park S, et al. Synthesis of noble molybdenum and tungsten complexes for hydrocracking catalyst of heavy oil. Journal of Industrial and Engineering Chemistry. 2019;**72**:408-413. DOI: 10.1016/j.jiec.2018.12.043

[56] Kim K-D, Lee Y-K. Active phase of dispersed MoS_2 catalysts for slurry phase hydrocracking of vacuum residue. Journal of Catalysis. 2019;**369**:111-121. DOI: 10.1016/j.jcat.2018.10.013

[57] Liu B, Zhao K, Chai Y, Li Y, Liu D, Liu Y, et al. Slurry phase hydrocracking of vacuum residue in the presence of presulfided oil-soluble MoS2 catalyst. Fuel. 2019;**246**:133-140. DOI: 10.1016/j.fuel.2019.02.114

[58] Ahmed HS, Shaban SA, Menoufy MF, El Kady FY. Effect of catalyst deactivation on vacuum residue hydrocracking. Egyptian Journal of Petroleum. 2013;**22**:367-372

[59] Rodríguez E, Félix G, Ancheyta J, Trejo F. Modeling of hydrotreating catalyst deactivation for heavy oil hydrocarbons. Fuel. 2018;**225**:118-133. DOI: 10.1016/j.fuel.2018.02.085

Chapter 14

A Model-Based Investment Assessment for Heavy Oil Processing in the Petroleum Refining Industry

Cheng Seong Khor

Abstract

The need for heavy oil processing has increased in recent years worldwide, backed by higher demands for petroleum products in the face of declining light crude oil resources. The situation encourages refineries to focus more on maximizing the production of high-value outputs from this lower-value heavier feedstock. This study purports to assess heavy oil processing potential in the refining industry through model-based economic evaluation. We formulate a refinery model suitable for preliminary investment decision making, which considers various cost elements for a number of conventional commercial heavy oil processing technologies. The formulated model is applied to a case study on the worldwide potential for heavy oil processing. This chapter demonstrates the application of a model-based approach to perform or assist with investment assessment.

Keywords: refinery design, optimization model, linear programming, residue fluid catalytic cracking (RFCC), fluid coking

1. Introduction

Heavy crude oil upgrading has gained the interests of refineries as demand for petroleum products increased in the face of declining lighter crude oil resources. In today's market, there are abundant heavier crude oils in the market as compared to conventional lighter ones. However, multiple competing technologies exist with a wide range of product yields and energy (or utility) requirements to refine these heavy oil resources [1, 2].

Heavy crude oils contain high fractions of residue and are generally classified by the density measure of API gravity of less than 20. The residue requires additional upgrading processes to break the complex molecular structure in obtaining valuable products. Residue upgrading processes include several thermal and catalytic processes, which can be categorized as carbon rejection or hydrogen addition. Examples of carbon rejection processes are delayed coking (DCK), visbreaking (VB), fluid coking (FCK), and solvent deasphalting, while hydrogen addition technologies include fixed bed hydroprocessing (e.g., Hyvahl F) and ebullated bed hydroprocessing (e.g., LC Fining) [3].

Integrating these technologies into refinery systems requires a systems approach-based economic evaluation instead of relying on, for example, monovariable decision making such as solely based on attaining the highest product yields. The increasing demand for high-value petroleum products and declining for that of bottom distillate products encourage refinery to give more focus on maximizing the yields on heavier crudes. In addition to that, the price for heavy crude oils is generally lower than lighter crude oils. Installing and operating heavy oil upgrading technologies enable refineries to buy cheaper feedstock and still produce high-value marketable products [4].

There are available refinery optimization models of various complexities in terms of time and space scales, which give rise to different computational requirements as based on the purpose and activity. For high-level decision making, linear programming (LP) models are suitable when only preliminary results are needed [5–8]. Nonlinear and/or mixed-integer models have been proposed for detailed refinery design [9–12] and for operation management [13–15]. A recent review on refinery optimization advances, which encompass developments in both academic and industrial settings, is available in [16].

The present work attempts to contribute toward assessing heavy oil processing potential in the petroleum refining industry by adopting a model-based economic evaluation approach. Using product demands and crude oil feed properties as base data, a refinery model can be developed to evaluate potentially profitable technologies including those for residue oil upgrading. For this purpose, we formulate an optimization model suitable for a preliminary high-level investment decision making with an appropriate economic objective function and a set of constraints that consider a number of conventional commercial technologies. A case study using available current data on market conditions is illustrated to carry out the intended assessment. A secondary goal of the study is to demonstrate the use of a standard business productivity tool (such as an Excel spreadsheet) to conduct such an assessment.

2. Problem statement

We consider the following investment decision-making problem for heavy oil processing in refineries. Given the (a) fixed market demand for desired refinery products and their prices, (b) available process technologies and their cost structures and capacities, and (c) cost of crude oil (single type or mixtures) and their nominal product yields, we wish to determine the optimal process technologies or units and their indicative processing capacities (flow rates) by minimizing the total operating cost, which mainly consist of utility requirements on energy demand for processing operations.

3. Optimization model formulation

A refinery model suitable for preliminary investment decision making is posed as a linear optimization (LP) model. The model admits process parameters for heavy oil processing including raw material availabilities, nominal product yields of a number of representative commercial technologies, market demands and prices for main product streams, and global processing or product capacities besides various cost-related economic parameters. An optimum solution is determined as a point in the solution space, which minimizes an economic-based objective function that stipulates the total operating cost for all heavy oil processing technologies considered that is feasible in satisfying all the associated constraints encompassing the aforementioned economic parameters.

This model uses the following notations:

Parameters

$\mathrm{yd}_{i,u}$ yield of component i from unit u

oc_u operating cost of unit u

nc_i capacity expansion cost of component i

cp_u capacity of process unit u

dm_i product demand of component i

Variables

$F_{i,u}$ inlet flow rate of component i to unit u

Z_i new capacity flow rate of component i

A compact representation of the optimization model formulation is presented and explained as follows:

$$\text{Minimize} \sum_{i,u} \mathrm{oc}_u F_{i,u} + \sum_i \mathrm{nc}_i Z_i \tag{1}$$

$$\text{Subject to} \sum_u \mathrm{yd}_{i,u} F_{i,u} = 0, \forall i \in I \tag{2}$$

$$\sum_i \mathrm{yd}_{i,u} F_{i,u} = 0, \ \forall u \in U \tag{3}$$

$$\sum_i F_{i,u} \le \mathrm{cp}_u, \forall u \in U \tag{4}$$

$$Z_i + \sum_u F_{i,u} \ge \mathrm{dm}_i, \forall i \in I \tag{5}$$

$$F_{i,u}, Z_i \ge 0, \forall i \in I \tag{6}$$

where the minimizing objective function shown in Eq. 1 caters for operating cost oc_u, which consists of raw material cost on crude oils and utility cost of process units as based on their inlet flow rates as well as capacity expansion cost nc_i to meet market demands dm_i. Eq. 2 describes component balances for each material i using fixed yield coefficients $\mathrm{yd}_{i,u}$ (on mass basis), which render linear relation between the feed inputs and product outputs of unit u that are implicitly dependent on the unit's operating conditions. On the other hand, Eq. 3 represents the total material balances for each unit u. Eq. 4 ensures that the total inlet flows into unit u does not exceed its maximum capacity cp_u in determining the required processing level. Eq. 5 stipulates that total processing rates for material i meet or exceed its demand dm_iincluding a provision for new capacity Z_i (or alternatively available product imports) to cover market requirements. Equation 6 enforces nonnegative values for all the decision variables.

4. Case study

We consider a case study of assessing the worldwide potential for heavy oil processing in the downstream petroleum processing sector by applying the foregoing model. Economic model parameters are estimated based on commercial data available in the literature as cited for **Tables 1–4**. The raw material is assumed to be a vacuum residue stream available from a vacuum distillation unit or, alternatively, a vacuum rerun unit with comparable processing capacity.

Product	Price ($/kg)	Demand (kg/hour)	Reference
Dry gas	0.0078	6000	[17, 18]
Total LPG	0.0020	30,000	[18, 19]
Gasoline	0.0097	150,000	[20, 21]
Diesel	0.0039	121,000	[18, 22]
Gas oil	0.0031	40,000	[18, 22]
Coke	0.0027	60,000	[23]

Table 1.
Model economic parameters for products used in this chapter.

The products of each process technology are categorized according to their cut temperatures. Product yields of process technologies are typically given in volume percentages in the literature. To make use of mass conservation principle, we convert them to weight percentages by assuming fixed densities of the product and feed components. The weight-based yields are then normalized as listed in **Table 2** according to the process technologies for use as input-output constants in the process unit material balances described by Eqs. (2) and (3).

Table 1 gives the product economic parameters in terms of selling prices and market demands. The operating cost data for the heavy oil process technologies are summarized in **Tables 3** and **4**. Annual operating time is taken to be 8150 hours per year corresponding to an onstream factor of about 93% (0.9304).

The technologies considered in this case study (with their associated abbreviations in parentheses as used in **Tables 3** and **4**) are delayed coking (DCK); fluid coking (FCK); fluid catalytic cracking (FCC); visbreaking (VB); ebullated bed hydrocracking technology of LC Fining (LCF); Cherry-P (CP) and fluid thermal cracking (FTC) technologies; residual fluid catalytic cracking (RFCC) technologies of heavy oil treating (HOT), heavy oil cracking (HOC), and R2R (roughly stands for residue cracking with two-step regeneration); solvent deasphalting technology of MDS; and other residue hydrotreating and hydroconversion technologies of asphaltenic bottoms cracking (ABC) and Hyvahl F (HF).

Further, we consider several assumptions in representing operating requirements of these technologies. Solvent deasphalting operation depends on the solvent type. Visbreaking is a relatively inexpensive mild thermal cracking process that is assumed to generate steam on a net basis, which can be sold (i.e., negative steam cost), while its cooling utility uses air instead of water. Delayed coking requires a furnace to heat the feed stream for coke removal; thereby, it uses a large fuel quantity as compared to other technologies. Fluid coking is a catalytic operation which uses steam for heating and air or water for cooling. Heavy oil cracking (HOC) is similar to fluid catalytic cracking (FCC) with the capability to remove heat from the generator, which can be recovered to produce steam, thus contributing as revenue (i.e., negative steam cost). Hydrotreating (HDT) heavy oil consumes hydrogen in the reaction scheme to decrease carbon-to-hydrogen ratio, in which the model considers the worst-case operating requirements for cycle oil feed. Due to limited literature data, cost parameters for certain technologies are approximated to similar ones (e.g., LC fining to FCK).

We use the model to conduct a general assessment of the probable technologies required to meet heavy oil processing capacity globally. The result obtained is graphically summarized in **Figure 1**. The objective value on total annualized cost of heavy oil processing is found to be about 164.2 million US$ with total utility cost for the selected units determined to make up 80%, while that of raw material cost only

Product	Technology													
	DCK	FCK	FCC	VB (low[a])	VB (high[b])	LCF	CP	FTC	HOT	HOC	MDS	ABC	HF	R2R
Dry gas	0.8	1.1	1.7	0.0	0.0	0.0	0.0	0.0	0.0	0.0	0.0	0.0	0.0	0.0
Total LPG	12.5	11.0	15.0	2.0	2.3	2.6	0.0	0.0	0.0	12.4	0.0	0.0	0.0	0.0
Naphtha	16.3	19.2	42.1	5.5	6.2	12.0	6.6	16.5	13.0	53.7	0.0	7.5	0.0	63.0
Middle distillate	14.9	22.9	20.6	10.5	11.8	41.2	30.7	38.2	9.7	14.6	61.2	17.9	4.2	18.1
Gas oil	33.6	13.1	10.1	82.0	79.7	44.2	62.7	24.2	28.7	8.8	38.8	74.5	25.5	4.8
Coke	21.5	32.7	10.4	0.0	0.0	0.0	0.0	21.1	48.6	10.6	0.0	0.0	70.3	14.1

[a]*Low API feed.*
[b]*High API feed.*

Table 2.
Product yields (in normalized weight percentages) for process units used in this work [24–26].

Technology	Electricity (kWh/b)	HPS (lb/b)	LPS (lb/bl)	Fuel (kBtu/b)	CW (gpm/b/hour)	Cat (lb/b)	H_2 (ft^3/b)	Total ($/(year·(kg/hour)))
MDS	0.133	60	—	80	—	—	—	266.4
VB	0.033	-50	—	80	—	—	—	265.8
DCK	0.239	—	40	120	0.6	—	—	403.5
FCK	0.865	200	100	—	30	—	—	273.9
FCC	0.067	—	20	80	400	0.3	—	312.2
HOC	0.017	—	-80	80	—	0.25	—	267.0
HDT	0.093	—	—	24	400	—	900	128.3

Notes: HPS = HP steam, LPS = LP steam, CW = cooling water, Cat = catalyst, H_2 = hydrogen, unit b = barrel (0.136 barrel = 1000 kg), gpm = gallon/minute.

Table 3.
Utility requirements (base data) and cost for heavy oil process technologies [21].

Technology	Parameter							
	Basis	Operating cost ($·year^{-1}(kg/hour)$^{-1}$)						
		SDA	VB	DC	FCK	FCC	HOC	HDT
Electricity tariff	0.060 $/kWh	0.133	0.033	0.239	0.865	0.067	0.017	0.093
HP[1] steam cost	0.0045 $/lb	0.299	−0.249	0.200	1.330	0.100	−0.399	0.036
LP[2] steam cost	0.003 $/lb							
Fuel cost	3.000 $/kBtu	266.0	266.0	399.0	0.0	266.0	266.0	79.8
Cooling water cost	0.10 $/gal	0.000	0	3.990	3.325	44.336	0.000	44.34
Catalyst cost	5.00 $/lb	0.000	0.000	0.000	0.000	1.663	1.386	0.000
Hydrogen cost	0.004 $/ft^3	0.000	0.000	0.000	0.000	0.000	0.000	3.990
Total	—	266.4	265.8	403.5	273.9	312.2	267.0	312.2
Capacity (ton/hour)	—	173.3	141.6	130.8	130.8	288.8	173.3	288.8

[1]*High pressure.*
[2]*Low pressure.*

Table 4.
Operating cost parameters [21].

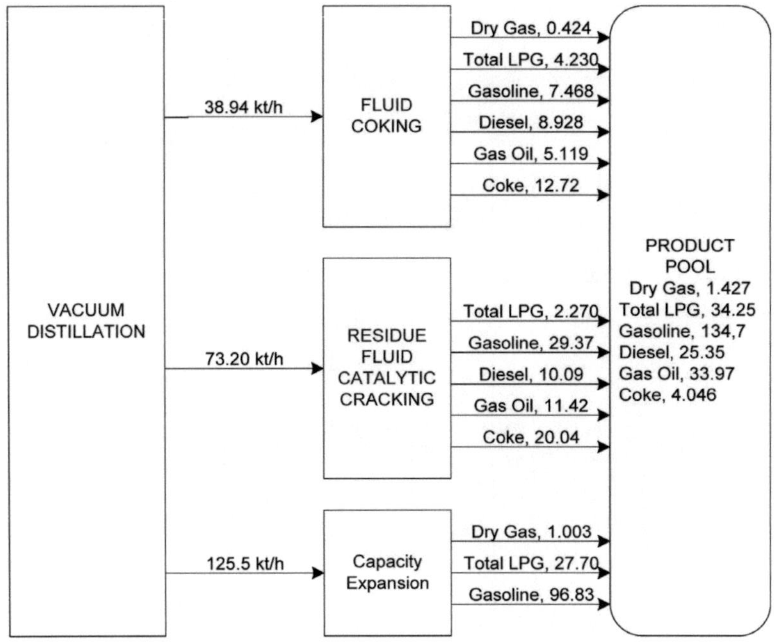

Figure 1.
Model solution for the case study (all flow rates in kiloton per hour).

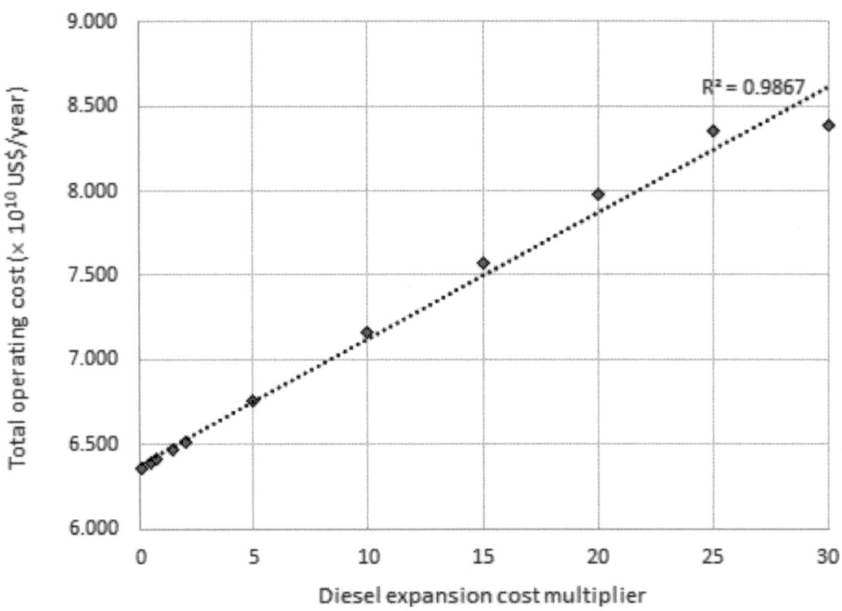

Figure 2.
Sensitivity analysis on effect of diesel capacity expansion cost on total operating cost.

17%. RFCC can account for nearly 31% of the available capacity, while the remaining can be met by a fluid coking technology. The potential RFCC technologies identified include HOC, HOT, and R2R [24]. It is also projected that there is demand for 53% of capacity expansion for heavy oil processing.

In general, RFCC can be designed compactly to produce high yields of valuable products with low maintenance cost as similar to the FCC technology that it is based upon. Fluid coker is reported to promote reactor heat transfer, which allows it to be operated at high temperature for high product yields with increased product separation into valuable products. It also uses burner operated with steam and air as utilities as opposed to an expensive fuel, which is reflected in its low operating cost [26, 27].

We conduct sensitivity analysis to examine how the model parameter values influence the solution. As an example, **Figure 2** shows the linear effect of varying the capacity expansion cost for diesel product output (in terms of a fixed multiplicative factor) on the total operating cost for heavy oil processing as is considered in our case study. Indeed, a trend of continuous high demand for distillate products (including diesel) necessitates correspondingly increased investment in the processing cost.

The model implemented in Excel (version for Microsoft Office 365) is freely available upon request from the author.

5. Concluding remarks

This chapter presents a model-based approach to conduct a preliminary assessment for investment decision making in heavy oil processing for refineries. The economic evaluation can be carried out using an Excel spreadsheet or other similar business productivity tools. The results provide an order of magnitude indication of refining capacity potential for this increasingly important resource in the hydrocarbon industry.

Author details

Cheng Seong Khor[1,2]

1 Chemical Engineering Department, Universiti Teknologi PETRONAS, Perak, Malaysia

2 Centre for Process Systems Engineering, Institute of Autonomous Systems, Universiti Teknologi PETRONAS, Perak, Malaysia

*Address all correspondence to: chengseong.khor@utp.edu.my; khorchengseong@gmail.com

References

[1] Frecon J, Le-Bars D, Rault J. Flexible upgrading of heavy feedstocks. In: Petroleum Technology Quarterly. Craven Arms, UK: Crambeth Allen Publishing; 2019. pp. 31-39

[2] Speight JG. Chapter 2: Refining processes. In: The Refinery of the Future. Boston: William Andrew Publishing; 2011. p. 57

[3] Ancheyta J, Speight JG. Hydroprocessing of Heavy Oils and Residua. Boca Raton, FL: CRC Press; 2007

[4] Elshout RV, Bailey J, Brown L, Nick P. Upgrading the Bottom of the Barrel, Hydrocarbon Processing. Houston, TX: Gulf Publishing Company; 2018

[5] Albahri TA, Al-Sharrah G, Khor CS, Elkamel A. Grassroots petroleum refinery configuration for heavy oil processing. Petroleum Science and Technology. 2019;**37**:275-281

[6] Hartmann JC. Interpreting LP outputs. Hydrocarbon Processing. 1999;**78**:64-68

[7] Leiras A, Ribas G, Hamacher S, Elkamel A. Tactical and operational planning of multirefinery networks under uncertainty: An iterative integration approach. Industrial & Engineering Chemistry Research. 2013;**52**:8507-8517

[8] Zhang J, Zhu XX, Towler GP. A simultaneous optimization strategy for overall integration in refinery planning. Industrial and Engineering Chemistry Research. 2001;**40**:2640-2653

[9] Albahri TA, Khor CS, Elsholkami M, Elkamel A. Optimal design of petroleum refinery configuration using a model-based mixed-integer programming approach with practical approximation. Industrial & Engineering Chemistry Research. 2018;**57**:7555-7565

[10] Albahri TA, Khor CS, Elsholkami M, Elkamel A. A mixed integer nonlinear programming approach for petroleum refinery topology optimisation. Chemical Engineering Research and Design. 2019;**143**:24-35

[11] Khor CS, Elkamel A. Superstructure optimization for oil refinery design. Petroleum Science and Technology. 2010;**28**:1457-1465

[12] Khor CS, Loh CY. Elkamel A. A logic-based mixed-integer superstructure optimization approach for the optimal design of petroleum refinery topology with environmental considerations, In: Bassett MH, Ierapetritou M, Pistikopoulos E, editors; 5th International Conference on Foundations of Computer-Aided Process Operations (FOCAPO); Cambridge, Massachusetts, USA; 2008. pp. 221-224

[13] Castillo Castillo P, Castro PM, Mahalec V. Global optimization algorithm for large-scale refinery planning models with bilinear terms. Industrial & Engineering Chemistry Research. 2017;**56**:530-548

[14] Ibrahim D, Jobson M, Li J, Guillén-Gosálbez G. Optimization-based design of crude oil distillation units using surrogate column models and a support vector machine. Chemical Engineering Research and Design. 2018;**134**:212-225

[15] Pinto JM, Joly M, Moro LFL. Planning and scheduling models for refinery operations. Computers & Chemical Engineering. 2000;**24**:2259-2276

[16] Khor CS, Varvarezos D. Petroleum refinery optimization. Optimization and Engineering. 2017;**18**:943-989

[17] Messick D. World sulphur outlook. 2019. Available from: http://www.firt.

org/sites/default/files/donmessick_sulphur_outlook.pdf [Accessed: 17 January 2019]

[18] Gary JH, Handwerk GE, Kaiser MJ. Petroleum Refining: Technology and Economics. 5th ed. New York: Marcel Dekker; 2007

[19] Temizer M. Global LPG demand to rise sharply in 10 years. 2019. Available from: https://www.aa.com.tr/en/energy/natural-gas/global-lpg-demand-to-rise-sharply-in-10-years/1334 [Accessed: 20 May 2019]

[20] Petrochemical Update. Global naphtha surplus forecast to grow; rising oil product exports split US-Mexico trade balance. 2019. Available from: http://analysis.petchem-update.com/engineering-and-construction/global-naphtha-surplus-forecast-grow-rising-oil-product-exports-split [Accessed: 21 January 2019]

[21] Maples RE. Petroleum Refinery Process Economics. 2nd ed. Oklahoma: Pennwell; 2000

[22] Fitzgibbon T, Ding C, Szabat P. Diesel Demand: Still Growing Globally Despite Dieselgate. 2019. Available from: https://www.mckinsey.com/industries/oil-and-gas/our-insights/petroleum-blog/diesel-demand-still-growing-globally-despite-dieselgate [Accessed: 15 January 2019]

[23] Savant A. Petroleum coke market report overview, global price trend, increasing demand, updated leading countries, industry analysis and future forecasts 2018-2025. 2019. Available from: https://www.reuters.com/brandfeatures/venture-capital/article?id=61034 [Accessed: 15 January 2019]

[24] Kamiya Y. Heavy Oil Processing Handbook. Japan: Research Association for Residual Oil Processing (RAROP); 1991

[25] Gray MR. Upgrading Petroleum Residues and Heavy Oils. New York: Marcel Dekker; 1994

[26] Speight JG, Özüm B. Petroleum Refining Processes. New York, Basel: Marcel Dekker; 2002

[27] Speight JG. Chapter 6: Catalytic cracking. In: The Refinery of the Future. Boston: William Andrew Publishing; 2011. pp. 181-208

Chapter 15

Environmental Challenges Associated with Processing of Heavy Crude Oils

Samuel O. Sojinu and Onome Ejeromedoghene

Abstract

The petroleum industry is one of the largest industries in the world and plays a pivotal part in driving a nation's economy. However, the exploration and exploitation of heavy crude oil have raised series of environmental challenges and caused increased concern for the communities where the oil refineries are cited. Activities such as gas flaring and oil spillage have led to the release of toxic organic and inorganic pollutants, which has resulted in acid rain, climate change, and contamination of soil, water, and air. These environmental hazards have caused adverse effects directly or indirectly to the ecosystem. This chapter offers a general overview of the processes involved in the processing and some of the potential environmental challenges associated with heavy crude oil processing.

Keywords: heavy crude oil, toxic pollutants, oil refinery, environment

1. Introduction

Heavy oil has been part of the natural environment for centuries. It is a substance believed to have formed over the years by the death and decomposition of plant and animal remains that have become incorporated in the sediments of shallow seas and later overlaid by a succession of strata of sedimentary rocks for millions of years. These organic residues are acted upon by intense heat and pressure into petroleum, migrating upwards, sometimes over extensive areas, either to reach the surface or be occasionally trapped in what are to become oil reservoirs [1].

Heavy oil is a naturally occurring, unrefined petroleum that is basically composed of hydrocarbon deposits and other organic materials. The heavy oil can be processed into more useful products like gasoline, kerosene, jet fuel, diesel, heating oil, and other allied products called petrochemicals by refining process.

Basically, crude oil undergoes various stages of exploration before getting to the end consumers or retailers. The overall well-to-consumer supply chain for petroleum products is often described as being segmented into three main segments:

- Upstream activities: This comprises the preliminary stages. It involves exploration activities of crude oil deposits leading to the production of crude oil. Companies in this category include ExxonMobil and Shell Petroleum; they own rights to drill. Also included in this category are companies providing support services such as Halliburton.

- Midstream activities: These involve the transportation of crude oil to refinery; the refining of crude oil into marketable products; and the onward distribution of these products to wholesalers and retailers. Included in this category are companies that transport oil by pipeline, truck, or barge (e.g., Magellan Pipeline, Tulsa, Oklahoma) and companies that refine crude oil (e.g., Tesoro, San Antonio, Texas).

- Downstream activities: Similarly, these categories of activities involve the retail end of petroleum industry. Gasoline stations are the main downstream companies; companies that supply heating oil or propane also fall into this category [2].

The American Petroleum Institute (API) classifies crude oils according to their contents, origins, and specific gravity. Light crude oils or sweet crude are liquid petroleum with low density and can flow freely at room temperature. They are characterized by low viscosity (a property that defines ease of flow), low specific gravity, and high API gravity due to the presence of a high proportion of light hydrocarbon fractions basically from C_6–C_{12}. However, heavy crude oils are highly viscous that cannot easily flow to production wells under normal reservoir conditions. They yield more heat upon burning but have lower API gravity [3].

2. Heavy oils

In the past decades, heavy oils were formed when certain algae species degrades hydrocarbon deposits which lead to the loss of its lighter hydrocarbon fractions, leaving behind the heavier hydrocarbon fractions. By definition, heavy crude oil is oil with high viscosity (approximately 10,000 centipoise). Heavy crude oil ranges from free flowing oil to bitumen and/or tar sands "ultraheavy oil" that is actually embedded in sand and located at depths less than 75 m [4].

Heavy oils are among the class of unconventional crude oil. They are one of the world's largest resources as well as a potential major contributor to the future of energy globally. They are found all over the world with Canada and Venezuela accounting for more than half of world deposits. Basically, the processing of heavy crude oil is faced with challenges such as cost of extraction/processing, haulage to refineries, and processing them into valuable products meeting market demands while adhering to environmental requirements. In addition, the market for selling these heavy crude oil directly is limited because only a few existing refineries are capable of receiving and/or processing such low-quality crude oils [5].

Heavy crude oil is usually characterized by a low content of lighter cuts and contains significantly higher contents of asphaltenes (altered fragments of organic chemical compounds) (**Tables 1** and **2**), which have been reported to greatly

Fraction	Weight percentage (%)	Elementary composition based on C_{20}+ (%)				
		C	H	N	O	S
Asphaltene	14.1	83.8	7.5	1.3	1.7	4.8
Resin	37.3	82.8	8.9	1.5	2	4.3
Aromatic	37.2	84.3	10	<0.3	1.1	4
Saturate	11.4	86.6	13	<0.3	<0.2	<0.1

Table 1.
Typical elemental composition for heavy oil [5].

Origin	Composition (%)					H/C ratio
	Carbon	Hydrogen	Nitrogen	Oxygen	Sulfur	
United States[a]	88.6	7.4	0.8	2.7	0.5	1
Kuwait[a]	82.4	7.9	0.9	1.4	7.4	1.44
Venezuela[a]	85.5	8.1	3.3	1.8	1.3	1.14
Mexico[a]	81.4	8	0.6	1.7	8.3	1.18
Brazil[b]	83	9	2	—	—	1.3
Italy[a]	78	8.8	Trace	3	10.2	1.35
Canada[a]	85.1	11.1	0.7	2.5	0.6	1.56

[a]*Data from Marcel Dekker, Inc. [8].*
[b]*Total content of oxygen and sulfur is 6%.*

Table 2.
Elemental composition of asphaltenes from several oil samples [7].

complicate the refining process. Subsequently, certain asphaltene requires that the heavy oil also undergo a special refining process called deasphalting [6]. The impurities present in heavy crude oil are in the form of compounds of sulfur, oxygen, hydrogen, nitrogen, carbon, and the heavy metals (nickel and vanadium) [4].

The first Canadian producers of heavy bitumen known as Athabasca oil sands used surface mining with massive equipment to mine the oil sands, separate the bitumen from the sand, and return the sand to the excavation site. More recent bitumen production uses the steam-assisted gravity drainage (SAGD) method developed by the Alberta Oil Sands Technology and Research Authority in the 1980s. SAGD uses steam to heat the bitumen, allowing it to flow by gravity to a reservoir where it is recovered [9].

Orinoco tar sand is the most common unconventional crude oil produced in Venezuela. Typical qualities for Venezuelan unconventional crude oils are 5–15° API gravity, 4–6 wt% sulfur, and 1–2 wt% nitrogen. There are also conventionally produced heavy crude oils such as Ku-Maloob-Zaap oil that is very similar in quality to unconventional heavy crudes [9].

More recently, there have been wide ranges of crude oil upgrading options that can allow a large selection of upgraded crude oil qualities. They range from the simple process of diluting with light sweet crude oil (naphtha or natural gas condensates) to produce Maya crude equivalent to complex flow schemes that include delayed coking and residue hydrocracking, as well as other various high-pressure hydrocracking technologies. Specifications for producing this crude type include a gravity target range of around 20–25° API and a target sulfur content of around 3–4 wt%. More so, there is currently high demand for this quality crude oil because the production of Maya and other similar heavy crude oils has been declining in recent years. Exports of Maya crude oil have decreased by about 1 million barrel per day over the last 7 years. Heavy oil processing capacity has increased significantly over the same period (notably at Reliance, Jamnagar, India; Motiva, Port Arthur, USA; and Marathon, Garyville, USA).

Creating a higher-quality, sweet synthetic crude oil with API gravity between 30 and 40° API opens up the potential market for upgraded crude oil by an order of magnitude, because most refineries are capable of processing crude oils within this gravity range. It may be difficult for the producer to justify the cost of the additional upgrading required. However, this level of upgrading could be phased in, if necessary, to accommodate potential market changes in the future. In the same vein, creating high-quality finished products directly from unconventional crude oils is possible but unlikely to be economically viable unless the refinery's location is near a large, high-value market for finished products, has economic logistical options available for product movement, or has a unique specification it can meet. For example, ultralow pour point diesel is a high-value product in Western Canada, near the upgrader site, because of the cold winter season at that location [9].

3. Crude oil exploration and exploitation

Oil exploration and exploitation is a major revenue earner in petroleum-producing countries [10] and serves as the driver of the economies of some of these countries. However, like most human activities, it results in environmental hazards that could be referred to as "slow poisons," in that they often take a long time before causing disease and, in extreme cases, death [11]. The usually unrecognized and slow action of the hazards created by oil exploration and exploitation makes it difficult to fully appreciate their contribution to the disease burden in a country like Nigeria, especially in the oil-bearing communities [12].

4. Crude oil processing

4.1 Separation

The first step of crude oil processing involves the separation of the complex mixtures in the heavy crude according to their molecular weight via atmospheric distillation at atmospheric pressure. During the process, which is also referred to as topping (refining), the oil is subjected to intense heated at the base of a 60 m distillation column at a temperature of 350–400°C, causing it to turn to vapor. The vapors rises inside the column, while the heaviest molecules, or residuals, remain at the bottom. As the vapors rise, the molecules condense into liquids at different temperatures along the length of the fractionating column. Only gases reach the top of the column, where the temperature has dropped to about 150°C. The liquids, which have become increasingly light, are collected on trays located at different heights of the column. Each tray collects a different petroleum fraction, with highly viscous hydrocarbons like asphalt (bitumen) at the bottom and gases at the top.

4.2 Conversion

The heavy fraction leftover after atmospheric distillation still retains many compounds of medium density. This fraction is transferred to another column where it undergoes a second round of distillation to recover middle distillates like heavy fuel oil and diesel.

In the conversion process, the residual heavy hydrocarbon molecules from the separation process are broken down into two or more lighter molecules. The conversion process also known as catalytic cracking is carried out at 500°C. This converts 75% of the heavy products into gas, gasoline, and diesel. The yield can be increased further by adding hydrogen, a process called hydrocracking, or by using deep conversion to remove carbon. The more complex the operation, the more it costs and the more energy it uses.

4.3 Treating

Treating entails deliberate removal of molecules that are corrosive or results in air pollution, especially sulfur. Commencing from January 1, 2009, gasoline and diesel sold in Europe cannot contain sulfur in excess of 10 parts per million (ppm) or 10 mg per kg. The purpose of these measures is to improve air quality and optimize the effectiveness of catalytic converters used to treat exhaust gas. For diesel fraction, desulfurization is done at 370°C and at a pressure of 60 bar. The hydrogen employed in the process reacts with the sulfur to form hydrogen sulfide (H_2S), which is then further treated to remove the sulfur, an important industrial material. Similarly, lighter fractions of kerosene, butane, and propane are washed in a caustic soda (sodium hydroxide) solution to remove thiols, also known as mercaptans. This process is referred to as sweetening [13].

5. Challenges associated with heavy crude oil processing

5.1 Gas flaring

Gas flaring has been one of the most challenging energy and environmental problems facing the world today [14]. As a matter of fact, this act has been condemned in different countries of the world, but the practice has not been totally abolished in some countries especially in the emerging economies [15]. The act of flaring has been allowed by petroleum-producing countries with insufficient fund and investment on structural infrastructure for the efficient utilization of the associated gases obtained from crude oil refining processes [16]. This is in-line with the definition of gas flaring, according to the Canadian Association of Petroleum Producers, as the controlled burning of natural gas that cannot be processed for sale or use because of technical or economic reasons (**Figure 1**) [17]. The World Bank estimates that the annual amount of associated natural gas being flared and vented is about 110 billion cubic meters (bcm), which is practically enough to provide the combined annual natural gas consumption of Germany and France, with Nigeria topping the list of highest gas flaring countries (**Table 3**). Also, estimates calculated from satellite images of flares (National Oceanic and Atmospheric Administration, (NOAA) data, reported by Global Gas Flaring Reduction Partnership (GGFR) indicate that global gas flaring in 2012 alone was 144 bcm. This represents massive resource wastage and a remarkable environmental problem, representing some 400 million tons in CO_2 release into the environment and being at the level of one third of EU's gas consumption [18]. In addition, the amount is twice the annual gas consumption of Africa and three quarters of the Russian gas export [19, 20].

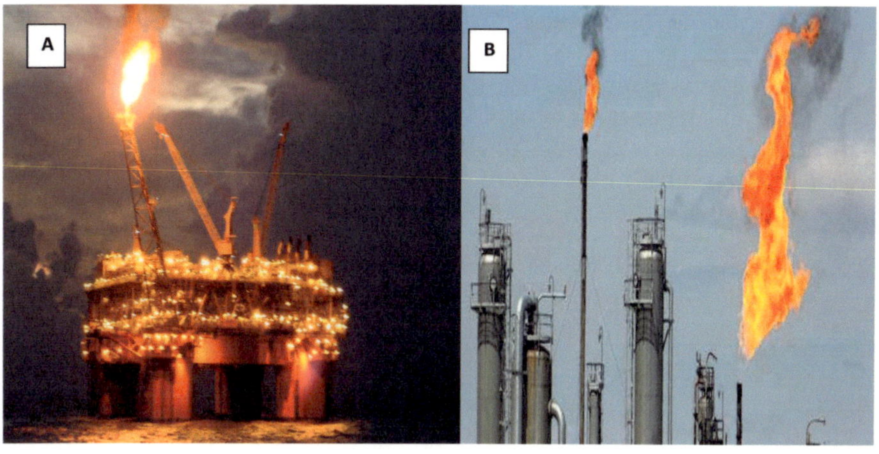

Figure 1.
Flaring (A) from start-up operations on the Deepwater Atlantis platform [15] (B) in Nigeria [16].

2004 rank	Country	Reported flaring, 2004	2004 rank	Country	Reported flaring, 2004
1.	Nigeria	24.1	11.	U.S.A.	2.8
2.	Russia	14.7	12.	Kazakhstan	2.7
3.	Iran	13.3	13.	Libya	2.5
4.	Iraq	8.6	14.	Azerbaijan	2.5
5.	Angola	6.8	15.	Mexico	1.6
6.	Qatar	4.5	16.	UK	1.6
7.	Algeria	4.3	17.	Brazil	1.5**
8.	Venezuela	3.7**	18.	Gabon	1.4
9.	Equatorial Guinea	3.6	19.	Cameroon	1.1
10.	Indonesia	3.5**	20.	Canada	1.0
	Total top 20	**107.5**			

**Source: GGFR (The Global Gas Flaring Reduction Public-Private Partnership).*
***These figures, expressed in billion cubic meters (bcm) may include some venting as well, due to unavailability of segregated data [19].*

Table 3.
Top 20 gas flaring countries in the world [19].*

A flare is normally visible and generates both noise and heat. During flaring, the burned gas generates mainly water vapor and CO_2. Generally, the gases being flared consist of a mixture of different gases; their composition depends on the source of the gas going to the flare system. Natural gas predominantly contains about 90% methane (CH_4) with ethane and small amounts of other hydrocarbons as well as variable amount of inert gases like N_2 and CO_2 [14, 21]. Efficient combustion process involves achieving good mixing ratios between the fuel gas and air (or steam) [22] and on the absence of liquids. Notably, low pressure pipe flares are not intended to handle liquids and do not perform efficiently when hydrocarbon liquids are released into the flare system [23].

Typical gas flaring systems are set up on onshore and offshore platforms of production fields, on transport ships, port facilities, storage tank farms, and along distribution pipelines to vent off trapped gas. The system consists of the flare stack or boom and pipes which collect the gases to be flared (**Figure 2**) [24]. Most flaring processes take place at the top of stack through combustion of gases with the visible flame. However, the height of the flame depends upon the volume of released gas, while brightness and color of the flame largely depend upon composition. The tip of the flare at the end of the stack or boom is designed to assist entrainment of air into the flare to improve burning efficiency. Seals located in the stack prevent flashback of the flame, and a vessel at the base of the stack removes and conserves any liquids from the gas passing to the flare [20].

5.1.1 Classification of flaring processes

5.1.1.1 Emergency flaring

Emergency flaring occurs in cases of fire outbreak, breakage of valves, or compressor failures. This leads to the burning of a large volume of gas within a very short time [20].

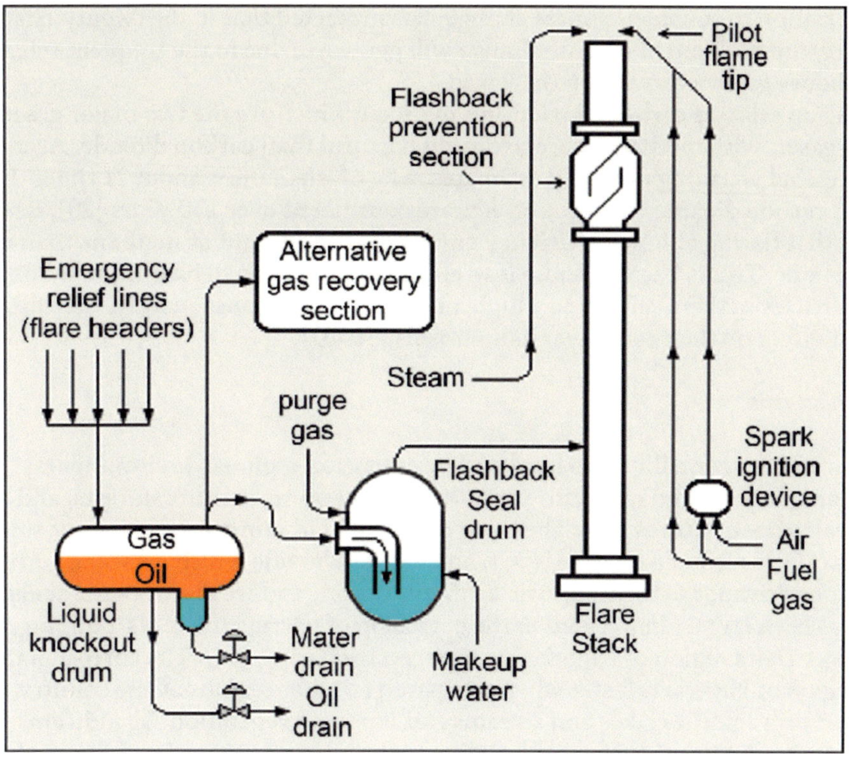

Figure 2.
Schematic flow diagrams of an overall vertical elevated flare stack system [25].

5.1.1.2 Process flaring

Process flaring usually comes with a lower rate, such as during petrochemical process; this involves the removal and subsequent flaring of waste gases in the production stream. The volumes of flared gas at such processes can vary during normal functionality and plant failures from a few m^3/h to thousands m^3/h, respectively [26].

5.1.1.3 Production flaring

Production flaring occurs in the exploration of crude oil. Large volumes of gas are usually combusted during the evaluation of a gas-oil potential test as an indication of the capacity of the well for production.

It has been reported that the flares contain recognized toxins, such as benzene, which pollute the environment drastically [27].

5.2 Environmental challenges associated with gas flaring

5.2.1 Climate change

Gas flaring contributes to climate change, which has serious environmental implications globally. The burning of fossil fuel gives off carbon dioxide, methane, and other gases which has led to global warming with more serious environmental challenges for developing countries, especially Africa which is highly vulnerable with limited ability to adapt [28]. The Intergovernmental Panel on Climate Change (IPCC), a scientific body set up in 1988 by the UN and the World Meteorological

Organization to consider climate change, has projected that in the twenty-first century, the problem of climate change will get worse due to the frequent release of greenhouse gases warming up the world.

Flaring releases carbon dioxide and methane which are the two major greenhouse gases, with methane more toxic and harmful than carbon dioxide. According to the global warming potential estimates, a kg of methane is about 21 times that of a kg of carbon dioxide when the effects are considered over 100 years [29]. Research shows that flaring at lower efficiency emits a higher amount of methane than carbon dioxide. This is because those less-efficient flares tend to have more moisture and particles in them reflecting a high amount of heat. Consequently, they produce similar effect on the ozone layer like aerosols [30–32].

5.2.2 Acid rain

Flaring also contributes to local and by extension regional environmental problems, such as acid rain with attendant impact on agriculture, forests, and other physical infrastructures [33]. The main causes of acid rain are emissions of sulfur oxides (SO_x), nitrogen oxides (NO_x), and carbon dioxide which combine with atmospheric vapor to form sulfuric acid, nitric acid, and weak carbonate acids. Acid rain has been largely implicated in the corrosion of corrugated roofs/buildings in the Niger Delta region of Nigeria due to excessive flaring [34]. The corrosion rate in this region of Nigeria is faster when compared to other regions of the country.

Acid rain acidifies lakes and streams and damages vegetation. In addition, acid rain accelerates the decay of building materials and paints. In addition, these contaminants acidify the soil, hence depleting soil nutrient, decreasing soil fertility, and reducing the nutritional value of crops cultivated within such flaring vicinity [35]. The effects of the changes in temperature on crops included stunted growth, scotched plants, and such other effects as withered young crops [36]. In some cases, there is no vegetation (**Figure 3**) in the areas surrounding the flare due partly to the tremendous heat that is produced and acidic nature of the soil pH [37].

5.2.3 Agriculture

Gas flaring has also been studied widely and reported to show negative effects on plant growth and wildlife [20]. Studies on the effects of waste gas flares on the surrounding vegetation in southeastern Nigeria show that the effect of flare extends

Figure 3.
Effect of acid rain on the environment [38].

well beyond a distance of 110 m from the stacks, with the exception of the suppression of the flowering of short-day plants [39]. It was therefore recommended that further studies were necessary to determine the effects of the flares on the productivity of crops grown in the region (**Table 4**) [41]. Soils of the study area are quickly losing their fertility and capacity for sustainable agriculture due to the acidification of the soils by the various pollutants associated with gas flaring in the area. This is in agreement with previous studies [35].

In one of our unpublished data, it was discovered that the carbon (^{13}C) isotopic composition for various categories of plants have been altered due to gas flaring in the region. Studies show that gas flaring significantly affects not only the microclimate but also the soil's physiochemical properties of the flare sites [42]. This was reflected in the study on the effect of gas flaring on maize yield size. The study reveals increased soil physiochemical properties (such as sand content of the soil, pH, bulk density, air and soil temperatures) toward the flare site. It was recommended that for an optimum yield of maize within the Niger Delta where gas flaring is taking place, the maize must not be cultivated within 2 km of the bund wall of the flare sites.

Furthermore, the destructive effect of gas flaring on wildlife has also been reported [43]. The study opined that gas flaring has been associated with the disruption of wildlife in the immediate surrounding area. As a result, the continuous bright light of gas flares scares wildlife causing them to migrate to more friendly territories or locations.

5.2.4 Health effects

Gas flaring causes the adjoining communities to suffer from increased health risks caused by exposure to those hazardous air pollutants emitted during incomplete combustion of gas flare. These pollutants are associated with a variety of adverse health impacts, including deformities in children, skin problems, cancer, lung damage, neurological and respiratory tract diseases, reproductive and developmental effects, etc. [44–46].

Information from the World Bank on the adverse health impact of particulate matter suggests that gas flaring from Bayelsa State, in Nigeria, would likely result to 49 premature deaths, 4960 respiratory illness among children, and 120 asthma attacks annually [47]. Also, the US Environmental Protection Agency (EPA) warns that exposure to benzene causes acute leukemia and a variety of other blood-related disorders in humans. Exposure to H_2S at concentrations below the permissible limit can cause spontaneous abortion [48].

5.3 Current trend in global gas flaring

Over the years, policies and regulations to tackle the menace of gas flaring have been stepped up considerably by various oil-producing countries with oil and gas companies taking active steps to plan and implement flare-reduction measures.

Distance of farmland from flare site	Percentage loss in yield of crops (%)
100 m	100
600 m	45
1 km	10

Table 4.
Loss in crop yield due to gas flaring [40].

Globally, a decline in the volume of associated gases flared from 2005 to 2010 of about 20% has been reported [18].

Azerbaijan's national oil company, State Oil Company of Azerbaijan Republic (SOCAR), had reduced her associated petroleum gas flaring and venting almost by half to less than 300 million cubic meters within 2 years after joining the GGFR in 2010. The key to this success was measurement (by relying on available data on how much, where, and how gas was being flared) and planning, with the GGFR contributing to both. The GGFR helped by providing measuring equipment and trained the company's staff in using it. Based on information gathered, SOCAR in collaboration with the British Petroleum Company drafted a gas recovery plan that put 1.6 bcm of natural gas to productive use in 5 years. This was used to power Azerbaijan's economy and development while ensuring a cleaner environment.

In the Al-Shaheen project, GGFR partnered with Qatar Petroleum and Maersk and worked together to capture and use 180 million cubic feet of gas per day for power generation. The gas, produced as a byproduct of pumping oil from 300 off-shore wells, had been flared since oil production started at the Al-Shaheen field in 1994. This hitherto wasted gas is now being used to provide about a third of Qatar's electricity, and the gas emissions have been drastically reduced at the completion of the project.

Mexico likewise reduced gas flaring by 66% in 2 years owing to collaboration among the country's Secretary of Energy, its national oil company, Pemex, regulators, and GGFR. Pemex is committed to reducing gas flaring while strengthening flaring and venting regulations, as well as investing in gas recovery strategies.

The government in partnership with operating companies in the Republic of Congo developed a 350 megawatt gas-to-power project feeding two power plants with gas from the M'Boundi oil field. The laudable initiative has reduced emissions of greenhouse gases and increased access to electricity for some 300,000 people.

Kuwait Oil Company and GGFR worked together on a plan to reduce flaring below 1% of the intake associated gas and reach a "technical limit" for gas flaring reduction.

In Nigeria, huge investments from GGFR partners have also helped in reducing gas flaring by 4 bcm in the past 5 years: from 18.6 bcm in 2006 to 14.6 bcm in 2011. GGFR is working with the government and the oil companies to continue building on this positive trend. The World Bank also provides partial risk guarantees to investors in gas-to-power projects [49].

5.4 Crude oil spillage

Oil spillage have been a global issue that has been occurring since the discovery of crude oil [50]; spillage have the potentials of adversely affecting the different spheres of the environment such as coastal and marine habitats, wildlife, fisheries, and human activities (**Figure 2**) [51–53]. Crude oil spillage is simply regarded as the unlawful release of liquid petroleum hydrocarbons within the environments, due to accident and/or intentional human factors. This phenomenon occurs in different forms, and it is attributed to several factors as discussed below.

5.4.1 Equipment failures

Thousands of barrels of oil have been spilt into the environment through failures of oil pipelines and tanks in the country. This spillage is a result of lack of regular maintenance of the pipelines, drilling rigs, and storage tanks and leakages/spills during processing and at refining plants. Some of these facilities have been in use for decades without replacement.

5.4.2 Human error/theft

Sabotage is another major cause of oil spillage in some countries like Nigeria. Some of the citizens of the country with connivance with people from other countries engage in oil bunkering. They damage and destroy oil pipelines in their effort to steal oil from them. Shell Petroleum Development Company (SPDC) in 1996 claimed that sabotage was responsible for more than 60% of all oil spilled at its facilities in Nigeria, stressing that the percentage has increased over the years because the number of sabotage incidents has increased adding that spills due to corrosion of pipelines have decreased with programs to replace oil pipelines (**Figure 4**, **Table 5**). Pirates are stealing Nigeria's crude oil at a phenomenal rate, carting away nearly 300,000 barrels per day, and selling it illegally to the international trade market [41, 54]. **Table 6** gives an example of the number of spills and the reported volumes of oil spilled into the marine environment from 1997 to 2014. Nigeria's largest spill till date was an offshore well blowout in January 1980, when an estimated 200,000 barrels of oil (8.4 million US gallons) spilled into the Atlantic Ocean from an oil industry facility, and that incidence damaged 340 hectares of mangrove [56]. In Russia, aging and leaking pipelines are responsible for the wastage of millions of tons of oil a year, causing widespread environmental damage that largely goes unreported, according to Greenpeace Russia. However, reports have it that the Russia's 4.5 million tons of leaked oil is seven times greater than the BP Deepwater Horizon spill in the Gulf of Mexico in 2010 [57].

Notably, in the highest oil-producing region in Russia (Khanty-Mansi region in Northwestern Siberia), the annual oil spills from pipelines amounted to 5781.4 tons on an area of 229.6 hectares in 2009. In 2010, the Russian Company, Rosneft, alone spilled 3738 tons, and in 2011 a total of 5289 tons of oil was spilled in the region [58, 59].

Oil spills have and are still posing major threat to the environment of the oil-producing areas, which if not effectively checked can lead to the total destruction of the ecosystems. During an oil spill incidence in water, the gaseous and liquid components evaporate with time. Some other components immediately get dissolved in the water and become oxidized, while some undergo bacterial changes and eventually sink to the bottom. Oil spills often result in contamination of surface water with hydrocarbons and trace metals (**Figure 5**) [10]. Also, the soil is contaminated with a wanton negative impact on the terrestrial inhabitants. As the continued evaporation of the volatile and semivolatile lower-molecular-weight compounds affects aerial life, the dissolution of the less-volatile components with the resulting emulsified water affects aquatic life [60].

In addition, oil spill destroys plants and animals in the estuarine habitat. Oil settles on beaches and kills organisms that live there; it also settles on ocean floor

Figure 4.
Oil pipelines in Okrika community, Niger Delta [55].

Causes	Number (N = 135)	Percentage
Mechanical failure	23	17.04
Corrosion	21	15.56
Operational error	17	12.59
Third party activity	28	20.74
Natural hazard	3	2.22
Unknown	43	31.85

Table 5.
The causes of oil pipeline failures between 1999 and 2005 [51].

Year	Total number of reported spills	Quantity in barrels	Source
1997	339	59,272.00	[49]
1998	390	—	
1999	319	—	
2000	637	84,072.00	
2001	412	120,967.00	
2010	537	17,658.10	[50]
2011	673	66,906.84	
2012	844	17,526.37	
2013	522	4066.20	
2014	1087	10,302.16	

Table 6.
Reported oil spillage in the Niger Delta region.

Figure 5.
Impact of oil spillage on (A) loss of vegetation in Ogoni land, Niger Delta [62]. (B) Russia wasting millions of tons of oil from leaking pipes [56].

and kills benthic (bottom-dwelling) organisms such as crabs. Oil poisons algae, disrupts major food chains, and decreases the yield of edible crustaceans. It also coats birds, impairing their flight or reducing the insulative property of their feathers, thus making the birds more vulnerable to cold. Oil spill also endangers fish hatcheries in coastal waters and as well contaminates the flesh of marketable fish. Studies show that the pollution caused by oil spillage does not end with the mopping up of the spilled oil [61].

It is now known that health risk is not averted by abstinence or nonconsumption of fish killed by spilled oil. Some of the fishes and animals that escape death from oil spill pollution are known to have taken in some of the toxic substances, which in turn get into human beings that eat them as a result of bioaccumulation of these substances. This will in turn cause infections on man coupled with other "side effects" such as genetic mutations [61]. In another study [63], the density of sandstone interlaced with shale, beach ridge sand, and medium-coarse sand samples was reduced by 17.7, 13.3, and 15.0% on the average, respectively, due to oil spillage. Furthermore, crude oil-rich beach ridge sand, sandstone interlaced with shale, and medium-coarse sand decreased averagely by 4.4, 9.9, and 15.2%, respectively, of the original value of the specific heat capacity of the unmixed samples, while thermal conductivity of the crude oil beach ridge sand, medium-coarse sand, and sandstone interlaced with shale derivative increased by 9.8, 2.6, and 12.3%, respectively, on the average.

For example, a Nigerian coastal environment housing an extensive mangrove ecosystem has been destroyed. The mangrove was once a source of both woods being used as fuel for the indigenous people and a habitat for the area's biodiversity, but it is now unable to survive the oil toxicity of its habitat. Oil spills in the Niger Delta have been a regular occurrence, and the resultant degradation of the surrounding environment has caused significant tension between the people living in the region and the multinational oil companies operating there [64].

5.5 Noise

Noise pollution is one of the environmental impacts associated with heavy crude processing. The main sources of noise during the production of heavy crude oil would include compressor and pumping stations, producing wells, and vehicular traffic [65]. It has been reported that compressor stations produce noise levels between 64 and 86 dBA close to the station to between 58 and 75 dBA at about 1 mile (1.6 km) away from the station [66]. The direct impacts from the noise pollution would be localized disturbance to wildlife and residents.

5.6 Ecological resources

The major impacts to ecological resources during production of heavy crude oil could occur from disturbance of wildlife from noise and human activity; exposure of biota to contaminants; and death of biota from accidental collision with ground facilities or vehicles [67]. The presence of production wells, ancillary facilities, and access road can also reduce the habitat quality, disturb the biota, and thus affect ecological resources [68]. The presence of an oil field could also interfere with the migratory and other behaviors of some wildlife. The inappropriate discharge of produced water onto the soil or surface water bodies can result in high salinity which will not be able to sustain plant growth. In locations where naturally occurring radioactive material (NORM)-bearing produced water and solid wastes are generated, mismanagement of these wastes can result in radiological contamination of soils or surface water bodies leading to more harmful effects on humans [69].

5.7 Hazardous materials and waste management

Improper handling of industrial/hazardous waste obtained from the processing of heavy crude oil could cause an adverse effect on the ecosystem when released into the environment. These wastes are usually produced during routine operations in process plants [65]. Chemicals contained in open pits used to store wastes may pose a threat to wildlife and livestock because of potential seepage of such chemicals into the underground water. Sand separated from produced water should be properly disposed as it is often contaminated with oil, trace amounts of metals, or other naturally occurring constituents. Production processes could also cause accumulation of large volumes of scale and sludge wastes inside pipelines and storage vessels [70]. These wastes may be transported to offsite disposal facilities. Produced water can become a significant waste stream during the production of crude oil.

5.8 Air quality

The primary emission sources during the production of heavy crude oil include compressor and pumping station operations, vehicular traffic, production well operations, separation of oil phases, and on-site storage of the crude oil. Toxic pollutants emitted during operations would include volatile organic compounds (VOCs), nitrogen oxides, sulfur dioxide, carbon monoxide, benzene, toluene, ethylbenzene, xylenes, polycyclic aromatic hydrocarbons (PAHs), hydrogen sulfide, particulates, ozone, and methane which normally pose as impurities in operation columns [71].

5.9 Health and safety

Potential impacts to general public health and safety during production include accidental injury or death to workers and, to a lesser extent, the public. Health impacts could also result from water contamination, air pollutions, noise, soil contamination, and stress (e.g., associated with living near an industrial zone) [65]. Potential fires and explosions could constitute hazards. Cavitation could ignite grass fires. Increased or reckless driving by oil or gas workers would also create hazards. In addition, health and safety hazards include working in extreme weather conditions and possible contact with natural hazards, such as uneven terrain and dangerous plants, animals, or insects [72].

6. Conclusion

The processing of heavy crude oil generates different types of toxic organic and inorganic pollutants which pose direct or indirect impacts on the environment. Gas flaring, oil spillage, and pipeline vandalization have been some of the major processes leading to the release of toxic pollutants into the environment. Equipment failure and sabotage/theft have been major causes of oil spillage. These incidences have destroyed the coastal vegetation, polluted ground/surface water, and led to ethnic and regional crises. Other environmental impacts associated with heavy crude oil processing include noise pollution from producing wells, interference with ecological resources, release of hazardous materials, reduction in air quality, and health and safety implication. It is therefore pertinent for the petroleum sector to adopt procedures that would provide a reasonable degree of protection to the environment. Such procedural techniques, if implemented, should be able to reduce or eradicate those toxic pollutants.

Conflict of interest

The authors declare no conflict of interest.

Author details

Samuel O. Sojinu* and Onome Ejeromedoghene
Department of Chemistry, Federal University of Agriculture, Abeokuta, Nigeria

*Address all correspondence to: sojinuok2000@yahoo.com

References

[1] Kingston PF. Long-term environmental impact of oil spills: Review paper. Spill Science and Technology Bulletin. 2002;**7**(1-2):53-61

[2] Wikipedia. API gravity. 2018 [Retrieved December 14, 2018]. Available from: https://en.wikipedia.org/wiki/API_gravity

[3] JBlumsac S. The process of crude oil refining. [Retrieved December 14, 2018]. Available from: https://www.e-education.psu.edu/eme801/node/470

[4] Halliburton.Heavy oil. 2018 [Retrieved December 14, 2018]. Available from: https://www.halliburton.com/en-US/ps/solutions/heavy-oil/default.html

[5] Gateau P, Hénaut I, Barré L, Argillier JF. Heavy oil dilution. Oil & Gas Science and Technology. 2000;**59**:503-509

[6] Santos RG, Loh W, Bannwart AC, Trevisan OV. An overview of heavy oil properties and its recovery and transportation methods. Brazilian Journal of Chemical Engineering. 2014;**31**(3):571-590

[7] Loh W, Mohamed RS, Santos RG. Crude oil asphaltenes: Colloidal aspects. In: Somasundaram P, editor. Encyclopedia of Colloid and Surface Science. New York: Taylor and Francis; 2007

[8] Speight JG. The Chemistry and Technology of Petroleum. New York: Marcel Dekker, Inc; 1991

[9] Alain-Yves HUC. Heavy Crude Oils From Geology to Upgrading: An Overview. 2010. Available from: http://www.editionstechnip.com/en/catalogue-detail/1044/heavy-crude-oils.html [Accessed: October 4, 2018]

[10] Digital Refining, Processing, Operations and Maintenance. 2012. Available from: http://www.digitalrefining.com/article/1000563,Challenges_of_heavy_crude_processing.html#.W7kGW9dKgdU [Accessed: October 4, 2018]

[11] Ordinioha B, Brisibe S. The human health implications of crude oil spills in the Niger Delta, Nigeria: An interpretation of published studies. Nigerian Medical Journal. 2013;**54**(1):10-16

[12] World Health Organization. The World Health Report 2002: Reducing Risks, Promoting Healthy Life. Geneva: World Health Organization; 2003. pp. 1-71

[13] WHO. Facing the Facts: The Impact of Chronic Disease in Nigeria. Geneva: WHO; 2005. Available from: http://www.who.int/chp/chronic_disease_report/en/

[14] Planete Energies. The Three Stages of Refining. 2015 [Retrieved December 4, 2018]. Available from: https://www.planete-energies.com/en/medias/close/three-stages-refining

[15] Eman AE. Gas flaring in industry: An overview. Petroleum and Coal. 2015;**57**(5):532-555

[16] Anslem OA. Negative effects of gas flaring: The Nigerian experience. Journal of Environment Pollution and Human Health. 2013;**1**(1):6-8. DOI: 10.12691/jephh-1-1-2

[17] JINN. Gas Flaring in Nigeria: An Overview, Justice in Nigeria Now. 2010. Available from: www.justiceinnigerianow.org

[18] Canadian Association of Petroleum Producers. Flaring and Venting. 2012. Available from: http://www.capp.ca/environmentCommunity/

airClimateChange/Pages/FlaringVenting.aspx [Accessed: October 10, 2012]

[19] Haugland T, Saunier S, Pederstad A, Holm T, Darani H, Kertesheva A. Associated petroleum gas flaring study for Russia, Kazakhstan, Turkmenistan, and Azerbaijan—Final Report. Carbon Limit. 2013;**13**(28):1-80

[20] Gervet B. Gas Flaring Emission Contributes to Global Warming. Luleå, Sweden: Renewable Energy Research Group Division of Architecture and Infrastructure, Luleå University of Technology; 2007

[21] Ismail OS, Umukoro GE. Global impact of gas flaring. Energy and Power Engineering. 2012;**4**:290-302. DOI: 10.4236/epe.2012.44039

[22] Peterson J, Tuttle N, Cooper H, Baukal C. Minimise facility flaring. Hydrocarbon Processing. 2007;**86**(6):111-115

[23] Gzar HA, Kseer KM. Pollutants emission and dispersion from flares: A gaussian case-study in Iraq. Journal of Al Nahrain University. 2009;**12**(4):38-57

[24] Duck B. Reducing emissions in plant flaring operations. Hydrocarbon World. 2011;**6**(1):42-45

[25] Wikipedia, The Free Encyclopedia. Gas flare. 2012. Available from: http://en.wikipedia.org/wiki/Gas_flare

[26] Aregbe AG. Natural gas flaring—Alternative solutions. World Journal of Engineering and Technology. 2017;**5**:139-153

[27] United States Environmental Protection Agency. Industrial Flares. 1991. United States Environmental Protection Agency. Available from: http://www.epa.gov/ttn/chief/ap42/ch13/final/c13s05.pdf [Accessed: October 13, 2012]

[28] Gas Flaring in Nigeria, Friends of the Earth Media Briefing. 2004. p. 1-2. Available from: www.foe.co.uk/resource/media_briefing/gasflaringinnigeria.pdf [Accessed: October 14, 2004]

[29] Uwem U, Akpan EB. Gas flaring in Nigeria: Problems and prospects. Global Journal of Politics and Law Research. 2017;**5**(1):16-28. Online ISSN: ISSN 2053-6593

[30] The International Association of Oil & Gas Producers (OGP). Flaring & Venting in the Oil & Gas Exploration & Production Industry: An Overview of Purpose, Quantities, Issues, Practices and Trends. Flaring and Venting Task Force Report, Report No. 2.79/288. 2000

[31] Bassey N. Gas flaring: Assaulting communities, jeopardizing the world. In: Proceedings at the National Environmental Consultation, the Environmental Rights Action in Conjunction with the Federal Ministry of Environment; December 10-11, 2008. Abuja; 2008

[32] Orubu CO, Odusola A, Ehwarieme W. The Nigerian oil industry: Environmental diseconomies, management strategies and the need for community involvement. Journal of Human Ecology. 2004;**16**(3):203-214

[33] Gobo AE, Richard G, Ubong IUJ. Health impact of gas flares on Igwuruta/Umuechem communities in Rivers state. Journal of Applied Sciences and Environmental Management. 2009;**13**(3):27-33

[34] Hassan A, Konhy R. Gas flaring in Nigeria: Analysis of changes in its consequent carbon emission and reporting. Accounting Forum. 2013;**37**(2):124-134

[35] Imevbore AAA, Adeyemi SA. Environmental monitoring in relation to pollution and control of oil

pollution. Proceeding at the Seminar on the Petroleum Industry and the Nigerian environment. 1981;**6**:135-142 (In: Ref. [16])

[36] Orimoogunje OI, Ayanlade A, Akinkuolie TA, Odiong AU. Perception on the effect of gas flaring on the environment. Research Journal of Environmental and Earth Sciences. 2010;**2**(4):188-193

[37] Ubani EC, Onyejekwe IM. Environmental impact analysis of gas flaring in the Niger delta region of Nigeria. The American Journal of Scientific and Industrial Research. 2013;**4**(2):246-252

[38] Packwood R. Dead spruce trees killed by 'acid rain'. [Retrieved December 14, 2018]. Available from: https://www.gettyimages.com/detail/photo/dead-spruce-trees-killed-by-acid-rainhigh-res-stockphotography/126541656

[39] Augustine OI, Sanford WW. The effects of waste gas flares on the surrounding vegetation in south-eastern Nigeria. Journal of Applied Ecology. 1976;**13**(1):177-187. DOI: 10.2307/2401936

[40] Adeyemo AM. The oil industry extraministerial institutions and sustainable agricultural development: A case study of Okrika L.G.A of Rivers state, in Nigeria. Journal of Oil and Politics. 2002;**2**(1):60-78

[41] Ejiba IV, Onya SC, Adams OK. Impact of oil pollution on livelihood: Evidence from the Niger Delta region of Nigeria. Journal of Scientific Research and Reports. 2016;**12**(5):1-12

[42] Odjugo PAO, Osemwenkhae EJ. Natural gas flaring affects microclimate and reduces maize (Zea mays) yield. International Journal of Agriculture and Biology. 2009;**11**(4):408-412

[43] Hutchful E. Disarmament and development: An African view. IDS Bulletin. 1985;**16**(4):61-67

[44] Ovuakporaye SI, Aloamaka CP, Ojieh AE, Ejebe DE, Mordi JC. Effects of gas flaring on lung function among residents of a gas flaring community in Delta state, Nigeria. Research Journal of Environmental and Earth Sciences. 2012;**4**(5):525-528

[45] Canadian Public Health Association. Background to 2000 Resolution No. 3: Gas Flaring. Canadian Public Health Association 2000 Position Paper; 2000

[46] Kostiuk LW, Thomas GP. Characterization of Gases and Liquids Flared at Battery Sites in the Western Canadian Sedimentary Basin. Technical Report. Edmonton: University of Alberta; 2004. 95 p

[47] Collins C, Oshodi O. Improper abandonment of oil. Nigerians in America. 2010. Available from: http://www.nigeriansinamerica.com/articles/4403/1/Improper-Abandonement-Of-Oil/Page1.html

[48] Argo J. Unhealthy Effects of Upstream Oil and Gas Flaring, Save Our Seas and Shores (SOSS). Nova Scotia; 2002

[49] World Bank Group. Global Gas Flaring Reduction Results. Fact Sheet. [Retrieved December 14 2018]. Available from: http://siteresources.worldbank.org/EXTSDNET/Resources/GGFR-Results-Fact-Sheet.pdf

[50] Kadafa AA. Oil exploration and spillage in the Niger Delta of Nigeria. Civil and Environmental Research. 2012;**2**(3):38-51

[51] Fattal P, Maanan M, Tillier I, Rollo N, Robin M, Pottier P. Coastal

vulnerability to oil spill pollution: The case of Noirmoutier Island (France). Journal of Coastal Research. 2010;**26**(5):879-887

[52] Beyer J, Trannum HC, Bakke T, Hodson PV, Collier TK. Environmental effects of the deepwater horizon oil spill: A review. Marine Pollution Bulletin. 2016;**110**(1):28-51. DOI: 10.1016/j. marpolbul.2016.06.027

[53] Ifelebuegu AO, Ukpebor JE, Ahukannah AU, Nnadi EO, Theophilus SC. Environmental effects of crude oil spill on the physicochemical and hydrobiological characteristics of the Nun River, Nigeria Delta. Environmental Monitoring and Assessment. 2017;**189**(173):1-12

[54] Saleh MA, Ashiru MA, Sanni JE, Ahmed TA, Muhammad S. Risk and environmental implications of oil spillage in Nigeria (Niger-Delta region). International Journal of Geography and Environmental Management. 2017;**3**(2):44-53

[55] Kashi EA. Girl walks over oil pipelines in Okrika, Rivers State, Nigeria with a Shell umbrella. 2016 [Retrieved December 14, 2018]. Available from: http://ukpuru.tumblr.com/post/138384416907/a-girl-walks-over-oil-pipelines-in-okrikarivers

[56] Achebe CH, Nneke UC, Anisiji OE. Analysis of oil pipeline failures in the oil and gas industries in the Niger delta area of Nigeria. In: Proceedings of The International Multi Conference of Engineers and Computer Scientists. 2012. pp. 1274-1279

[57] Sharkov D. Russia Wasting Millions of Tonnes of Oil From Leaking Pipes. 2015 [Retrieved December 14, 2018]. Available from: https://www.newsweek.com/2015/04/17/russia-need-not-drill-arctic-if-itrepairs-oil-pipelineleaks-321372.html

[58] Moskovchenko DV. Oil products in bottom sediments of water bodies in Khanty-Mansi autonomous Okrug. Water Resources. 2005;**32**(1):79-83

[59] Vaver OY. The analysis of social conflicts of environmental management in the Khanty-Mansi autonomous Okrug—Ugra. Geographical Sciences. 2012;**11**:533-537

[60] Akpofure EA, Efere ML, Ayawei P. The Adverse Effects of Crude Oil Spills in the Niger Delta. Urhobo Historical Society. 2000;**1**:21-23

[61] Atubi AO. Effects of oil spillage on human health in producing communities of Delta state, Nigeria. European Journal of Business and Social Sciences. 2015;**4**(8):14-30

[62] Greyl L. Multinational Oil Companies on the Niger Delta, Nigeria. 2018 March 15 [Retrieved December 14, 2018]. Available from: https://ejatlas.org/conflict/multinational-oil-companieson-the-niger-delta-nigeria

[63] George NJ, Akpabio GT, Udofia KM. The implication of oil spillage on the thermal properties of soil samples in the Niger Delta, southern Nigeria. Archives of Physics Research. 2010;**1**(4):64-72

[64] Nwilo CP, Badejo TO. Impacts and Management of Oil Spill Pollution along the Nigerian Coastal Areas. Lagos, Nigeria: Department of Survey and Geoinformatics, University of Lagos; 2005. pp. 567-570. Available from: www.fig.net/pub.figpub36/chapters/chapter_8.pdf

[65] Ngene S, Tota-Maharaj K, Eke P, Hills C. Environmental and economic impacts of crude oil and natural gas production in developing countries. International Journal of Economy Energy & Environment. 2016;**1**(3):64-73

[66] Davorin M. Risk Analysis for Prevention of Hazardous Situations in Petroleum and Natural Gas Engineering. IGI Global; 2013. 433 p. ISBN: 10:9781466647770

[67] Ouren DS, Haas C, Melcher CP, Stewart SC, Ponds PD, Sexton NR, et al. Environmental Effects of Off-Highway Vehicles on Bureau of Land Management Lands: A Literature Synthesis, Annotated Bibliographies, Extensive Bibliographies, and Internet Resources. US Geological Survey Open File Report 1353. 2007. 225 p

[68] Macdonal C, Lockhart L, Gilman A. Effects of oil and gas activities on the environment and human health. In: AMAP Assessment. 2007

[69] Burton GA, Basu N, Ellis BR, Kapo KE, Entrekin S, Hydraulic NK. "Fracking": Are surface water impacts an ecological concern? Environmental Toxicology and Chemistry Journal. 2014;**33**(8):1679-1689

[70] Attallah MF, Awwad NS, Aly HF. Environmental Radioactivity of TE-NORM Waste Produced from Petroleum Industry in Egypt: Review on Characterisation and Treatment. Rijeka, Croatia: Intech; 2012

[71] Oil and Gas Production Phase Impacts. Available from: www.teeic.indianaffairs.gov/er/oilgas/impact/prod/index.html [Accessed: August 31, 2016]

[72] McMichael AJ, Campbell-Lendrum DH, Corvalan CF, Ebi KL, Githeko AK, Scheraga JD, et al. Climate Change and Human Health: Risks and Responses. World Health Organisation; Geneva. 2003;1-338. ISBN: 924156248x